QUANTUM THEORY

Quantum Theory

An Information Processing Approach

Jochen Rau

RheinMain University of Applied Sciences

OXFORD
UNIVERSITY PRESS

OXFORD
UNIVERSITY PRESS

Great Clarendon Street, Oxford, OX2 6DP,
United Kingdom

Oxford University Press is a department of the University of Oxford.
It furthers the University's objective of excellence in research, scholarship,
and education by publishing worldwide. Oxford is a registered trade mark of
Oxford University Press in the UK and in certain other countries

First Edition published in 2021

Impression: 3

Published in the United States of America by Oxford University Press
198 Madison Avenue, New York, NY 10016, United States of America

British Library Cataloguing in Publication Data
Data available

Library of Congress Control Number: 2021932516

ISBN 978–0–19–289630–8 (hbk.)
ISBN 978–0–19–289631–5 (pbk.)

DOI: 10.1093/oso/9780192896308.001.0001

Printed and bound by
CPI Group (UK) Ltd, Croydon, CR0 4YY

Preface

Roughly one hundred years after its original inception, quantum theory is undergoing a paradigm shift. In its original form, quantum theory was developed as a theory of microscopic *matter*. By the early twentieth century, experimental evidence had mounted that classical physics breaks down on the microscopic scale. For instance, it had been observed that matter sometimes behaves like a wave; that, conversely, electromagnetic radiation sometimes behaves like a particle; and that there exist atoms, composed of nuclei and electrons, which are stable and, when excited, emit light at characteristic, distinct frequencies. All these phenomena eluded a classical explanation and called for a radically different approach, leading eventually to a whole new body of theory known as quantum theory. Its early focus on the description of microscopic matter has yielded rich rewards both in scientific terms—culminating in the erection of an impressive theoretical edifice known as the standard model of elementary particles—and in technological terms: there is nowadays hardly any modern device that does not make use, in one way or the other, of quantum effects; just think of a laser or the ubiquitous transistor.

When one focuses on the description of particles and their interactions, two concepts in particular are of paramount importance: energy and time. *Energy* is the quantity which characterizes the possible stationary states of matter, both its ground state—the state of lowest energy—and the various excitations above that ground state. The energies of these excitations are related, for example, to the distinct frequencies of the light emitted by excited atoms. *Time* is the parameter which governs the evolution of a system, and is needed whenever one wants to describe a dynamical process, such as a scattering or a decay process. In fact, energy and time are intimately related. The additional constant of nature that was introduced by quantum theory—*Planck's constant*—relates the physical unit of energy to the physical unit of time; and the central equation of this original version of quantum theory—the *Schrödinger equation*—establishes a connection precisely between the time evolution of a system and its possible energies.

Yet quantum theory is more than a theory of matter. On a deeper level it is also a theory of *measurements*, their compatibility or incompatibility, the probabilities of their respective outcomes, and the interrelationships between these probabilities. On this level, quantum theory deviates not just from classical mechanics but also from classical *logic* and the rules of classical *probability theory*. One of the most intriguing consequences of quantum theory is the possibility of peculiar statistical correlations, dubbed *entanglement*, which cannot be explained classically and appear to contradict our intuitive conception of reality. While these issues attracted great minds early on and animated famous debates between Niels Bohr, Albert Einstein, and others, they took a back seat for much of the twentieth century. One reason for this neglect was the practical difficulty of investigating these phenomena in the laboratory. In contrast to the quantum properties of matter, which may be probed using large assemblies of particles, the study of these more exotic quantum effects requires the careful preparation, manipulation, and measurement of *individual* quantum systems—which for a long time was beyond physicists' experimental capabilities.

Since about the last decade of the twentieth century, several developments have gained momentum that are now bringing this deeper level of quantum theory to the fore. Technology has made huge progress, which was recognized in 2012 with the Nobel Prize (to David Wineland and Serge Haroche) 'for ground-breaking experimental methods that enable measuring and manipulation of individual quantum systems'. At the same time, theorists came to realize that peculiar quantum phenomena like entanglement can be exploited for *information processing* and, strikingly, that this novel form of information processing allows one to tackle problems which are intractable or even impossible classically. One prototypical example is Shor's algorithm for the factorization of large integers, which requires computational resources that grow only polynomially with the size of the problem, as opposed to exponentially in all known classical algorithms. Another example is the secure transmission of information via a quantum channel, protected from eavesdropping by the fundamental laws of physics. While some of these new possibilities are still theoretical, others have been implemented experimentally or even started to be commercialized. We see the dawn of a new 'quantum industry', based upon the systematic exploitation of those peculiar quantum effects which were long at the margins. Some people call this the 'second quantum revolution'.

Unfortunately, the teaching of quantum theory has not kept pace with these developments. While it is true that quantum computing and, more broadly, quantum technology is now being taught in many universities, and quite a few specialized training programmes have sprung up, these offers are mostly supplementary and targeted at graduate students of physics who wish to specialize in that field. The mandatory introductory quantum theory courses at the undergraduate level are, for the most part, still being taught in the old way. This is a pity for several reasons:

1. Much of conventional quantum mechanics is dominated (and, arguably, obscured) by rather complicated mathematics, such as functional analysis or the theory of partial differential equations. This makes it sometimes difficult to disentangle the physical concepts from the mathematical methods. Introducing quantum theory under the aspect of information processing, on the other hand, would require only minimal mathematics: essentially, just finite-dimensional complex vector spaces. It would thus offer a much simpler, more direct route to quantum physics, all the way to some of the most fascinating conceptual issues.

2. Learning quantum theory the traditional way requires a solid grounding in classical physics. It is therefore taught relatively late in the undergraduate curriculum, and rarely outside the physics department. By contrast, the theory of quantum information processing may be formulated without caring about the underlying material substrate. This cuts away a lot of the complexity and requires virtually no prior knowledge of physics. Teaching quantum theory with that

focus would make the subject more accessible to younger students and also to students in other disciplines such as computer science, mathematics, chemistry, or engineering.

3. The latter aspect is particularly salient in view of the growing demand of the emerging quantum industry for graduates skilled in quantum information processing. Meeting this greater demand will require bringing quantum information processing into the mainstream of teaching and making it a central part of every introductory course on quantum mechanics. Moreover, it will be necessary to open these courses to students of other disciplines, as well as to professionals wishing to retrain later in their career.

It is high time to try a new approach to teaching quantum theory—especially at the introductory level.

The present book is intended to support that change. It grew out of lecture notes for a pair of courses which I have taught for many years at Goethe University: a course on 'probability in quantum physics', focused on the basic concepts of measurement, probabilities, and statistical correlations, leading all the way to the Bell and Kochen–Specker theorems and the fundamental issues of locality, non-contextuality, and realism; and a second course, on 'quantum computing', which aims to give an overview of various quantum technologies—quantum computing proper, quantum simulation, high-precision metrology, and quantum communication—and to provide the essential tools for understanding and designing the pertinent protocols. In combination, the two courses cover both the conceptual, physical aspects and the key applications of quantum information processing. By merging the respective notes into a single book, I have sought to create one coherent story, told in a unified language, that fits the new information processing paradigm. I limit myself to the theoretical aspects of the basic physics and algorithms and do not discuss specific hardware implementations, a vast area of—mainly experimental—research which would merit a course of its own. To follow the story in the present book, no prior physics background and no complicated mathematics are required; some familiarity with linear algebra, complex numbers, and basic probability theory will suffice. Therefore, it may be taught equally well in a computer science, mathematics, chemistry, or engineering department.

The book begins with a discussion of selected experimental evidence that eludes a classical explanation and forces us to consider an alternative. The experiments chosen for this discussion are simple and go straight to the heart of the matter: in the quantum realm, the order of measurements is no longer irrelevant; sequences of measurements can steer a system from one state to another; and there are statistical correlations which cannot be explained classically. Next, I dwell quite extensively on basic desiderata of logical consistency and operational meaning that an alternative theory should satisfy. Even before introducing the full framework of quantum theory, this already anticipates many of its salient features and will make its mathematical structure appear more natural later on. This part of the story—Chapter 2—is a bit more formal and may be skipped by the impatient who want to advance quickly to practical applications. Then comes the framework proper of quantum theory, set in complex Hilbert space. Here I introduce all requisite mathematical tools, define the key objects in Hilbert space—states, observables, transformations—and spell out the rules for doing calculations with them. All of this is discussed only to the extent that it is relevant for quantum information processing. Therefore, some topics that one would find in a traditional, 'theory of matter' textbook are conspicuously absent: energy, time, the Schrödinger equation, and even (except for a brief guest appearance) Planck's constant. Much space is devoted instead to the simplest possible quantum system, with just two basis states, which constitutes the elementary building block of quantum information processing: the *qubit*. And since quantum protocols typically process multiple qubits in parallel, there is also an important section on the theory of composite quantum systems. For the more conceptually minded, this is followed by an investigation of some foundational issues: locality, non-contextuality, realism, and the classical limit. The more practical types may skip that part and forge ahead to the next chapter, where quantum information processing begins in earnest. First, I introduce the basic building blocks: in addition to qubits, these are the quantum logic and measurement gates, pieced together in a quantum circuit. There is then another little detour for the theory-minded, regarding the universality of such circuits, as well as a special type of circuit consisting of measurements only; again, this may be skipped by the impatient. After that point the path is clear for considering practical applications. Since there are so many of them, I had to make some tough choices. Rather than striving for completeness, I give a panoramic view of quantum technology, with just one or two representative examples taken from each major area that illustrate the basic working principle. Readers might miss some well-known protocols, like the aforementioned Shor's algorithm, but find others instead that are rarely included in an introductory text; for instance, the variational quantum eigensolver or entanglement-assisted metrology. In my view, this rebalancing renders the examples more representative of the diversity of current quantum technology, and it also reflects more faithfully their practical importance. For instance, the variational quantum eigensolver is closer to being realized with present-day hardware and applied to real-life computational problems than Shor's algorithm, impressive as the latter may be on paper.

In this book I try to strike a balance between two competing goals. On the one hand, I aim to lower the barrier of entry to quantum information processing and guide the student to practical applications as quickly as possible. For this reason, I assume virtually no prior physics knowledge and keep the mathematics as simple as possible. Yet, on the other hand, I want this book to be more than a mere collection of wondrous recipes; I also want the student to develop a deeper understanding of the underlying physical concepts. I am convinced that in the long run this will give the student a more solid base for developing their own ideas and novel applications. Since the basic concepts of quantum theory are so at odds with our intuition, they cause confusion all too often. Therefore, I allot considerable space to the careful definition, motivation, and interpretation of these concepts—probably more than other authors. The result is a book with two interwoven tracks, one more conceptual and one more practical, which can be used in a flexible way, depending on the respective emphasis one wants to place on either of these tracks. A complete course covering the entire material would require about three hours a week for one semester, or one to two hours per week if spread out over two semesters. A more applications-oriented course might leave out some of the discussion of logic and probability in Chapter 1, all of Chapter 2, the last two sections of Chapter 3, the universality analysis in Chapter 4, and perhaps the information-theoretic part of Chapter 5. Such a course could be taught in one to two hours per week in a single semester only. As could its mirror image, a course focused on the theoretical foundations, which would include all that material and leave out most of the practical applications instead.

Each section of the book roughly corresponds to a one- to two-hour teaching unit, and the exercises at the end of each chapter are appropriate for the weekly assignments. Some exercises are marked as 'Project' and may be assigned as student projects. Also, at the end of each chapter I provide suggestions for further reading. Since the literature on quantum theory and quantum information processing is vast, this can only be a tiny selection that reflects my own reading history and personal taste, and is in no way meant to be complete.

Acknowledgements

Quantum theory has intrigued me since my student days. But my professional interests, both inside and outside academia, initially took a different course. It was only much later that I returned to quantum theory and made it the subject of my research and teaching. Such late career changes are never easy, and I am grateful to those who extended a helping hand and welcomed me into the community of quantum theorists by sharing their insights, offering generous invitations, and engaging in stimulating discussions. I wish to mention in particular Chris Fuchs, Lucien Hardy, and Robert Spekkens, who took notice of my early work and invited me to spend time at Perimeter Institute, thus opening for the first time the door to that new community; Martin Plenio and Renato Renner, who took the trouble to scrutinize my *Habilitation* thesis on the role of probability in physics—part of which has found its way into this book—and offered their kind hospitality in Ulm and Zurich; and Gernot Alber, who gave me a foothold in the form of a first job in the new field; as well as Caslav Brukner, Otfried Gühne, Rüdiger Schack, and Philip Goyal, with whom it was always a pleasure to discuss.

The present book grew out of a pair of courses on quantum information processing— one more conceptual, one more applied—that I have taught for many years at Goethe University Frankfurt. By entrusting these courses to the novice that I once was, my colleagues at Goethe University offered me a great opportunity to learn by teaching. Indeed, over time I have learnt invaluable lessons from the many penetrating questions, approving nods, or empty stares of my students, forcing me to keep adapting the structure and contents of my course. Should some explanations in this book feel particularly clear and simple, they are the fruit of this continuing interaction with my students; if others still look weak or cumbersome, they are, of course, entirely due to my own shortcomings.

My special thanks go to María García Díaz, who went out of her way to check my manuscript carefully and suggested many valuable improvements. I would also like to thank Michail Skoteiniotis for suggesting some of the more unconventional exercises. And, once again, it has been a pleasure to work with the professional team at Oxford University Press, who supported me all along and saw the final product smoothly through the press.

The final parts of this book were written in the spring of 2020 during a sabbatical visit at Scuola Internazionale Superiore di Studi Avanzati (SISSA) in Trieste, Italy. I thank Stefano Liberati for his kind invitation and Oxana Mishina for offering her generous help with getting settled. Alas, what should have been an opportunity for lively scientific exchange turned out to be a frightening experience for people all over the world, and in

Italy in particular, after the outbreak of the Covid-19 pandemic. To make matters worse, I also had to deal with some health issues of my own. Thankfully, during this period I found the people in Trieste to conduct themselves with exemplary seriousness, calm, and consideration for others. And I am most grateful to my family and friends—and first and foremost, my wonderful daughter Milena—who were there when I needed them.

Contents

1

Introduction

1.1 Some Experimental Evidence

The quantum realm exhibits phenomena that run counter to conventional logic.
I illustrate this for a particularly simple class of experiments: measurements or
combinations of measurements which each have only two possible outcomes. In
particular, I highlight the importance of the order in which measurements are
performed.

The quantum world exhibits plenty of peculiar phenomena; many more than we can
possibly cover in this brief introduction. We shall limit ourselves to the simplest of these
phenomena, those which involve measurements with just two outcomes: 'up' or 'down',
'left' or 'right', 'pass' or 'no pass', 'click' or 'no click'—like Heads or Tails in a coin
flip. Measurements of this kind are termed *binary*. More specifically, we shall consider
two kinds of physical systems on which such binary measurements may be performed:
atoms and photons. In the first case we deal with silver atoms, vaporized in an oven and
emitted from the latter in some definite spatial direction. A measurement then consists in
sending this beam of silver atoms through a so-called *Stern–Gerlach apparatus*, a pair of
magnets which generates an inhomogeneous magnetic field, and subsequently blocking
the atoms with a screen. It turns out that the apparatus will split the beam in two, and,
hence, that each individual silver atom will hit the screen at one of just two possible
locations (Fig. 1.1). For the purposes of the present introduction, the precise physical
mechanism behind this—the interaction of the atom's quantized magnetic moment with
the inhomogeneous field—need not concern us; it suffices to note that this measurement
is binary. The pair of magnets can be rotated about the beam axis, yielding a continuum
of possible measurement configurations parametrized by the rotation angle, θ (Fig. 1.2).
Regardless of the angle, the measurement continues to be binary.

The other physical system that we shall take as an example is a photon. Photons are
the elementary quanta of light. Their existence was first hypothesized at the beginning
of the twentieth century as an explanation of the photoelectric effect (an explanation
which earned Albert Einstein his Nobel Prize). Nowadays, photons can be made visible
with the help of strongly dimmed, or 'attenuated', lasers (Fig. 1.3). Photons may be
subjected to binary measurements in various ways. One specific example is the passage
through a so-called *polarizing beam splitter*. This is typically a plate or a cube (whose

Quantum Theory: An Information Processing Approach. Jochen Rau, Oxford University Press (2021). © Jochen Rau.
DOI: 10.1093/oso/9780192896308.003.0001

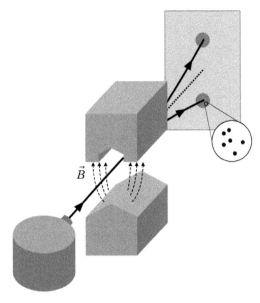

Fig. 1.1 *Stern–Gerlach experiment. The pair of magnets splits a beam of silver atoms in two. Under a microscope, one can resolve the hits by individual atoms.*

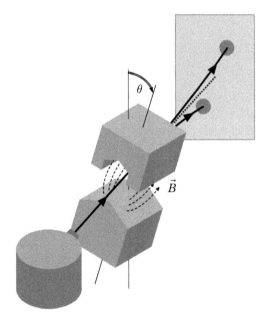

Fig. 1.2 *Rotated Stern–Gerlach apparatus. The pair of magnets is rotated about the beam axis. There is a continuum of possible configurations, parametrized by the rotation angle, θ.*

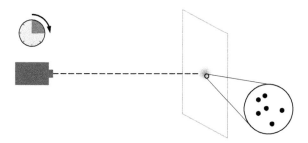

Fig. 1.3 *Quanta of light. A dimmed (or 'attenuated') laser shines on a photosensitive screen for a certain period of time. Viewed from afar, the beam leaves on the screen a dot of non-zero, albeit small, size whose darkness varies on a continuous scale. Viewed more closely, however, this dot is made up of many smaller spots which, individually, all have the same darkness. There is either a spot or not; on this microscopic level there is no continuous grey scale. Each spot corresponds to a hit by an elementary quantum of light, a photon. The appearance of a continuous grey scale on the macroscopic level is created by the varying density of these spots.*

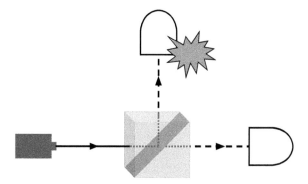

Fig. 1.4 *A polarizing beam splitter placed in a laser beam. The beam splitter, represented by the cube, either lets an incoming photon pass or deflects it in another direction. Which way the photon went is registered by the two detectors, which 'click' when they are hit by a photon.*

precise composition and inner workings again need not concern us) which either lets a photon pass or deflects it in some different direction. Thus, an incident light beam, which is composed of many photons, will be effectively split in two. Placed behind the beam splitter are two detectors, one in each direction, which 'click' whenever they register a photon (Fig. 1.4). Like the Stern–Gerlach apparatus, a beam splitter may be rotated about the beam axis, yielding a continuum of possible configurations (Fig. 1.5). (In practice, it is usually not the splitter itself which is rotated. Rather, the splitter is placed in between so-called 'half-wave plates', and it is these which are rotated. For our purposes, this technicality is not relevant.) Again, these configurations can be parametrized by the rotation angle, which—for reasons that we will discuss shortly—we shall denote by $\theta/2$ rather than θ.

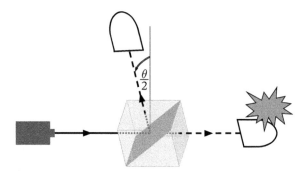

Fig. 1.5 *Rotated polarizing beam splitter. The polarizing beam splitter can be rotated about the beam axis, yielding a continuum of possible measurement configurations, parametrized by the rotation angle, $\theta/2$. (The reason for the factor one half is explained in the text.)*

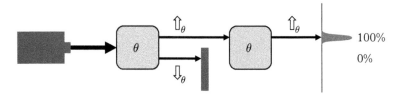

Fig. 1.6 *Reproducibility of binary measurements. Two identical measurement devices (Stern–Gerlach apparatuses or polarizing beam splitters), oriented at the same angle θ, are placed in sequence. Of the two beams exiting the first device, one is blocked out (say, the lower or deflected beam, respectively, labelled by $\Downarrow$) and the other (upper or straight) beam, $\Uparrow$, is allowed to continue. When this remaining beam is sent through the second, identically configured device, all of it will exit as a single beam, corresponding to the same outcome (here: $\Uparrow$) as before. In other words, the second measurement confirms with certainty the outcome of the first.*

In order to qualify as genuine measurements rather than just random splittings, the experiments described above must yield outcomes that are *reproducible*. If an atom or photon takes the upper or straight path, respectively, then upon passing through a second, identical measurement device, it must always take the same path again. The same must hold for the lower or deflected path, respectively. This is indeed the case, as can be verified experimentally with the setup illustrated in Fig. 1.6. Such reproducibility ceases to be guaranteed as soon as the two devices are rotated against each other and hence are no longer identical. Then the second device will generally split the beam again (Fig. 1.7). A particle which exited the first device in the upper beam, $\Uparrow_\theta$, will end up in the upper beam of the second device, $\Uparrow_{\theta+\Delta\theta}$, only with some probability, $\mathrm{prob}(\Uparrow_{\theta+\Delta\theta} \mid \Uparrow_\theta)$. While reproducibility demands that for $\Delta\theta = 0$ it is

$$\mathrm{prob}(\Uparrow_\theta \mid \Uparrow_\theta) = 1, \tag{1.1}$$

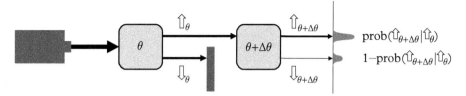

Fig. 1.7 *Binary measurements at different angles. Two measurement devices (Stern–Gerlach apparatuses or polarizing beam splitters) are placed in sequence, with the second rotated against the first in parameter space by $\Delta\theta$. In the case of Stern–Gerlach apparatuses this corresponds to an actual spatial rotation by $\Delta\theta$, whereas in the case of beam splitters it corresponds to a spatial rotation by $\Delta\theta/2$. Only one beam from the first device (here: $\Uparrow_\theta$) is sent through the second device. In general, the second device will split this beam in two, labelled $\Uparrow_{\theta+\Delta\theta}$ and $\Downarrow_{\theta+\Delta\theta}$, with respective intensities depending on the parameter offset, $\Delta\theta$. The intensity of the upper beam, $\Uparrow_{\theta+\Delta\theta}$, is proportional to the conditional probability, prob($\Uparrow_{\theta+\Delta\theta}$ | $\Uparrow_\theta$), that a particle will exit the second device in $\Uparrow_{\theta+\Delta\theta}$, given that it left the first device in $\Uparrow_\theta$. The intensity of the other beam is proportional to the complementary probability, $[1 - prob(\Uparrow_{\theta+\Delta\theta}$ | $\Uparrow_\theta)]$.*

for other values of $\Delta\theta$ this probability may be smaller. In fact, when a Stern–Gerlach apparatus is turned upside down, $\Delta\theta = \pi$, what was formerly 'up' will become 'down', and vice versa. Thus, an atom from the upper beam of the first device will end up with certainty in the *lower* beam of the second device; the probability of it exiting in the upper beam of the second device has fallen to zero:

$$\mathrm{prob}(\Uparrow_{\theta+\pi} \mid \Uparrow_\theta) = 0. \tag{1.2}$$

Experimentally, it turns out that the same happens in a sequence of polarizing beam splitters when the second beam splitter is rotated in space by $\pi/2$—rather than π—relative to the first. This explains why in the case of polarizing beam splitters we parametrized spatial rotations by $\theta/2$ rather than θ: in this manner, we have ensured that the last equation holds for polarizing beam splitters as well.

For offsets somewhere in between the two extreme cases, $0 \leq \Delta\theta \leq \pi$, we would expect that the conditional probability in question drops monotonically as the parameter offset increases; and indeed, for both types of devices this is precisely what one observes experimentally (Fig. 1.8). The function is universal: it is the same for Stern–Gerlach apparatuses and polarizing beam splitters; it does not depend on any specifics of the respective source; nor does it depend on the initial parameter setting, θ. Aside from its two endpoints and its monotonicity in between, the function exhibits some interesting symmetries. To begin with, for the conditional probability it makes no difference whether the second device is rotated clockwise or counterclockwise; the probability is an even function of the parameter offset:

$$\mathrm{prob}(\Uparrow_{-\Delta\theta} \mid \Uparrow) = \mathrm{prob}(\Uparrow_{\Delta\theta} \mid \Uparrow). \tag{1.3}$$

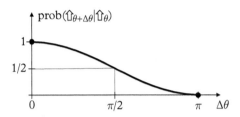

Fig. 1.8 *Conditional probability of a particle exiting 'up' from the second device, given that it exited 'up' from the first, as a function of the parameter offset, $\Delta\theta$.*

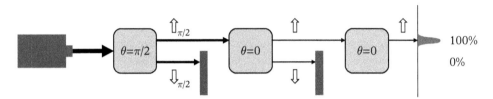

Fig. 1.9 *Example of a sequence composed of three binary measurements. In this particular sequence the screen or detectors behind the last device register only one beam.*

(To simplify notation, we chose the initial parameter value $\theta = 0$.) Secondly, we already noted that an offset by π amounts to swapping the outcomes (Eq. (1.2)). More generally, this means that we have

$$\text{prob}(\Uparrow_{\Delta\theta+\pi} \mid \Uparrow) = 1 - \text{prob}(\Uparrow_{\Delta\theta} \mid \Uparrow). \tag{1.4}$$

Replacing $\Delta\theta$ with $-\Delta\theta$ and exploiting the previous symmetry, Eq. (1.3), this further implies that

$$\text{prob}(\Uparrow_{\pi-\Delta\theta} \mid \Uparrow) = 1 - \text{prob}(\Uparrow_{\Delta\theta} \mid \Uparrow). \tag{1.5}$$

Consequently, at the midpoint, $\Delta\theta = \pi/2$, the probability must equal one half,

$$\text{prob}(\Uparrow_{\Delta\theta=\pi/2} \mid \Uparrow) = \frac{1}{2}. \tag{1.6}$$

Again, this agrees with the experimental evidence.

Now we place a third device into our sequence. Two specific examples are shown in Figs 1.9 and 1.10. They feature the same set of measurement devices but arranged in different orders. Interestingly, the different ordering alone leads to very different final outcomes. In the first example, the initial splitter-cum-block tests whether a particle goes 'up' when the device is oriented at $\theta = \pi/2$. All particles which pass this initial test then enter the second device, oriented at a different angle, $\theta = 0$. Once again, only particles which go 'up' with respect to this new orientation are selected and passed on to the

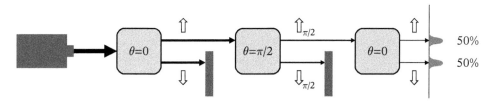

Fig. 1.10 *Alternative measurement sequence that results from swapping the first two devices in Fig. 1.9. In this alternative arrangement the screen or detectors register two beams of equal intensity rather than just one.*

third device. This last device, however, is identical to the preceding one. Therefore, by reproducibility (Fig. 1.6), it has no effect; it simply confirms the previous measurement outcome. As a result, the screen or detectors will be hit by one beam only. In the second example, the first two devices have been swapped. The third splitter is then no longer an identical copy of the second, and thus may no longer be expected to be harmless. Indeed, due to the universality of the conditional probability function (Fig. 1.8), the splitting effected (or not) by the third device depends solely on its orientation relative to its immediate predecessor; the fact that its orientation coincides with that of another device further away in the sequence is irrelevant. Since the parameters of the last two devices differ by $\Delta\theta = -\pi/2$, Eqs (1.3) and (1.6) imply that the last device will split the incoming beam into two of equal intensity. As a result, the screen or detectors will be hit by two beams of equal intensity rather than just one.

We may also compare the setup in Fig. 1.10 to the earlier setup in Fig. 1.6 (with $\theta = 0$), which comprised only two measurements. The three-step sequence differs from its two-step counterpart by the insertion of an intermediate device that is rotated by $\pi/2$ relative to the other two. This intermediate, rotated device spoils the reproducibility observed in the original setup. When, in the third and final step, the initial measurement is repeated, the initial outcome is no longer confirmed with certainty. Rather, both conceivable outcomes occur with equal probability. It is as if the intermediate measurement has erased all memory of, or 'overwritten', the outcome of the first measurement; the first measurement might as well never have happened. This is quite disturbing, especially if viewed on the level of a single particle. In the course of its passage through the experimental setup (assuming that it gets through) the particle is subjected to the same test twice: does it go 'up' in a splitter oriented at $\theta = 0$? As the evidence shows, with probability one half the two identical tests on the identical particle will give contradictory answers. This forces two rather radical conclusions upon us:

1. Measurements may entail not only a gain but also a loss of information. There are situations where a measurement will invalidate (wholly or in part) the outcomes of earlier measurements. This loss is not due to a lack of accuracy on the experimenter's part but is an inevitable consequence of the types of measurements involved.

2. As a result, the outcome of a combined experiment (or, more precisely, the probabilities assigned to its various possible outcomes) generally depends on not just which measurements are being combined, but also *in which order*. The probability that 'device A will reveal x *and then* device B will reveal y' may differ from the probability that 'device B will reveal y *and then* device A will reveal x'. This order dependence is once again a matter of principle and not of a lack of accuracy.

These conclusions clearly contradict our everyday experience. We are used to gathering new information without fear of losing the information we already have, and we feel free to collect pieces of new information in arbitrary order. For instance, if we want to learn about a playing card, it makes no difference whether we first ascertain the suit to which it belongs (Diamonds, Clubs, Hearts, or Spades) and subsequently the face within the suit (King, Queen, Jack, etc.), or vice versa; and we need not fear that learning about the one will invalidate what we have learnt about the other. Now the experimental evidence is telling us that in the quantum domain we can no longer be so sure.

The experimental setups considered so far comprised a varying number of splitters, each (except for the last) followed by blocking out one of the two outgoing beams. Every such combination of splitter plus blocking-out amounts to a *filter*: it performs a measurement and lets through only those particles for which the measurement yielded a particular result. Consequently, each filter reduces the intensity of the beam. One would expect, therefore, that the more filters there are, the weaker the beam which will eventually hit the screen or detectors. However, sometimes the opposite is the case. An extreme example is illustrated in Figs 1.11 and 1.12. Initially there are just two filters,

Fig. 1.11 *Two filters arranged such that all particles are blocked.*

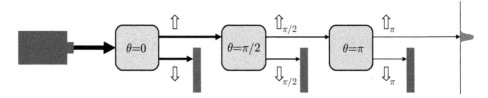

Fig. 1.12 *Insertion of a third filter in between the two filters of Fig. 1.11. Thanks to the additional filter, some particles can now pass.*

with the second filter rotated in parameter space by π relative to the first. Thanks to Eq. (1.2), every particle which has passed the first filter will necessarily fail to pass the second. Hence, these two filters together block all particles. Then a third filter with an intermediate orientation is inserted in the middle. In the resultant sequence each successive filter is rotated in parameter space by $\pi/2$, rather than π, relative to its predecessor. The extent to which the first filter reduces the intensity of the beam is not known; this may depend on the properties of the source. But for all subsequent filters, we know from Eq. (1.6) that each of them will cut the intensity in half. Therefore, there will be a beam left to hit the screen or detectors. Its intensity will be one quarter of the intensity right after the first filter. In other words, on average, one in four particles which have passed the first filter will pass all filters.

The procedure laid out in the preceding paragraph can be iterated. In a first iteration we insert two more filters, one in each gap in the existing sequence, so as to create a new sequence where each successive filter is rotated in parameter space by $\pi/4$, rather than $\pi/2$, relative to its predecessor. After the nth iteration we will have a sequence where the parameter difference between any two successive filters equals $\pi/2^{n+1}$. Then the probability that a particle will pass through all the 2^{n+1} filters behind the first filter will be

$$\text{prob}(\Uparrow_\pi \leftarrow \ldots \leftarrow \Uparrow_{3\pi/2^{n+1}} \leftarrow \Uparrow_{2\pi/2^{n+1}} \leftarrow \Uparrow_{\pi/2^{n+1}} \mid \Uparrow) = \left[\text{prob}(\Uparrow_{\pi/2^{n+1}} \mid \Uparrow)\right]^{2^{n+1}}. \quad (1.7)$$

On the left-hand side we used the harpoon ($\leftarrow$) to indicate the order—which we know to be relevant!—in which the necessary measurement results are obtained. Since we take the passage of the first filter as a given, this is denoted as a condition on the very right, behind the vertical bar. Starting from this condition, the sequence of measurement results then runs from right to left. On the right-hand side of the equation we employed the product rule,

$$\text{prob}(\ldots \leftarrow \Uparrow_{\theta_{k+1}} \leftarrow \Uparrow_{\theta_k} \mid \Uparrow_{\theta_{k-1}}) = \text{prob}(\ldots \leftarrow \Uparrow_{\theta_{k+1}} \mid \leftarrow \Uparrow_{\theta_k}) \text{prob}(\Uparrow_{\theta_k} \mid \Uparrow_{\theta_{k-1}}), \quad (1.8)$$

multiple times, as well as the fact that every single factor depends on the parameter offset only,

$$\text{prob}(\Uparrow_{\theta_k} \mid \Uparrow_{\theta_{k-1}}) = \text{prob}(\Uparrow_{(\theta_k - \theta_{k-1})} \mid \Uparrow). \quad (1.9)$$

Without exact knowledge of the conditional probability function (Fig. 1.8), the probability of passage is impossible to calculate for arbitrary n, but a few generic assumptions will allow us to say something about the limit $n \to \infty$. We assume that the conditional probability function is sufficiently smooth so that it can be Taylor expanded around $\Delta\theta = 0$. Since the function is even, Eq. (1.3), there will be no linear term in the Taylor expansion; the first non-trivial contribution will be of second order or higher,

$$\text{prob}(\Uparrow_{\Delta\theta} \mid \Uparrow) \approx 1 - O\left((\Delta\theta)^2\right). \quad (1.10)$$

Inserting this expansion into the right-hand side of Eq. (1.7) and using

$$\left[1 - O\left(\frac{1}{x^2}\right)\right]^x \approx 1 - O\left(\frac{1}{x}\right)$$

for large x, we find that for large n it is

$$\text{prob}(\Uparrow_\pi \leftarrow \cdots \leftarrow \Uparrow_{3\pi/2^{n+1}} \leftarrow \Uparrow_{2\pi/2^{n+1}} \leftarrow \Uparrow_{\pi/2^{n+1}} \mid \Uparrow) \approx 1 - O\left(\frac{1}{2^{n+1}}\right). \tag{1.11}$$

Thus, with each iteration the probability of passage increases, until, in the limit $n \to \infty$, it approaches unity. In other words, by adding more and more filters at suitable orientations we are able to build a sequence that lets virtually all particles pass—despite the fact that the first and the last filter in the sequence impose contradictory constraints which, in isolation, would be impossible to overcome. In effect, the intermediate filters 'steer' a particle away from the situation where it has just passed the first filter towards a situation where it will satisfy the opposite constraint imposed by the last filter, and they do so with a success probability equal to one. This peculiar possibility is known as *lossless steering*. Indeed, such steering can be realized experimentally. It is easiest to demonstrate with photons. For photons, a filter may be realized directly, without the two separate steps of splitting and blocking, with an absorptive linear *polarization filter*, or *polarizer*, of the kind used in photography or polarizing sunglasses. Two polarizers can be rotated against each other in such a way that they block all light. By placing an intermediate polarizer in the middle, light can be made to pass again; and when more and more polarizers are added which reduce the incremental angle between any two successive polarizers, the intensity of the transmitted beam continues to increase (Fig. 1.13). In practice, however, there will be a limit. Real filters are never perfect, so at some point the intensity loss due to imperfections of the filters is going to outweigh the intensity gain due to the reduction of the incremental angle.

So far we have only considered measurements or sequences of measurements on individual particles. Now we turn to combined measurements on *pairs* of particles. These exhibit features that are even more peculiar. We shall focus on pairs of photons, which are fairly straightforward to produce experimentally; the basic setup is illustrated schematically in Fig. 1.14. The two members of a photon pair produced in this way are each measured independently by a polarizing beam splitter and subsequent detectors. Both beam splitters are oriented at the same angle in parameter space, θ. When the nonlinear crystal used to produce the photon pair is of a particular kind (technically, a 'type-II parametric down-converter'), one observes that the measurement outcomes for the two photons are perfectly anticorrelated: whenever one device measures 'up' the other measures 'down', and vice versa, *regardless of the parameter value, θ*. If such perfect anticorrelation were observed for one specific parameter value only, say, $\theta = 0$, this could be easily understood. One would conclude that one member of a pair is always produced in a '$\Uparrow$ state' (meaning that a measurement in the $\theta = 0$ direction will yield 'up' with certainty) and the other, in a '$\Downarrow$ state' (meaning that said measurement will certainly yield

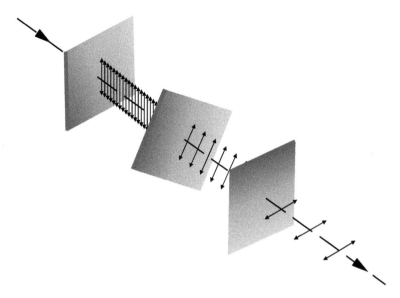

Fig. 1.13 *Steering. Light passes through a sequence of linear polarizers which are rotated against each other. When further intermediate polarizers are placed in between the first and the last polarizers, reducing the incremental angle between any two successive polarizers, the intensity of the transmitted beam increases.*

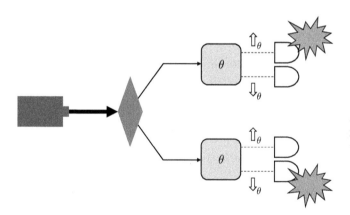

Fig. 1.14 *Entanglement. A strong laser shines on a nonlinear crystal. By a process known as 'spontaneous parametric down-conversion', the crystal splits some of the incoming photons into pairs of lower-energy photons. Both members of a pair are then separately subjected to measurement (polarizing beam splitter plus detectors), with identical parameter setting,* θ*. For a special type of crystal ('type II') the two members of a pair are perfectly anticorrelated, in the sense that the measurements on the two photons will always yield opposite results; when one photon hits the upper detector, the other photon will invariably hit the lower detector, and vice versa. This perfect anticorrelation is exhibited for all values of* θ*.*

'down'); only that it might be uncertain which member enters which device. While this would explain the perfect anticorrelation for $\theta = 0$, it could not possibly explain why the same perfect anticorrelation continues to be observed as the devices are rotated to, say, $\theta = \pi/2$. For if a $\Uparrow$ photon is measured by such a rotated device, the outcome will be 'up' or 'down' with equal probability (Eq. (1.6)); likewise for a $\Downarrow$ photon. Consequently, a $\Uparrow\Downarrow$ pair subjected to measurements at $\theta = \pi/2$ will produce identical outcomes 50 per cent of the time—a far cry indeed from the observed anticorrelation. As a matter of fact, the perfect anticorrelation for arbitrary values of θ has no straightforward explanation. It is a signature of a genuine quantum phenomenon termed *entanglement*.

In sum, we have seen that in the quantum realm already the simplest of measurements—binary measurements—reveal features that run counter to our everyday experience. They include:

1. the order dependence of measurements
2. the possibility of lossless steering
3. entanglement, as evidenced in universal perfect anticorrelation.

These and many other counterintuitive phenomena call for a thorough theoretical explanation. At the same time, they might open up possibilities for powerful practical applications in areas such as metrology, communication, and computing. To elucidate both the conceptual underpinning and the potential applications of these unusual quantum effects shall be the focus of the present book.

1.2 Logic

The phenomena discussed in Section 1.1 have challenged some of our classical notions of logic. I briefly review these classical notions and pin down where exactly they fail in the quantum realm. I do so using a language and a notation that will facilitate as much as possible the changes that will be required in the quantum realm.

When discussing an experiment in logical terms, the basic objects of interest are *propositions* (or 'hypotheses', or 'events') pertaining to physical systems, such as 'this die shows "4"'. Such propositions exhibit a logical structure. A part of this structure may be characterized as follows:

1. Some (but not all) propositions are related by *logical implication*. For instance, whenever the proposition x: 'this die shows "4"' is true, the proposition y: 'this die shows an even number' must also be true. One says that 'x implies y', or 'x refines y', and writes $x \subseteq y$. Logical implication induces an ordering in the set of propositions. This order is
   (a) transitive:

$$x \subseteq y, \; y \subseteq z \quad \Rightarrow \quad x \subseteq z \tag{1.12}$$

(b) reflexive:

$$x \subseteq x \tag{1.13}$$

(c) antisymmetric:

$$x \subseteq y, \; y \subseteq x \quad \Rightarrow \quad x = y. \tag{1.14}$$

In particular, two propositions which merely use different words—such as 'this die shows "2"' and 'this die shows an even number smaller than three'—but are equivalent in their meaning are considered identical.

However, the order is only partial, as there may be pairs of propositions which are not related by logical implication in either direction. It is important not to confuse logical implication with causation. While a car accident may cause injury, and so the proposition *a*: 'there was a car accident' may render the proposition *b*: 'someone was injured' more plausible, the former does not logically imply the latter.

2. There exists one particular proposition called the *absurd proposition*, denoted by $\emptyset$. It is the unique proposition which is always false; and, since from a false premise anything can be derived—*ex falso quodlibet*—it is also the unique proposition which implies all others:

$$x = \emptyset \quad \Leftrightarrow \quad x \subseteq y \; \forall y. \tag{1.15}$$

3. If two propositions, x and y, contradict each other—in the sense that whenever one of them is true, the other must be false—then it is possible to formulate the proposition '*either x or y*'; we shall denote the latter by $x \oplus y$. This logical operation is 'binary' in the sense that it combines two elements, x and y, to produce a third one, $x \oplus y$. It is important that $\oplus$ means 'either ... or ...' rather than an ordinary 'or'. For this reason, it is not defined for arbitrary pairs of propositions but only for those which are mutually exclusive. For instance, while we may combine x_1: 'this die shows "4"' and x_2: 'this die shows "6"' into $x_1 \oplus x_2$: 'this die shows either "4" or "6"', we cannot combine x_1 and y: 'this die shows an even number'; the proposition $x_1 \oplus y$ does not exist. The binary operation $\oplus$ has the following properties:

(a) It is commutative. If $x \oplus y$ is defined, then $y \oplus x$ is defined, and the two are equal:

$$x \oplus y = y \oplus x. \tag{1.16}$$

(b) It is associative. If $y \oplus z$ and $x \oplus (y \oplus z)$ are defined, then so are $x \oplus y$ and $(x \oplus y) \oplus z$, and it is

$$x \oplus (y \oplus z) = (x \oplus y) \oplus z. \tag{1.17}$$

Consequently, the brackets may be dropped altogether, and we may simply write $x \oplus y \oplus z$.

(c) The binary operation has the absurd proposition as its unique neutral element,

$$x \oplus y = x \quad \Leftrightarrow \quad y = \emptyset. \tag{1.18}$$

(d) For every $x \subseteq a$ there exists a unique proposition $\neg_a x$ such that $x \oplus (\neg_a x)$ is defined, and $x \oplus (\neg_a x) = a$. This proposition means '*a but not x*'; the two propositions, x and $\neg_a x$, are said to be 'complementary' refinements of a. Where a is some overarching proposition which is *a priori* given as true, the complement $\neg_a x$ may be regarded as the *negation* of x. Complementation exhibits two basic properties: it reverses the direction of logical implication,

$$x, y \subseteq a, \ x \subseteq y \quad \Rightarrow \quad \neg_a y \subseteq \neg_a x, \tag{1.19}$$

and, when applied twice, it returns the original proposition,

$$\neg_a \neg_a x = x. \tag{1.20}$$

These three features provide only a partial description of classical logic. For a complete specification additional properties are needed, which we will spell out later in this section. Before doing so, however, we shall elaborate briefly on some consequences of this first batch of properties.

We begin with two useful definitions. Given a proposition, y, a set of refinements, $\{x_i | x_i \subseteq y\}$, is called a *partition* of y if these refinements:

- mutually contradict each other, in the sense that whenever one of them is true, all others must be false

- collectively exhaust all possibilities, in the sense that whenever y is true, one of them must be true.

This is the case if and only if it is possible to write

$$y = \bigoplus_i x_i. \tag{1.21}$$

For example, the proposition y: 'this die shows an even number' is refined by the propositions x_1: 'this die shows "4"'; x_2: 'this die shows "6"'; and x_3: 'this die shows one of "2" or "4"'. The two latter refinements, $\{x_2, x_3\}$, constitute a partition of y, $y = x_2 \oplus x_3$. The set $\{x_1, x_2\}$, on the other hand, does not, because it does not exhaust all possibilities; nor do the sets $\{x_1, x_3\}$ (not mutually contradictory, not exhaustive) and $\{x_1, x_2, x_3\}$ (not mutually contradictory). The maximum number of non-absurd refinements into which a proposition can be partitioned,

$$d(y) := \max \# \left\{ x_i \,\middle|\, y = \bigoplus_{i=1}^{k} x_i, \, x_i \neq \emptyset \right\},\tag{1.22}$$

is a measure of its 'broadness'; we shall call it the *dimension* of that proposition. Propositions with dimension one cannot be partitioned; they are *most accurate*. In the die example, we find $d(x_1) = d(x_2) = 1$, $d(x_3) = 2$, and $d(y) = 3$. Indeed, x_1 and x_2 are among the most accurate propositions that one can make about a die, whereas x_3 and y are successively broader. The dimension satisfies a sum rule,

$$d\left(\bigoplus_i x_i \right) = \sum_i d(x_i).\tag{1.23}$$

Indeed, in our example we have $d(y) = d(x_2 \oplus x_3) = d(x_2) + d(x_3)$. You will investigate the sum rule more closely in Exercise (1.1).

The set of all possible refinements of a proposition a, $\mathcal{L}_a := \{x | x \subseteq a\}$, is closed under the binary operation $\oplus$: whenever two elements of the set, $x, y \in \mathcal{L}_a$, are mutually exclusive and hence can be combined with $\oplus$, the result is again contained in the set, $x \oplus y \in \mathcal{L}_a$. Therefore, for arbitrary a, the set $\mathcal{L}_a$ itself exhibits a logical structure, with the exact same properties as listed above. When the dimension of a is small, it may be convenient to visualize this logical structure of $\mathcal{L}_a$ with a graph, a so-called *Hasse diagram*. One such graph, pertaining to the Monty Hall problem, is shown in Fig. 1.15. In the Monty Hall problem, named after a popular American TV show of the 1970s, the contestant in a TV show is told by the host that behind one of three doors is a fancy car, whereas behind the other two there are goats. If the contestant chooses the correct door, she will win the car. The overarching proposition a: 'behind one of the doors is a fancy car' is given as true, so all deliberations will centre around more precise hypotheses as to the car's location. In total there are six distinct, non-absurd refinements of a which, together with a and $\emptyset$, are represented as nodes of a directed graph. These nodes are arranged in horizontal layers, ordered by the dimension of the respective propositions. At the very bottom is the node representing the absurd proposition, $\emptyset$, with dimension zero; immediately above, the nodes representing most accurate propositions, with dimension one; and so on, until, on the uppermost layer, there is the node representing the overarching proposition, a, with dimension $d(a)$. The latter dimension is also referred to as the dimension of the logical structure. Whenever a proposition logically implies another proposition on the next layer, the associated nodes are connected by an arrow, where the direction of the arrow indicates the direction of the logical implication. (Actually, the direction is already fixed by the ordering of the layers. An arrow must always point upwards.) Thanks to the transitivity of logical implication, Eq. (1.12), two propositions on layers which are further apart are related by logical implication if and only if they are connected by a sequence of arrows. For instance, in the Monty Hall problem the propositions r and a do not lie on adjacent layers, so they are not connected directly by an arrow. That the former implies the latter is nevertheless apparent

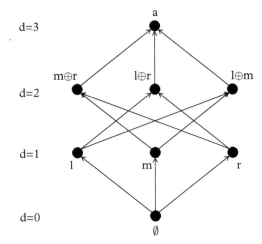

Fig. 1.15 *Hasse diagram for the Monty Hall problem. a denotes the overarching proposition, given as true, that there is a fancy car somewhere, and l, m, r, the most accurate propositions that it hides behind the left, middle, or right door, respectively. The arrows mark logical implications between members of successive layers. Layers are indexed by the dimension, d, of the propositions contained in them.*

from the fact that they can be connected by a sequence of arrows, either via $m \oplus r$ or via $l \oplus r$. Thus the Hasse diagram contains information about *all* logical implications among members of $\mathcal{L}_a$. As for the Monty Hall problem, there is more to it than just the simple logical structure shown in Fig. 1.15, but I will save the rest of the story for later.

The rules of logic laid out so far are not in conflict with the experimental evidence discussed in Section 1.1. What will become problematic, however, is one additional assumption that is made—explicitly or implicitly—in classical logic, with far-reaching consequences. In classical theory, one presumes that the object to which the propositions refer—a coin, a die, an arrangement of cars and goats—is in some possibly unknown, yet definite configuration; and that this configuration uniquely determines the truth values of all propositions. If this is the case, propositions and their logical relationships can be visualized by means of sets, or *Venn diagrams*. Given the truth of some overarching proposition, a, all object configurations compatible with a form a set, called the *sample space*. For instance, in the case of a die roll, given that a: 'there is a die', the sample space comprises the six configurations 'ith face up', with $i = 1, \ldots, 6$. Each proposition about the die then corresponds to some subset of this sample space, namely the subset containing all those configurations which yield truth value 1 ('true') for the proposition in question. Conversely, each subset of the sample space corresponds to a proposition. For instance, the subset $\{$'ith face up'$|i = 2, 4, 6\}$ corresponds to the proposition 'this die shows an even number'. In other words, the correspondence between propositions and subsets of sample space is one to one. A most accurate proposition corresponds to a single-element set, or 'singleton'; it is tantamount to specifying the full configuration. The absurd proposition corresponds to the empty set.

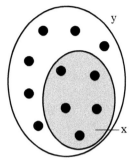 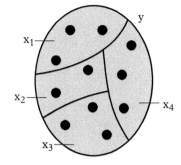

Fig. 1.16 *Logical implication (left) and partition (right) in the classical set picture. Logical implication corresponds to the subset relation. A proposition logically implies another, $x \subseteq y$, if and only if the set associated with the former is contained in the set associated with the latter. A partition of a proposition, $y = \oplus_i x_i$, corresponds to a partition of the associated set. The mutually exclusive refinements, $\{x_i\}$, are represented by pairwise disjoint subsets. Collectively, these subsets cover the entire set associated with y.*

In this set picture, logical implication and partition are represented by set inclusion and set partition, respectively (Fig. 1.16). Two propositions, x and y, are related by logical implication, $x \subseteq y$, if and only if the subset corresponding to the former is included in the subset corresponding to the latter. They are mutually exclusive, or contradict each other, if and only if the corresponding subsets are disjoint. Since two different singletons are always disjoint, this entails, in particular, that two different most accurate propositions always contradict each other. Indeed, two different most accurate propositions about a die—say, 'this die shows "2"' and 'this dies shows "5"'—are obviously mutually exclusive. A collection of mutually exclusive propositions, $\{x_i\}$, constitutes a partition of some broader proposition, y (in the sense of Eq. (1.21)), if and only if the associated disjoint subsets cover the set associated with y. Finally, the dimension of a proposition, Eq. (1.22), equals the number of elements in, or 'cardinality' of, the associated subset. In particular, the dimension of the logical structure as a whole, $d(a)$, coincides with the cardinality of the sample space. The latter, in turn, is identical to the total number of most accurate propositions.

The reason why this classical set picture is problematic in the quantum realm can be seen by considering, for example, the experiment depicted in Fig. 1.10. If indeed every particle were in a definite configuration throughout the experiment and the measurements simply revealed the truth values determined by that configuration, then after the first filter (splitter plus block) the beam should contain only particles whose configurations are compatible with 'up' in the $\theta = 0$ direction; and after the second filter the beam should be further reduced to particles whose configurations are *also* compatible with 'up' in the $\theta = \pi/2$ direction. Thus, the remaining particles that enter the final splitter should be in configurations for which the propositions 'up in the $\theta = 0$ direction' and 'up in the $\theta = \pi/2$ direction' are *both* true. However, as the experiment shows, this is not the case. The final measurement reveals that the proposition 'up in the $\theta = 0$ direction' is true for only half the particles. This discrepancy might have two different explanations. One rather harmless explanation is that, while there is still a definite underlying configuration at any given time, measurements can cause uncontrollable changes of this configuration, thereby invalidating wholly or in part the results of previous measurements. Another, more radical explanation is that the whole notion of a definite underlying configuration is misguided. It is still too early to make a definitive judgement on this matter. Yet, whatever the precise explanation, it has already become clear that in the quantum realm the classical set picture will have to be modified, if not abandoned altogether. In fact, this is the reason why we have so far avoided introducing the binary AND and OR operations. In the set picture, the logical AND and the logical OR correspond to taking the intersection and union of sets, respectively. In the absence of a set picture, however, there is no longer an obvious way to define them. We will dwell on these issues extensively throughout Chapter 2, and will present a more general definition of the binary AND and OR operations only in Section 2.3. Definitive conclusions about the existence or non-existence of an underlying configuration will have to wait until after we have introduced the mathematical apparatus of quantum theory; some formal results relating to this matter will be discussed in Section 3.6.

1.3 Probability

The experimental evidence calls for a critical review of not just the rules of logic but also those of probability theory. I show that the two are in fact intimately related. As in the case of logic, I identify the rules which continue to hold in the quantum realm and those which need to be modified. Moreover, I introduce the pivotal concept of a 'state'.

Logic stipulates how to reason consistently when propositions are either true or false. Probability theory extends this to situations where some or all truth values may be undetermined—for whatever reason. In the words of Bruno de Finetti (de Finetti (1990)):

Probabilistic reasoning [...] merely stems from our being uncertain about something. It makes no difference whether the uncertainty relates to an unforeseeable future, or to an unnoticed past, or to a past doubtfully reported or forgotten; it may even relate to something more or less knowable (by means of a computation, a logical deduction, etc.) but for which we are not willing or able to make the effort.

What is a probability? It is, first of all, a number between zero and one,

$$\text{prob}(x) \in [0,1],$$

assigned to some proposition (or 'hypothesis', or 'event'), x. The widespread assumption is that this number represents the limit of a relative frequency: when a random experiment is repeated infinitely many times, the relative frequency with which a proposition comes out as true converges to some limit, which is then identified with the probability of that proposition. Unfortunately, this definition of probabilities relies on an idealization—infinitely many repetitions—that can never be realized in practice; sometimes a random experiment cannot even be repeated once. Nevertheless, we do routinely assign probabilities to statements like 'It will rain on Monday', 'The economy will go into recession next year', or 'I will pass tomorrow's exam', which clearly pertain to one-time events that cannot be repeated, at least not under perfectly identical conditions. Therefore, for the time being, we shall deliberately refrain from identifying probabilities with relative frequencies. Rather, we shall consider probability merely as a quantitative measure of the 'degree of plausibility', or 'degree of belief', that a test will reveal a given proposition as true. This induces an ordering of propositions: the greater the probability assigned to a proposition, the greater the belief in its experimental confirmation (and the more money you would bet on it if it were a game of chance). The connection with measurable relative frequencies will be established only later, in Section 2.7. By postponing this issue, we will be able to distinguish clearly between those parts of probability theory which are a matter of mere logical consistency and those parts which have to do with the measurability of probabilities in terms of relative frequencies. Even without the explicit link to relative frequencies, we will be able to formulate a number of consistency requirements that any quantitative measure of a 'degree of plausibility' should satisfy. This will lead to a first set of rules which govern the assignment of, and calculations with, probabilities.

Closely related to probabilities is the notion of a *state*. In any given setting, whether a physics experiment, a die roll, or a game such as the Monty Hall problem, we shall mean by 'state' the entirety of our expectations as to the outcomes of measurements. These expectations are quantified by probabilities, so the state is nothing but an exhaustive collection of probabilities, pertaining to all conceivable propositions which one might wish to test. Our expectations, and hence the state, are influenced by the history of past preparation procedures and measurement results, to the extent that these are known. For instance, in the Monty Hall problem we know that the initial preparation (arrangement of cars and goats behind closed doors) is constrained by a certain rule (one car, two goats). Perhaps we also performed one measurement already; for example, we opened the first door, revealing a goat. These two pieces of information shape our expectations as to the outcome of any future measurement; for example, the opening of the second door. Thus, on the one hand, the state summarizes our expectations as to the outcomes of future measurements; on the other hand, it encapsulates our knowledge about past interventions, insofar as they are relevant for inferences about the outcomes of future measurements. The state constitutes the link between past data and future expectations. This dual role of the state is illustrated in Fig. 1.17. For the time being, we shall treat the state as an abstract mathematical object, which we denote by the Greek letter ρ.

Being the representative of our collected expectations, the state, ρ, assigns to each proposition, x, a probability:

$$\rho : x \rightarrow \text{prob}(x|\rho) \in [0,1]. \tag{1.24}$$

Here we made explicit that, strictly speaking, this probability of x is a *conditional* probability, conditioned on the state, ρ. The probability scale is calibrated such that a proposition which is certainly false has probability zero; so, in particular, it is

Fig. 1.17 *The role of the state.*

$$\mathrm{prob}(\emptyset|\rho) = 0. \tag{1.25}$$

By contrast, a proposition which is certainly true has probability one. Propositions whose truth values are undetermined have probabilities somewhere in between. The probability assignment must be consistent with logical implication:

$$x \subseteq y \quad \Rightarrow \quad \mathrm{prob}(x|\rho) \le \mathrm{prob}(y|\rho). \tag{1.26}$$

If a probability assignment satisfies this monotonicity property, then so does every other assignment,

$$\mathrm{prob}'(x|\rho) := \Phi\left[\mathrm{prob}(x|\rho)\right], \tag{1.27}$$

which is related to the former by a strictly monotonic function, $\Phi : [0,1] \to [0,1]$, with $\Phi(0) = 0$ and $\Phi(1) = 1$. All these assignments yield the same qualitative ordering of propositions in terms of belief in their experimental confirmation. In this sense, they are equivalent. They just differ in the choice of quantitative scale.

There are further properties that we demand of a consistent probability assignment. We begin by considering some proposition, a, and its partition into two complementary refinements, $a = x \oplus y$. The probabilities of these three propositions are not independent. Rather, the probabilities of, say, a and x should uniquely determine that of y. For instance, the probabilities of a: 'this die shows an even number' and its refinement x: 'this die shows one of "2" or "4"' should uniquely determine that of the complementary refinement y: 'this die shows "6"'. Consequently, there must exist a function $f : [0,1] \times [0,1] \to [0,1]$ such that, for arbitrary mutually exclusive x and y,

$$\mathrm{prob}(y|\rho) = f\left[\mathrm{prob}(x \oplus y|\rho), \mathrm{prob}(x|\rho)\right].$$

This function must be monotonically increasing in its first argument and monotonically decreasing in its second. Moreover, applying the function once again to $\mathrm{prob}(y|\rho)$ (together with the same $\mathrm{prob}(x \oplus y|\rho)$) must return the probability of the refinement complementary to y—that is, the probability of x. Therefore, the function must satisfy the condition

$$f(\alpha, f(\alpha, \beta)) = \beta.$$

As Cox (1946) and Jaynes (2003) have shown, every probability assignment which satisfies these additional requirements is equivalent, in the sense of Eq. (1.27), to an assignment where it is $f(\alpha,\beta) = \alpha - \beta$ and hence the probabilities obey the sum rule,

$$\text{prob}(x \oplus y | \rho) = \text{prob}(x | \rho) + \text{prob}(y | \rho). \tag{1.28}$$

In other words, it is always possible to set the probability scale in such a way that this sum rule holds. Thus, the sum rule is a matter of logical consistency and convenient choice of scale. Its derivation does *not* require the identification of probabilities with limits of relative frequencies. By induction, we obtain the *law of total probability*,

$$\text{prob}\left(\bigoplus_i x_i \,\middle|\, \rho\right) = \sum_i \text{prob}(x_i | \rho), \tag{1.29}$$

for arbitrary partitions, $\{x_i\}$.

Whenever a measurement yields new data, expectations change and the probabilities must be updated. Let ρ be the state prior to measurement, and let the measurement reveal the truth of some proposition, y. Upon this measurement, the probability of a proposition, x, generally changes from its prior value, $\text{prob}(x | \rho)$, to some updated, posterior value, $\text{prob}(x | y, \rho)$. Only in the special case where the measurement confirms the truth of a proposition which was already known to be true beforehand, nothing changes:

$$\text{prob}(y | \rho) = 1 \quad \Rightarrow \quad \text{prob}(x | y, \rho) = \text{prob}(x | \rho). \tag{1.30}$$

In all other cases, the posterior probabilities will differ from their prior values. The updated state, $\rho_{|y}$, defined via

$$\text{prob}(x | \rho_{|y}) := \text{prob}(x | y, \rho) \; \forall x, \tag{1.31}$$

must still satisfy all the consistency requirements discussed earlier, to wit, the monotonicity property and the law of total probability, Eqs (1.26) and (1.29). Furthermore, the probability that one finds *first* y *and then* x as true, $\text{prob}(x \leftharpoonup y | \rho)$, must be uniquely determined by the probability of y, $\text{prob}(y | \rho)$, and the posterior probability of x given y, $\text{prob}(x | y, \rho)$. In other words, there must exist another function $F : [0,1] \times [0,1] \to [0,1]$ such that

$$\text{prob}(x \leftharpoonup y | \rho) = F\left[\text{prob}(x | y, \rho), \text{prob}(y | \rho)\right] \; \forall x, y.$$

(The direction of the harpoon ($\leftharpoonup$) indicates the order of the measurements. It points from right to left so that x and y appear in the same order on both sides of the equation.) This function must be monotonically increasing in both arguments. Moreover, the function must be such that for longer measurement sequences it will not matter how these are broken down into subsequences. In concrete terms, by applying the function iteratively, it is possible to calculate the probabilities of longer sequences such as the probability of revealing *first z* as true *and then y and then x*. This can be done in two different ways: either by considering the partial sequence $x \leftharpoonup y$ and extending it to the right by z,

$$
\begin{aligned}
\mathrm{prob}(x \leftharpoonup y \leftharpoonup z|\rho) &= \mathrm{prob}((x \leftharpoonup y) \leftharpoonup z|\rho) \\
&= F[\mathrm{prob}(x \leftharpoonup y|z,\rho), \mathrm{prob}(z|\rho)] \\
&= F[F\{\mathrm{prob}(x|y \leftharpoonup z,\rho), \mathrm{prob}(y|z,\rho)\}, \mathrm{prob}(z|\rho)],
\end{aligned}
$$

or by considering the partial sequence $y \leftharpoonup z$ and extending the latter to the left by x,

$$
\begin{aligned}
\mathrm{prob}(x \leftharpoonup y \leftharpoonup z|\rho) &= \mathrm{prob}(x \leftharpoonup (y \leftharpoonup z)|\rho) \\
&= F[\mathrm{prob}(x|y \leftharpoonup z,\rho), \mathrm{prob}(y \leftharpoonup z|\rho)] \\
&= F[\mathrm{prob}(x|y \leftharpoonup z,\rho), F\{\mathrm{prob}(y|z,\rho), \mathrm{prob}(z|\rho)\}].
\end{aligned}
$$

Both approaches must yield the same final result. Consequently, the function F must be associative:

$$
F(F(\alpha,\beta),\gamma) = F(\alpha, F(\beta,\gamma)).
$$

According to Cox and Jaynes, these constraints ensure that it is always possible to set the probability scale in such a way that $F(\alpha,\beta) = \alpha\beta$. Then probabilities satisfy not only the law of total probability but also the *product rule*,

$$
\mathrm{prob}(x \leftharpoonup y|\rho) = \mathrm{prob}(x|y,\rho) \cdot \mathrm{prob}(y|\rho). \tag{1.32}
$$

Once again, the derivation of this rule does not presuppose that probabilities are limits of relative frequencies. Like the law of total probability, the product rule merely reflects logical consistency and a convenient choice of scale.

   The product rule ensures that posterior probabilities are consistent with logical relationships, in the following sense. A proposition x logically implies all broader propositions of the form $x \oplus y$, for arbitrary complementary y. Consequently, when a measurement has revealed the truth of x, an additional test of $x \oplus y$, either before or afterwards, is superfluous; ascertaining the truth of both x and $x \oplus y$, in whatever order, is equivalent to just ascertaining the truth of x. Symbolically, we may write

$$
x \leftharpoonup (x \oplus y) = (x \oplus y) \leftharpoonup x = x. \tag{1.33}
$$

Inserting both variants into the product rule yields, on the one hand,

$$
\mathrm{prob}(x|\rho) = \mathrm{prob}(x|x \oplus y,\rho) \cdot \mathrm{prob}(x \oplus y|\rho) \tag{1.34}
$$

and, on the other,

$$\text{prob}(x \oplus y | x, \rho) = 1. \tag{1.35}$$

The latter equation means that once x has been found to be true, a subsequent test of $x \oplus y$ will certainly reveal that to be true, too. Thus, the posterior probability correctly reflects the logical implication $x \subseteq x \oplus y$. In the special case $y = \emptyset$ we have $\text{prob}(x | x, \rho) = 1$: testing a proposition x once again after its truth has already been ascertained will confirm the previous result with certainty. This reflects the basic requirement that measurements be reproducible. Finally, in combination with the sum rule, Eq. (1.28), the last equation entails

$$\exists x \oplus y \quad \Rightarrow \quad \text{prob}(y | x, \rho) = 0. \tag{1.36}$$

Whenever two propositions can be joined with $\oplus$, they must necessarily contradict each other. Indeed, the posterior probability correctly shows that once one of them has been found to be true, the other has zero probability of being true.

For two states, ρ and σ, we define a *convex combination*, $t\rho + (1-t)\sigma$ with $t \in [0, 1]$, to be that state which yields the probabilities

$$\text{prob}(x | t\rho + (1-t)\sigma) := t\,\text{prob}(x|\rho) + (1-t)\,\text{prob}(x|\sigma) \tag{1.37}$$

for arbitrary x. If the two states individually satisfy the monotonicity property, Eq. (1.26), and the sum rule, Eq. (1.29), then so does this convex combination. Hence it, too, yields consistent probabilities. One says therefore that the states which are logically consistent form a *convex set*. The process of forming a convex combination of two states is also referred to as *mixing*, the resultant state as a *mixture*, and the coefficients, t and $(1-t)$, as the respective weights. Like the constituent states, a mixture must be updated after a measurement. In order for this update to be consistent, it must respect the product rule, Eq. (1.32). This applies to both the individual states and the mixture. Therefore, we require, on the one hand,

$$\text{prob}(x \leftarrow y | t\rho + (1-t)\sigma) = t\,\text{prob}(x \leftarrow y|\rho) + (1-t)\,\text{prob}(x \leftarrow y|\sigma)$$
$$= t\,\text{prob}(x|y, \rho) \cdot \text{prob}(y|\rho) + (1-t)\,\text{prob}(x|y, \sigma) \cdot \text{prob}(y|\sigma)$$

and, on the other hand,

$$\text{prob}(x \leftarrow y | t\rho + (1-t)\sigma) = \text{prob}(x|y, t\rho + (1-t)\sigma) \cdot \text{prob}(y|t\rho + (1-t)\sigma).$$

Equating the right-hand sides yields

$$\text{prob}(x|y, t\rho + (1-t)\sigma)$$
$$= \frac{t\,\text{prob}(y|\rho)}{\text{prob}(y|t\rho + (1-t)\sigma)}\,\text{prob}(x|y, \rho) + \frac{(1-t)\,\text{prob}(y|\sigma)}{\text{prob}(y|t\rho + (1-t)\sigma)}\,\text{prob}(x|y, \sigma).$$

According to Eq. (1.31), we can glean from these posterior probabilities the update of the mixture:

$$[t\rho + (1-t)\sigma]_{|y} = \frac{t\,\mathrm{prob}(y|\rho)}{\mathrm{prob}(y|t\rho + (1-t)\sigma)}\rho_{|y} + \frac{(1-t)\mathrm{prob}(y|\sigma)}{\mathrm{prob}(y|t\rho + (1-t)\sigma)}\sigma_{|y}, \qquad (1.38)$$

with $\rho_{|y}$ and $\sigma_{|y}$ being the respective updates of the two constituent states. In other words, the update of the mixture is a mixture of the updates. However, the new weights generally differ from those of the original mixture.

In order to illustrate how the various rules can be applied in practice, we return to the unfinished story of the Monty Hall problem. It proceeds as follows. After the candidate has made an initial choice of door, the host opens one of the two other doors, revealing a goat. The candidate is then given the opportunity to either stick to her initial choice, or else switch to the other door that is still closed. Should the candidate switch or stay? Suppose the candidate initially picked the left door (L), and the host subsequently opened the middle door (M), revealing a goat and, thus, demonstrating that the car must be behind either the left or the right door ($l \oplus r$). Given this sequence of events, we seek the posterior probability that the car is behind a specific door; say, the left door, $\mathrm{prob}(l|(l\oplus r) \leftharpoonup M \leftharpoonup L, \rho)$. If this probability is strictly larger than one half, the candidate should stick to the left door; otherwise, it is advantageous (or at least neutral) to switch to the right door. With the help of the product rule, Eq. (1.32), we may first write

$$\mathrm{prob}(l|(l \oplus r) \leftharpoonup M \leftharpoonup L, \rho) = \frac{\mathrm{prob}(l \leftharpoonup (l\oplus r) \leftharpoonup M|L,\rho)}{\mathrm{prob}((l\oplus r) \leftharpoonup M|L,\rho)},$$

which, thanks to Eq. (1.33), is the same as

$$\mathrm{prob}(l|(l \oplus r) \leftharpoonup M \leftharpoonup L, \rho) = \frac{\mathrm{prob}(l \leftharpoonup M|L,\rho)}{\mathrm{prob}((l\oplus r) \leftharpoonup M|L,\rho)}.$$

Here the state, ρ, represents the candidate's expectations before the start of the game, which are shaped merely by knowledge about the rules of the game. We can apply the product rule one more time to the denominator on the right-hand side,

$$\mathrm{prob}((l \oplus r) \leftharpoonup M|L,\rho) = \mathrm{prob}(l \oplus r|M \leftharpoonup L, \rho) \cdot \mathrm{prob}(M|L,\rho).$$

Provided the rules compel the host to always open a door that differs from the one indicated by the candidate, the candidate's initial choice, L, leaves only two options for the host, M and R, which are a priori equally likely; hence it is $\mathrm{prob}(M|L,\rho) = 1/2$. As far as the probability in the numerator is concerned, the same indifference stipulates that it should not make a difference whether the middle (M) or the right (R) door was opened. Moreover, it will not hurt to add $\mathrm{prob}(l \leftharpoonup L|L,\rho)$, which, by the rules of the game, is equal to zero. Thus we find

$$\text{prob}(l \gets \mathcal{M}|L,\rho) = \frac{1}{2}\left[\text{prob}(l \gets \mathcal{M}|L,\rho) + \text{prob}(l \gets \mathcal{R}|L,\rho) + \text{prob}(l \gets \mathcal{L}|L,\rho)\right].$$

Since the host always has to open *some* door, the three alternative sequences, $(l \gets \mathcal{M})$, $(l \gets \mathcal{R})$, and $(l \gets \mathcal{L})$, constitute a partition of l. Then the law of total probability applies, yielding

$$\text{prob}(l \gets \mathcal{M}|L,\rho) + \text{prob}(l \gets \mathcal{R}|L,\rho) + \text{prob}(l \gets \mathcal{L}|L,\rho) = \text{prob}(l|L,\rho) = \frac{1}{3};$$

in the last step we assumed that the candidate has no advance knowledge about the location of the car. Altogether, we obtain

$$\text{prob}(l|(l \oplus r) \gets \mathcal{M} \gets L,\rho) = \frac{1}{3\,\text{prob}(l \oplus r|\mathcal{M} \gets L,\rho)}.$$

The numerical value of this posterior depends on the precise rules of the game (presumed to be known to the candidate). If the rules compel the host to always open a door with a goat behind it, then $\mathcal{M}$ logically implies $l \oplus r$, so $\text{prob}(l \oplus r|\mathcal{M} \gets L,\rho) = 1$. If, on the other hand, the host is free to open one of the two remaining doors at random, even if it means revealing the car, or if the host perhaps does not even know the location of the car himself, then the sequence $\mathcal{M} \gets L$ by itself says nothing about the location of the car, and it is $\text{prob}(l \oplus r|\mathcal{M} \gets L,\rho) = 2/3$. In the first case, the posterior probability that the car is behind the left door equals one third, like it was before the host opened the door. This is not surprising. After all, when the candidate initially attributed the probability one third to the left door, she knew full well that in the next step the host would reveal a goat behind one of the other doors; so the fact that he actually did does not convey any new information as regards the left door and therefore does not justify a change of probability assignment. In the second case, on the other hand, the posterior probability increases to one half. Thus, depending on the precise rules of the game, switching doors might improve the odds or not. In either case, it will not hurt.

In the Monty Hall problem we encountered several instances where the assignment of probabilities was guided by a symmetry consideration. In the first instance we assigned $\text{prob}(\mathcal{M}|L,\rho) = 1/2$ because, in a state of complete ignorance, there was no reason to prefer one of the two allowed doors over the other. For the same reason, we set $\text{prob}(l \gets \mathcal{M}|L,\rho) = \text{prob}(l \gets \mathcal{R}|L,\rho)$. Likewise, we assigned $\text{prob}(l|L,\rho) = 1/3$ because, again, based on the available information, all three possible locations of the car were equally likely; there was no justification for preferring one specific location over the others. These instances reflect a more general principle, known as the *principle of indifference*. This principle refers to the assignment of probabilities on some fixed layer of the logical structure; that is, to some set of propositions which have the same dimension, d. In a situation of 'indifference', the available information is such that it cannot justify any discrimination between these propositions. The principle of indifference stipulates that all propositions in the set must then be assigned the same probability. In the last instance mentioned above, a more careful line of reasoning would proceed as follows. The initial choice of door (here: L) is completely arbitrary and does not reveal any information about the car's actual location; had the candidate initially chosen the middle or right door instead, this would not have affected our pertinent expectations. The initial state, ρ, which encodes nothing more than the rules of the game, does not provide any clue either. Consequently, neither the initial choice of door, L, nor the state, ρ, can justify any discrimination between the rival most accurate hypotheses l, m, and r (which all have the same dimension, $d = 1$). Hence, according to the principle of indifference, these three propositions must be assigned the same prior probability, $\mathrm{prob}(x|L,\rho)$, for all $x \in \{l, m, r\}$. Since the three propositions constitute a partition of the overarching proposition, $a = l \oplus m \oplus r$, the sum rule, Eq. (1.29), implies

$$\mathrm{prob}(a|L,\rho) = \mathrm{prob}(l \oplus m \oplus r|L,\rho) = \sum_{x=l,m,r} \mathrm{prob}(x|L,\rho) = 3\,\mathrm{prob}(x|L,\rho)\ \forall x = l,m,r.$$

From the rules of the game we know the overarching proposition, 'behind one of the doors is a fancy car', to be true, $\mathrm{prob}(a|L,\rho) = 1$. So indeed, it must be

$$\mathrm{prob}(x|L,\rho) = \frac{1}{3}\quad \forall x = l,m,r.$$

It is tempting to also apply the principle of indifference to the posterior probabilities, $\mathrm{prob}(x|(l \oplus r) \leftarrow \mathcal{M} \leftarrow L,\rho)$, for $x \in \{l,r\}$. (By then the third possibility, m, has been ruled out.) However, this is not warranted. In contrast to the candidate, the host might know the location of the car. So by opening the middle door, $\mathcal{M}$, rather than the right door, the host might convey a non-trivial piece of information which may well have a bearing on the candidate's expectations. In other words, we must allow for the possibility that this piece of information does discriminate between the two remaining options, l and r. Consequently, we are not permitted to invoke the principle of indifference here.

   All the rules of classical probability theory laid out so far continue to hold in the quantum realm. For instance, we had used the product rule, Eq. (1.32), to calculate the transmission probability for a sequence of rotated filters, Eq. (1.8); this was in perfect agreement with the experimental evidence. As in the case of classical logic, it is just one innocent-looking, and often implicit, extra assumption that will get us into trouble. Here the additional assumption is that the order of measurements never matters: it makes no difference whether I first check the suit to which a playing card belongs and then the face within the suit, or the other way round. If this is the case, then the probability of a sequence is independent of the order of measurements, $\mathrm{prob}(x \leftarrow y|\rho) = \mathrm{prob}(y \leftarrow x|\rho)$. Applying the product rule, Eq. (1.32), to both sides of this equation then yields the so-called *Bayes' rule*,

$$x \leftarrow y = y \leftarrow x \quad \Rightarrow \quad \mathrm{prob}(x|y,\rho) = \frac{\mathrm{prob}(y|x,\rho)\,\mathrm{prob}(x|\rho)}{\mathrm{prob}(y|\rho)}. \tag{1.39}$$

In classical probability theory, this rule is often used to describe how the probability of some 'hypothesis', x, must be updated from the 'prior' $\text{prob}(x|\rho)$ to the 'posterior' $\text{prob}(x|y,\rho)$ in the light of new 'data', y; in this sense, it encapsulates the process of *learning*. The conditional probability on the right-hand side, $\text{prob}(y|x,\rho)$, where the roles of hypothesis and data are interchanged, is commonly called the 'likelihood function'. We shall come back to this idea of learning in the context of measuring a state in Section 2.7.

Alas, the experimental evidence shows that in the quantum domain the order of measurements sometimes *does* matter. Consequently, Bayes' rule is sometimes violated. A case in point is once again the experiment depicted in Fig. 1.10. If Bayes' rule were valid in this setting, it would stipulate that the probability of 'up' in the last splitter satisfy

$$\text{prob}(\Uparrow \mid \Uparrow_{\pi/2}, \Uparrow) = \frac{\text{prob}(\Uparrow_{\pi/2} \mid \Uparrow, \Uparrow)\,\text{prob}(\Uparrow \mid \Uparrow)}{\text{prob}(\Uparrow_{\pi/2} \mid \Uparrow)}.$$

Thanks to the reproducibility of measurements (Fig. 1.6), it is $\text{prob}(\Uparrow \mid \Uparrow) = 1$ and $\text{prob}(\Uparrow_{\pi/2} \mid \Uparrow, \Uparrow) = \text{prob}(\Uparrow_{\pi/2} \mid \Uparrow)$, so the right-hand side is equal to one. The observed probability, however, is merely one half. Thus, in order to describe experiments in the quantum realm, we will need a more general version of probability theory that does not presuppose the order independence of measurements. We shall set out to develop such a theory in Chapter 2.

Chapter Summary

- Binary measurements are measurements with two possible outcomes, like a coin flip. In the quantum realm they are realized, for instance, by a Stern–Gerlach apparatus (for atoms) or a polarizing beam splitter (for photons) in combination with a screen or detectors. These measurements have a continuum of possible orientations.

- When binary measurements are performed in sequence, their order matters. In particular, a later measurement may invalidate the outcome of an earlier measurement. This is not due to a lack of accuracy on the experimenter's part but is rather a characteristic feature of the quantum realm.

- A filter performs a measurement and lets the particle pass only if there was a particular outcome. Counter to intuition, adding more filters may increase the probability of passage. Infinitely many filters can steer a particle from passing one filter to passing a different filter, even if the latter imposes the opposite condition.

- It is possible to produce particle pairs such that identical measurements on the two particles always yield opposite results, whatever the orientation of the device. This universal anticorrelation has no classical explanation.

- Propositions about measurements obey basic rules of logic. However, in the quantum realm the logical structure is weaker than familiar classical logic. In particular, it is no longer possible to identify propositions with subsets of a sample space. Instead, the logical structure may be visualized as a Hasse diagram.

- The dimension of the logical structure equals the maximum number of mutually contradictory propositions.

- In general, the outcomes of measurements are not certain, so one must resort to probabilities.

- Probabilities are numbers between zero and one, assigned to propositions. To be consistent, they must satisfy the law of total probability and the product rule. In the product rule, the order of measurements matters.

- In the quantum realm probabilities need not obey Bayes' rule. The latter presupposes that the order of measurements is irrelevant, which is no longer guaranteed.

- The complete catalogue of probabilities constitutes the state. It encapsulates knowledge about past preparation procedures and measurement results and, on this basis, shapes expectations as to the outcomes of future measurements. States that assign logically consistent probabilities form a convex set.

- Whenever a measurement yields new data, expectations change and the state must be updated. The update of a mixture of states is a mixture of the updates.

Further Reading

Most traditional textbooks motivate the need for a non-classical theory not with binary measurements but with experimental evidence that highlights the wave nature of matter, such as the famous double-slit experiment. A notable early exception is the book by Sakurai (1985). Lately, starting with binary measurements has become more widespread. You will find discussions of binary experiments, for instance, in the fine books by Schumacher and Westmoreland (2010), Scarani *et al.* (2010), and Susskind and Friedman (2014). Generalizations of classical logic which allow for propositions that cannot be tested jointly have been developed in various contexts. The approach that I have taken in this chapter resembles most closely the idea of 'operational statistics' introduced by Foulis and Randall (1972), Randall and Foulis (1973), and Foulis and Randall (1981). The mathematical structure underlying this generalized logic is that of an 'orthomodular partially ordered set' or 'orthoalgebra'. If you want to delve deeper into these structures, you may consult Foulis *et al.* (1992), Foulis *et al.* (1993), or Wilce (2000). There is a related kind of generalized logic, dubbed 'quantum logic', which we will briefly encounter in Chapter 2, and for which I will provide additional reading suggestions there. By introducing probabilities without an immediate link to relative frequencies, I have followed the modern paradigm of 'Bayesian probability'. As a concise introduction to Bayesian probability theory, I recommend the book by Sivia (1996). The desiderata of logical consistency underlying the basic rules of probability theory go back to an article by Cox (1946) and are explained compellingly in the classic book by Jaynes (2003).

EXERCISES

1.1. Sum rule for the dimension

    a. Consider the sum rule for the dimension, Eq. (1.23). Does it follow from the other properties of the logical structure, or does it constitute a separate assumption? In the former case, provide a proof. In the latter case, give an example of a logical structure that is compatible with all other properties but violates the sum rule.

    b. In the classical set picture the dimension of a proposition coincides with the cardinality of the associated subset of sample space. Show that in this classical setting the dimension satisfies a more general sum rule,

$$d(x \vee y) = d(x) + d(y) - d(x \wedge y),$$

where $x \vee y$ is the proposition associated with the union, and $x \wedge y$ is the proposition associated with the intersection, of the subsets corresponding to x and y. Show that this rule includes Eq. (1.23) as a special case.

1.2. Hasse diagrams

    a. Draw the Hasse diagram representing the logical structure of a coin flip. How many nodes and how many layers does it have?

    b. How many nodes and how many layers does the Hasse diagram for a die roll have?

    c. Consider the Hasse diagrams depicted in Fig. 1.18. Do they have the properties of a logical structure, as formulated in Section 1.2? If not, which requirements are violated? Imagine experiments or games to which these Hasse diagrams might correspond.

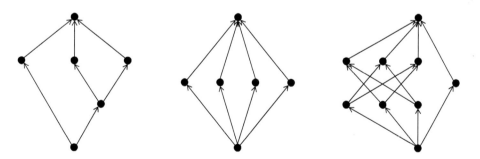

Fig. 1.18 *Examples of Hasse diagrams. The nodes at top and bottom represent the overarching and absurd propositions, respectively.*

1.3. Medical test

A medical test is used to screen for a disease that occurs with a frequency of about 1 per cent of the population. The test is 95 per cent accurate, meaning that if a person has the disease, the test result is positive with 95 per cent probability, and if the person is healthy, the test result is negative with 95 per cent probability. Given a positive test result, what is the probability that the person actually has the disease?

1.4. Monty Hall problem

Let $x_1,\ldots,x_4$ denote four unspecified propositions pertaining to the actual location of the car in the Monty Hall problem. For these propositions, two complete sets of conditional probabilities are given in Tables 1.1 and 1.2, corresponding to two different states ρ and σ. Based on the probability tables alone, assign the four propositions to nodes in Fig. 1.15. Calculate the probabilities and conditional probabilities of the remaining nodes for both choices of the state. What do these probabilities reveal about the setup of the game?

Table 1.1 *Conditional probabilities of the Monty Hall problem in the state ρ. The number in the ith row and jth column is the conditional probability $prob(x_i|x_j,\rho)$.*

Proposition	x_1	x_2	x_3	x_4
x_1	1	1	1	1
x_2	1/2	1	1	0
x_3	1/2	1	1	0
x_4	1/2	0	0	1

Table 1.2 *Conditional probabilities of the Monty Hall problem in the state σ. The number in the ith row and jth column is the conditional probability $prob(x_i|x_j,\sigma)$.*

Proposition	x_1	x_2	x_3	x_4
x_1	1	1	1	1
x_2	1/3	1	2/3	0
x_3	1/2	1	1	1/4
x_4	2/3	0	1/3	1

2

Reasoning About Measurements

2.1 Testable Propositions

Especially when dealing with counterintuitive phenomena, it is imperative to define concepts and to use language in a very careful way. I explain—and adopt—the approach known as 'operationalism', according to which a theory may admit only concepts that are associated with a concrete experimental procedure. I investigate what this means, in particular, for propositions about measurements and their logical structure. I illustrate the approach with a number of toy examples where the ability to perform measurements is limited by design, and consider their relation to phenomena observed in the quantum realm.

In the classical theory, the act of measurement is usually not given much thought. That all propositions can be tested in some way, at least in principle, is taken for granted; and so is the possibility of joining arbitrary propositions together via logical AND or OR operations. It is tacitly assumed that these combinations are again valid propositions amenable to testing, and that the order in which the requisite measurements are performed does not matter. However, the experimental evidence cautions us that in the quantum realm we have to reason much more carefully. Not every proposition which might perhaps be formulated in the abstract ('the particle is "up" in both the $\theta = 0$ and $\theta = \pi/2$ directions') is automatically meaningful, let alone susceptible to testing. We should make it clear from the beginning that such abstract, yet never verifiable, propositions—regardless of whether they might make hypothetical sense—shall be outside the scope of our theory; we shall be dealing exclusively with propositions that may be tested experimentally. Every admissible proposition must pertain to the outcome of some specific reproducible measurement that we can actually perform. (For our purposes, it shall suffice that the measurement could be done if only there were enough resources, even if in a particular situation those resources are not readily available.) This entails, in particular, that we must start wording propositions more carefully than we have done so far, spelling out explicitly the experimental setup with which they are tested. For instance, we'd better say: 'a Stern–Gerlach apparatus oriented at $\theta = 0$ deflects this atom upwards' rather than just 'the atom is "up"'. For ease of reading, and as long as the context is clear, we will sometimes still use the shorter version; but in the back of our minds there should always be an experimental setup.

Quantum Theory: An Information Processing Approach. Jochen Rau, Oxford University Press (2021). © Jochen Rau.
DOI: 10.1093/oso/9780192896308.003.0002

More formally, we are adopting an approach known as *operationalism*, an approach that has played an important role in the conceptual understanding of both relativity and quantum theory. It goes back to work by Percy Williams Bridgman, who at the time was deeply impressed with Einstein's—still young—theory of relativity. Prior to relativity, it had been considered self-evident that space and time were absolute. Einstein managed to overthrow this established view and revolutionize our understanding of space and time, chiefly because he investigated very carefully the respective experimental procedures by which length and time are measured. Bridgman (1927) posited that the same rigorous analysis must be applied to *every* concept employed in a physical theory. A concept has a place in a physical theory if and only if it can be given a purely operational definition, to wit, a concrete description of the experimental operations by which it is measured:

> [...] what do we mean by the length of an object? We evidently know what we mean by length if we can tell what the length of any and every object is, and for the physicist nothing more is required. To find the length of an object, we have to perform certain physical operations. The concept of length is therefore fixed when the operations by which length is measured are fixed: that is, the concept of length involves *as much as and nothing more than the set of operations* by which length is determined. In general, we mean by any concept nothing more than a set of operations; the concept is synonymous with the corresponding set of operations.

Accordingly, in our approach the concept of 'proposition' is inextricably linked to the set of operations by which a proposition can be tested.

We illustrate this general idea and its possible ramifications with a number of toy examples. In these examples our ability to measure is limited by design. We begin with the following—admittedly artificial—generalization of the Monty Hall problem. Rather than three doors, there is now a 3×3 array of doors. The host explains that hidden behind the doors are three fancy cars and six goats, and that the cars are arranged such that there is exactly one car in each row and column. This background information constitutes the overarching proposition, a. The candidate tries to localize the cars. She is allowed to proceed in rounds. In each round, up to three doors may be opened simultaneously. Where several doors are opened, these must be either in the same row or in the same column. If the candidate makes full use of her allowance, she will be able to localize in a single round one of the three cars: the one in the chosen row or column. At the end of the round the doors are closed again. In the following round the candidate is free to choose a different row or column, in which, again, she may open up to three doors. And so on. However, there is a catch: the opening of one or several doors may trigger a random reshuffle of the cars and goats behind all the other doors which remain closed. (The reshuffle respects the constraints imposed by the overarching proposition, a.) Whenever the candidate localizes a car by opening up to three doors in a given round, the ensuing reshuffle behind the other doors will be so thorough that any insights gleaned during previous rounds are completely obliterated; it is as if these previous rounds had never taken place. Thus, localizing a new car comes at the cost of losing sight of any previously localized car. In effect, the rules of the game preclude that at any given moment the candidate learns the locations of more than one car.

In view of these rules, opening all three doors in a row or column, which can be done in a single round, is already the most accurate measurement that the candidate can perform. There is no way to improve on this measurement by opening further doors in subsequent rounds. Such a most accurate measurement reveals the location of exactly one car, while the locations of the two other cars remain undetermined. Let e_{ij} denote the proposition that there is a car at position ij, where $i, j = 1, \ldots, 3$ label the rows and columns of the array. The $\{e_{ij}\}$ are not most accurate in the sense that they specify the complete configuration of all three cars, but they are most accurate in the operational sense, in that there are no other non-absurd propositions that would imply them and whose truth values could be ascertained by a measurement which is practically feasible. Those propositions whose truth values are amenable to an allowed measurement, as well as their logical interrelations, are represented by the Hasse diagram in Fig. 2.1. The logical structure has dimension $d(a) = 3$, since all possible most accurate partitions of the overarching proposition consist of three propositions:

$$a = \bigoplus_{i=1}^{3} e_{ki} = \bigoplus_{i=1}^{3} e_{ik} \quad \forall k = 1, \ldots, 3.$$

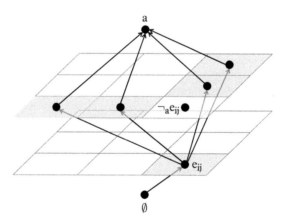

Fig. 2.1 *Hasse diagram for the 3×3 problem. The nodes at top and bottom represent the overarching and absurd propositions, respectively. In between, there are two layers, each comprising nine nodes arranged on a 3×3 chequerboard. For simplicity, not all nodes are drawn. The nodes on the lower layer represent the propositions $\{e_{ij}\}$ that there is a car at position ij. The nodes on the upper layer represent their respective complements, 'a but not e_{ij}', denoted by $\neg_a e_{ij}$. Each node on the lower layer is complementary to its mirror image on the upper layer. It logically implies all other nodes on the upper layer that occupy the same row or column.*

In this 3×3 problem, measurements are restricted. While each proposition in Fig. 2.1 may be tested in isolation, it is not allowed to test two propositions jointly if they pertain to different rows and different columns. For instance, the proposition e_{11} alone can be tested by opening the doors of the first row or column, and the proposition e_{22} alone can be tested by opening the doors of the second row or column; but because in total at most one row or column of doors may be opened simultaneously, the two propositions cannot be tested jointly. (Performing the two measurements in sequence would not help because, by the rules of the game, the second measurement would obliterate the result of the first.) Operationally, this also makes it impossible to combine e_{11} and e_{22} into a joint proposition via a logical AND operation; the proposition 'e_{11} and e_{22}' is *not* defined. While it might be formulated in the abstract, this proposition can never be tested, and hence is not part of our theory. Here we have imposed these restrictions artificially, via the rules of the game. Later we will discover that in quantum theory such knowledge constraints are, in fact, a fundamental aspect of Nature. In quantum theory, too, we shall encounter pairs of propositions which cannot be tested jointly.

The imposition of knowledge constraints necessitates modifications to the classical set picture laid out in Section 1.2. There we argued that, without any knowledge constraints, there is a one-to-one correspondence between propositions and subsets of sample space; in particular, between most accurate propositions and subsets with a single element. The sample space is the set of underlying configurations that uniquely determine the truth values of all propositions. In the 3×3 problem the rules of the game permit six distinct underlying configurations of cars (indicated by black dots) and goats (open circles):

$$
\begin{array}{ccc|ccc|ccc|ccc|ccc|ccc}
\bullet & \circ & \circ & \bullet & \circ & \circ & \circ & \bullet & \circ & \circ & \bullet & \circ & \circ & \circ & \bullet & \circ & \circ & \bullet \\
\circ & \bullet & \circ & \circ & \circ & \bullet & \bullet & \circ & \circ & \circ & \circ & \bullet & \bullet & \circ & \circ & \circ & \bullet & \circ \\
\circ & \circ & \bullet & \circ & \bullet & \circ & \circ & \circ & \bullet & \bullet & \circ & \circ & \circ & \bullet & \circ & \bullet & \circ & \circ
\end{array}
\tag{2.1}
$$

It is still the case that testable propositions are represented by subsets of this sample space; for example, the testable proposition 'there is a goat at position 13 (first row, third column)' by the subset comprising the first four configurations. However, due to the knowledge constraints, the correspondence between testable propositions and subsets of sample space is no longer one to one. For instance, among the testable propositions shown in Fig. 2.1, there is none that would correspond to a subset containing an odd number of configurations. In particular, since the knowledge constraints prevent us from ever uncovering the full configuration, there is no testable proposition that would correspond to a subset with only a single element. Those testable propositions which are most accurate in the operational sense, $\{e_{ij}\}$, merely specify the position of one of the cars. They are represented by subsets that contain two elements rather than just one; for example, the most accurate proposition e_{11} is represented by the subset comprising the first two configurations. In the absence of a correspondence with single-element sets, it is also no longer the case that two different most accurate propositions necessarily contradict each other. For instance, e_{11} and e_{22} are different but not mutually exclusive; the associated subsets, which each contain two configurations, partly overlap. In fact, this happens with all pairs of most accurate propositions that, like e_{11} and e_{22}, cannot be tested jointly. Finally, the total number of most accurate propositions need no longer coincide with the cardinality of the sample space, nor with the dimension of the logical structure. In the 3×3 problem there are nine different most accurate propositions, more than the cardinality of the sample space (six) and the dimension of the logical structure (three). In sum, we see that in the presence of knowledge constraints the links between testable propositions and subsets of sample space, and between logical and set relations, are loosened in multiple ways. In the present example the set picture survives insofar as testable propositions may still be thought of as selected subsets of some set of hypothetical configurations. Yet we shall soon discover that in other examples, not least in quantum theory itself, even this weaker set picture becomes highly problematic.

We move on to a second example, developed by Spekkens (2007) and known as the 'Spekkens toy model'. It is simpler than the 3×3 problem, and at the same time designed in such a way that it bears somewhat more resemblance to an actual quantum system. For this reason, it will accompany us not only in this section but also later in this chapter. Following in the footsteps of the previous example, the Spekkens toy model describes a game of chance with artificial knowledge constraints (Fig. 2.2). There are four boxes, one of which contains a present, a precious gem; this background information constitutes the overarching proposition, a. The information that can be accessed experimentally is limited in the following way. The only type of measurement which is allowed is to select two out of the four boxes and have a machine empty these two into a container, while hiding from view the contents of the individual boxes. Inspection of the container then yields one of two possible outcomes: either the container is empty, indicating that the two selected boxes were both empty; or the container contains the present, indicating that the present must have been in one of the two selected boxes. However, in the latter case it is impossible to trace back which of the two selected boxes it came from. There are $\binom{4}{2} = 6$ ways of choosing two out of four boxes, yet some of these choices are equivalent in terms of the insight they provide. For instance, emptying boxes 1 and 2 will provide the same insight about the prior location of the present as emptying boxes 3 and 4. Thus, in total, there are only three inequivalent measurements of this kind, each with two possible outcomes (Fig. 2.3). Every outcome, which in the figure corresponds to one end of an axis, constitutes a testable proposition. This proposition is most accurate—not in the

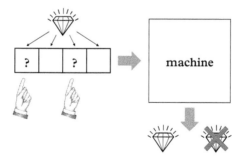

Fig. 2.2 *The game of chance behind the Spekkens toy model.*

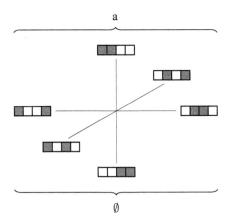

Fig. 2.3 *Logical structure of the Spekkens toy model. Between the overarching and the absurd propositions there is just a single layer. It contains six propositions, symbolized by different permutations of white and grey boxes and arranged at the ends of three distinct axes. The three axes correspond to the three inequivalent allowed measurements, and the two ends of an axis, to the respective possible outcomes. Every measurement allows one to determine whether a chosen pair of boxes was empty (white) or contained the present (grey).*

sense that it specifies the exact location of the present, but in the operational sense that there is no practical way to refine it any further. The two propositions at opposite ends of the same axis may be tested jointly in the same experiment. They are mutually exclusive, and together they constitute a partition of a. Thus, the logical structure has dimension $d(a) = 2$. Propositions pertaining to different axes, on the other hand, cannot be tested jointly.

In the Spekkens toy model, too, we may allow multiple rounds of the game. Two different measurements that cannot be performed jointly may then be performed sequentially. For instance, one might first have boxes 3 and 4 emptied *and then*, in a subsequent experiment, the two outer boxes, 1 and 4. Alas, like in the 3×3 problem, there is a catch. Unless the second measurement is just a repetition of the first (merely double-checking the contents of the container), it requires putting the contents of the container (if any) back into one of the boxes, closing all boxes, and resubmitting them to the machine. The rules of the game are such that this process inevitably entails a random relocation of the gem, rendering the result of the previous measurement totally useless. So once again, sequential measurements cannot be used to circumvent the prohibition of joint measurements. Even after two measurements, one still does not know more than that the gem must have been in one of two boxes, each with equal probability. An example of such sequential measurements is illustrated in Fig. 2.4.

The Spekkens toy model mimics several qualitative aspects of the experimental evidence discussed in Section 1.1. To begin with, like the Stern–Gerlach or photon

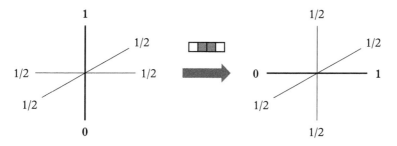

Fig. 2.4 *Sequential measurements in the Spekkens toy model. A first measurement has found boxes 3 and 4 empty. Accordingly, the proposition at the top of the vertical axis is true with certainty (probability one), and its complement at the bottom is false (probability zero). As to the potential outcomes of other measurements, nothing is known; so, by the principle of indifference, they are assigned probability one half. In a subsequent measurement the two outer boxes are found empty. This leads to an update of the probabilities, symbolized by the arrow. It is now the proposition on the right of the horizontal axis which is true with certainty, and its complement on the left which is false. By the rules of the game, the second measurement has completely obliterated the result of the first. Therefore, with respect to the vertical axis, one is back to a state of total ignorance; it is as if the first measurement had never taken place.*

polarization measurements, all measurements are binary. Secondly, two different measurements cannot be performed jointly, but only sequentially. And thirdly, each measurement completely 'overwrites' the outcomes of its predecessors; it is as if previous measurements had never happened. To a certain extent, the Spekkens toy model can even reproduce some of the quantitative observations. If we associate the vertical axis with a beam splitter oriented at $\theta = 0$ and the horizontal axis with a splitter oriented at $\theta = \pi/2$, then the probabilities shown on the left-hand side of Fig. 2.4, after the initial measurement along the vertical axis, are consistent with Eq. (1.6); and the probabilities on the right-hand side, after the subsequent measurement along the horizontal axis, are consistent with the 50/50 splitting seen in Fig. 1.10. The analogy could be taken still further to also encompass combined measurements of the kind shown in Fig. 1.14. To this end, we would have to extend the rules of the Spekkens toy model to two parallel games, so to speak, in a manner that resembles as closely as possible the phenomena seen in the quantum domain. However, at this point this would lead us too far astray. Suffice it to say that there are some striking parallels between actual quantum systems and toy models with artificially imposed knowledge constraints.

One of the experimental observations which the Spekkens toy model *fails* to reproduce is the lossless steering indicated in Fig. 1.13 and described mathematically by Eq. (1.11). Such lossless steering depends on there being a *continuum* of different measurements, parametrized by the angle θ—not just three different measurements, as in the Spekkens toy model. We can try to accommodate lossless steering in a toy model by simply replacing Spekkens's three measurement axes with a continuum. This is done in the 'circle' and 'sphere' models. In both models the logical structure has dimension $d = 2$, like in the Spekkens toy model; hence, casting the overarching and absurd propositions aside,

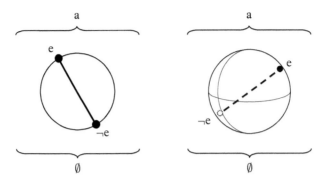

Fig. 2.5 *Logical structure of the circle and sphere models. The logical structure comprises three layers: at the bottom, the absurd proposition, ∅; at the top, the overarching proposition, a; and in between, a continuum of most accurate propositions with the topology of a circle (left) or the surface of a sphere (right), respectively. A most accurate proposition, e, and its complement, ¬ₐe, constitute antipodes. An allowed measurement corresponds to a diameter of the circle or sphere, respectively. It has two possible outcomes, indicated by black dots (or a dotted circle, if hidden), which correspond to the two intersections of the diameter with the circle or sphere, respectively. Two most accurate propositions which are neither identical nor complementary and hence do not lie on the same diameter, cannot be tested jointly.*

the differences lie solely in the intermediate, most accurate level of the logical structure. While in the Spekkens toy model this level comprised six discrete most accurate propositions (Fig. 2.3), there is now a continuum of most accurate propositions with the topology of a circle or a sphere, respectively (Fig. 2.5). The allowed measurements, too, now form a continuum rather than a discrete set of three. Still, several key features of the Spekkens toy model are maintained. Like in the Spekkens toy model, all measurements are binary. Each measurement corresponds to a line—a diameter—whose ends represent its two possible outcomes; and, like in the Spekkens toy model, measurements which correspond to different lines cannot be performed jointly. Sequential measurements are possible and entail consecutive probability updates. Again, the pertinent rules closely follow those of the Spekkens toy model (Fig. 2.6). It will turn out that one of these two continuous models—the sphere model—is already very close to the description of an actual two-dimensional quantum system, a 'qubit'. We will investigate this correspondence in detail in Section 3.3.

While some weaker version of the classical set picture could still be maintained for the 3×3 problem (and also for the Spekkens toy model, although we did not discuss this explicitly), this possibility becomes ever more remote in the continuous circle and sphere models. A sample space for the latter would have to comprise a continuum of underlying configurations. A most accurate proposition (in the operational sense) would be associated with a subset of this continuum; this subset, too, would contain a continuum of configurations. In fact, since the dimension of the logical structure equals two, a subset

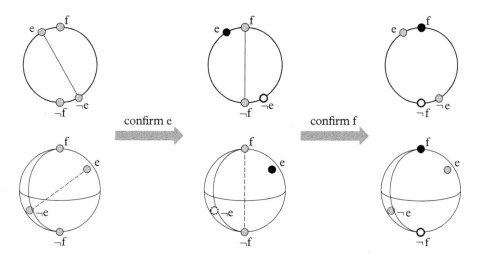

Fig. 2.6 *State updates in the circle (top) and sphere (bottom) models. In a given state, propositions are assigned probabilities which can be one (black dots), zero (open circles), or in between (grey dots). (Different grey dots need not represent the same numerical value of the probability.) Prior to any measurement, in a state of total ignorance, all propositions have a probability which is neither zero nor one (left). Once a measurement (diameter) reveals the truth of a most accurate proposition, e, these probabilities are updated. The updated state assigns to e, whose truth was just confirmed, probability one; to its complement, ¬_a_e, probability zero; and to all other most accurate propositions—which cannot be tested jointly with the one just confirmed—probabilities which are neither zero nor one (centre). Then a subsequent measurement along a different axis yields the truth of another most accurate proposition, f. This triggers yet another update to a posterior state in which f is true (right). The confirmation of f supersedes all prior history; it is as if the prior confirmation of e had never taken place. Hence, in the posterior state the proposition which was true previously, e, is no longer true with certainty but is assigned a probability which is neither zero nor one.*

associated with a most accurate proposition would contain half the sample space. So we would have to imagine the sample space as some continuous manifold which can be cut in two halves in infinitely many ways. Each cut would correspond to a partition into two complementary most accurate propositions. To give a concrete example, the sample space might be a circle. This circle can be cut along any diameter, yielding two opposite semicircles, which would represent complementary most accurate propositions. This continuous version of the set picture would have to allow a consistent assignment of probabilities. In the case of the circle, the posterior probability of a most accurate proposition, given the truth of another, would have to be some function of the arc length of the overlap of the respective semicircles. It may well be that such a function can still be found, consistent with all the probability rules and with the experimental evidence— but it should be evident that we are now reaching the limits of the set picture as a useful visualization of propositions and their logical relationships. It is time to leave it behind.

2.2 Reachable States

Operationalism also has ramifications for the set of states. I introduce the distinction between states which would be logically consistent but can never be reached experimentally, and those that can actually be reached. The latter form a convex set or cone, respectively. I determine the shapes of these sets or cones for the various toy models introduced in Section 2.1.

Not only the concept of 'proposition' but also the concept of 'state' must be operational. It is not sufficient that a state merely satisfies all the consistency requirements laid out in Section 1.3. In order for a state to be admissible, there must be a concrete description of both how it can be *reached*, or *prepared*, in practice and how it can be *measured*. In this section we shall begin discussing the first aspect: which states can actually be reached in an experiment. This discussion will only be preliminary. For a full answer, we will need an understanding of composite systems and exchangeable assemblies, to be provided in Sections 2.5 and 2.6. We will return to the issue of state preparation and complete our discussion in Section 2.8. The other aspect of the operational definition of the state—its measurability—will be investigated in Section 2.7.

Among the admissible states there is certainly the *state of total ignorance*, often called the *totally mixed state*. It is the starting point for all subsequent learning about a game of chance or a physical system. All that is known in a state of total ignorance is the truth of some overarching proposition, a, which asserts nothing more than the existence of the object that we are ignorant about—say, 'there is a car behind one of the doors' or 'there is a photon'—plus perhaps some background information about the rules of the game or the laws of physics. This overarching proposition is assigned probability one. As to its possible refinements—in a Hasse diagram, the propositions on the layers below a—we are totally ignorant; on any given layer, we have no information that would justify preferring one proposition over the other. Thus, by the principle of indifference, all propositions on the same layer must be assigned the same probability. This probability may depend on the layer, to wit, on the dimension of the propositions, but nothing else. In other words, there must exist a function $f : \mathbb{N} \to [0,1]$ such that

$$\text{prob}(x|\iota) = f[d(x)] \quad \forall x \subseteq a.$$

Here we have denoted the state of total ignorance by the Greek letter 'iota', ι. Both the dimension and the probability satisfy respective sum rules, Eqs (1.23) and (1.29). These imply, on the one hand,

$$f\left[d\left(\bigoplus_i x_i\right)\right] = f\left[\sum_i d(x_i)\right],$$

and, on the other hand,

$$f\left[d\left(\bigoplus_i x_i\right)\right] = \text{prob}\left(\bigoplus_i x_i \middle| \iota\right) = \sum_i \text{prob}(x_i|\iota) = \sum_i f[d(x_i)].$$

So, in effect, it is

$$f\left[\sum_i d(x_i)\right] = \sum_i f[d(x_i)],$$

which means that f must be linear: $f[d(x)] \propto d(x)$. Taking the normalization of the state into account, $\text{prob}(a|\iota) = 1$, we thus find

$$\text{prob}(x|\iota) = \frac{d(x)}{d(a)} \quad \forall x \subseteq a. \tag{2.2}$$

Starting from the state of total ignorance, we acquire knowledge (about the location of a car, or about a physical system) through a sequence of measurements, thereby ascertaining the truths of various propositions; say, first of x_1, then of x_2, and so on. We assume that in total some finite number, n, of propositions is confirmed, and denote the pertinent sequence, 'first x_1 is found to be true, *and then* x_2 is found to be true, *and then* ...' (in this order), by

$$s := x_n \leftarrow \dots \leftarrow x_2 \leftarrow x_1. \tag{2.3}$$

Whenever we have knowledge constraints, the order within this sequence generally matters. After such a sequence of ascertainments, we update the prior state of total ignorance, ι, to the posterior,

$$\iota_{|s} := (\dots((\iota_{|x_1})_{|x_2})\dots)_{|x_n}. \tag{2.4}$$

However, in practice, the sequence of ascertainments is not always known with certainty; there are various sources of error. For instance, an experiment might take place inside a black box, with limited control of the observer. The black box might be set up in such a way that it performs one of several possible measurement sequences, each with some non-zero probability. Each possible measurement sequence, in turn, may yield a variety of different outcomes, again with an associated probability distribution. Afterwards, the black box delivers an output to the observer. The output is a function of the measurements performed and the results obtained. Yet, once again, this function might be probabilistic: a given sequence of measurements and associated results need not produce a unique output but may produce one of several possible outputs, governed by yet another probability distribution. The net effect is that the output of the black box no longer specifies a unique sequence of propositions that was ascertained during the experiment. Rather, it allows for a range of possibilities, $\{s'\}$, with associated probabilities, $\{\lambda(s')\}$. Consequently, the observer must update the prior state not to a unique $\iota_{|s}$ but to a mixture,

$$\iota \to \rho = \sum_{s'} \lambda(s') \iota_{|s'}. \tag{2.5}$$

This general form of the update includes the simple update $\iota \to \iota_{|s}$ as a special case; it is then $\lambda(s) = 1$ and $\lambda(s') = 0$ for all $s' \neq s$.

To make things more concrete, we consider once again the 3×3 problem. Here the 'black box' is an assistant who volunteers to perform the following—admittedly peculiar—experiment on the candidate's behalf. Hidden from the candidate's view, he opens *either* the door at position 11 *or* the door at position 21, with respective probability one half. In the event that he opens the door at 11, he will raise a green flag if he finds a car and a red flag if he finds a goat; whereas in the event that he opens the door at 21, the flag colour will have the opposite meaning: he will raise a green flag if he finds a goat and a red flag if he finds a car. Initially, the candidate is still completely ignorant, so her expectations are described by the state of total ignorance, ι. By the principle of indifference, this state assigns to all most accurate propositions the same probability,

$$\{\mathrm{prob}(e_{ij}|\iota)\} = \begin{pmatrix} \frac{1}{3} & \frac{1}{3} & \frac{1}{3} \\ \frac{1}{3} & \frac{1}{3} & \frac{1}{3} \\ \frac{1}{3} & \frac{1}{3} & \frac{1}{3} \end{pmatrix}.$$

The candidate then accepts the assistant's offer, and eventually sees him raise a red flag. How will that change her expectations? How should she update the state? The red flag can mean either a goat at 11 or a car at 21. The former is twice as likely as the latter, simply because finding a goat is twice as likely as finding a car, regardless of the door chosen. Thus, with probability two thirds the red flag amounts to a confirmation of the proposition $\neg_a e_{11}$, whereas with probability one third it amounts to a confirmation of e_{21}. As a result, the candidate should update the state to a mixture of the respective posteriors,

$$\iota \to \rho = \frac{2}{3} \iota_{|\neg_a e_{11}} + \frac{1}{3} \iota_{|e_{21}}. \tag{2.6}$$

By the principle of indifference, the two posteriors assign the following probabilities on the most accurate level:

$$\{\mathrm{prob}(e_{ij}|\neg_a e_{11}, \iota)\} = \begin{pmatrix} 0 & \frac{1}{2} & \frac{1}{2} \\ \frac{1}{2} & \frac{1}{4} & \frac{1}{4} \\ \frac{1}{2} & \frac{1}{4} & \frac{1}{4} \end{pmatrix}, \quad \{\mathrm{prob}(e_{ij}|e_{21}, \iota)\} = \begin{pmatrix} 0 & \frac{1}{2} & \frac{1}{2} \\ 1 & 0 & 0 \\ 0 & \frac{1}{2} & \frac{1}{2} \end{pmatrix}.$$

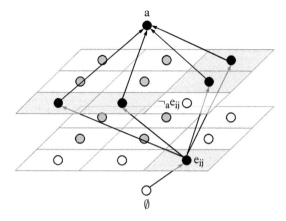

Fig. 2.7 *Probability assignments to propositions about the 3×3 problem, after the truth of some most accurate proposition, e_{ij}, has been ascertained. Black dots indicate probability one, open circles indicate probability zero. By the principle of indifference, all other propositions, marked with grey dots, are assigned probability one half.*

A posterior of the latter type is shown pictorially in Fig. 2.7. Mixing them, with respective weights two thirds and one third, then yields

$$\{\text{prob}(e_{ij}|\rho)\} = \begin{pmatrix} 0 & \frac{1}{2} & \frac{1}{2} \\ \frac{2}{3} & \frac{1}{6} & \frac{1}{6} \\ \frac{1}{3} & \frac{1}{3} & \frac{1}{3} \end{pmatrix}.$$

These probabilities on the most accurate level completely specify the updated state, ρ. Compared to a real physics experiment, the example presented here is not as crazy as it sounds. For example, many experiments employ photon or other particle detectors, which are black boxes, too. They are supposed to 'click' every time a particle hits, but they are far from perfect. Sometimes they fail to click even though a particle hits, or they click even though there is no particle. So strictly speaking, a click is not tantamount to a particle hit—just as a red flag is not tantamount to a goat.

When multiple experiments of this more general kind are performed, the ensuing state updates must be iterated. For instance, a first experiment might lead to an update as in Eq. (2.5) with some set of compatible sequences, $\{s'\}$, and associated probabilities, $\{\lambda(s')\}$. A subsequent experiment might allow for another set of compatible sequences, $\{s''\}$, and associated probabilities, $\{\tau(s'')\}$. The second update, then, has the form

$$\sum_{s'} \lambda(s')\, \iota_{|s'} \rightarrow \sum_{s''} \tau(s'') \left[\sum_{s'} \lambda(s')\, \iota_{|s'} \right]_{|s''}.$$

Thanks to the update rule for mixtures, Eq. (1.38), the measurement update of a mixture is a mixture of the measurement updates:

$$\left[\sum_{s'}\lambda(s')\,\iota_{|s'}\right]_{|s''} \propto \sum_{s'}\lambda(s')\,\mathrm{prob}(s''|\iota_{|s'})\,\iota_{|s''\leftharpoonup s'}. \tag{2.7}$$

(You will verify in Exercise (2.4) that Eq. (1.38) applies just as well after ascertaining a whole sequence, s'', rather than just an individual proposition, y.) The arrow in $s'' \leftharpoonup s'$ indicates the order in which the sequences have been ascertained: first s', and then s''. Thus, in effect, the two experiments combined entail a state update of the form

$$\iota \to \sum_{s'',s'} \kappa(s'',s')\,\iota_{|s''\leftharpoonup s'},$$

with some new weights, $\{\kappa(s'',s')\}$. This form is exactly the same as in Eq. (2.5) for a single experiment. In terms of their effect on the state, therefore, two (or more) consecutive experiments may be treated like one big experiment.

All the information which is encoded in a state originates from experiments of the kind just discussed, starting, ultimately, from the state of total ignorance. (Where there is prior information available, that information will be the fruit of earlier experiments, performed by somebody else.) Therefore, it is always possible to represent a state *which can actually be reached* in the form of Eq. (2.5). A concrete example is the state described by Eq. (2.6). Conversely, every state of this form is reachable via some suitably designed experiment. The reachable states form a convex set: whenever two states, individually, can be represented in the form of Eq. (2.5), so can any convex combination. Moreover, upon measurement, reachable states are mapped to reachable states (Eq. (2.7)); so this convex set is closed under measurement updates. In classical probability theory, the convex set of reachable states coincides with the convex set of logically consistent states, introduced previously in Section 1.3. (You will verify this in Exercise (2.3).) In the presence of knowledge constraints, however, the former might be strictly smaller than the latter. While every reachable state is a logically consistent state, the converse might no longer be true; there might be logically consistent states which can never be reached.

For sufficiently simple models, the convex set of reachable states may be visualized geometrically. Here we choose a graphical representation which also includes subnormalized states; that is, states where the overarching proposition, a, need not be true with certainty but may have any probability between zero and one. This allows for the possibility that the object of inquiry—a coin, a car, a photon—might be lost or discarded. The set of reachable states is then also closed under an arbitrary rescaling $\rho \to s\rho$, defined as

$$\mathrm{prob}(x|s\rho) := s\,\mathrm{prob}(x|\rho) \tag{2.8}$$

for $s \in [0,1]$, and is referred to as a *convex cone*. The convex cones for two classical examples without knowledge constraints, a coin flip experiment and the Monty Hall problem, are depicted in Figs 2.8 and 2.9, respectively. The figures also show the locations

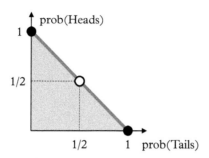

Fig. 2.8 *Convex cone of reachable states for a coin flip. As the coin might get lost during the experiment, the probabilities for Heads and Tails need not add up to one; rather, their sum equals* $\mathrm{prob}(a|\rho)$, *where a is the proposition that 'the coin landed in a visible place'. The shaded triangle indicates all reachable probability assignments, with* $\mathrm{prob}(a|\rho)$ *taking arbitrary values between zero and one. Thus, an arbitrary, not necessarily normalized, state is specified by two real parameters. Its diagonal boundary (thick line) is the subset of states that are normalized to* $\mathrm{prob}(a|\rho) = 1$. *The centre of this diagonal (open circle) represents the state of total ignorance. Its endpoints (black dots) are the only two pure states.*

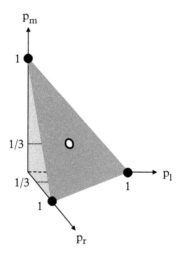

Fig. 2.9 *Convex cone of reachable states for the Monty Hall problem. The convex cone of the Monty Hall problem has the shape of a triangular pyramid. Points on its boundary or in its interior correspond to reachable probability assignments, with* $\mathrm{prob}(a|\rho)$ *taking arbitrary values between zero and one. Thus, an arbitrary, not necessarily normalized, state is specified by three real parameters. The triangular base of the pyramid (shaded dark grey) comprises all states that are normalized,* $\mathrm{prob}(a|\rho) = 1$. *The centre of this base (open circle) represents the state of total ignorance. Its corners (black dots) are the only three pure states.*

of the respective states of total ignorance. In Fig. 2.8 the state of total ignorance is situated at the centre of the thick diagonal line, where prob(Heads) = prob(Tails) = 1/2; whereas in Fig. 2.9 it is situated at the centre of the triangle, where all three probabilities are equal to one third. Finally, the figures indicate the so-called 'pure' states. This is a concept which we have yet to discuss; we will do so in Section 2.3. In these two examples there is no difference between logically consistent states and reachable states.

Now we turn to the Spekkens toy model, our simplest example with knowledge constraints. An arbitrary normalized state of the Spekkens toy model may be parametrized by Cartesian coordinates, x, y, and z. These determine the respective outcome probabilities of the three allowed measurements (Fig. 2.10). Prior to any measurement, the state is that of total ignorance. The latter assigns an equal probability of one half to all outcomes, corresponding to Cartesian coordinates $(0,0,0)$. After a measurement along one of the three axes, the posterior state will feature a coordinate value of $+1$ or -1 on the measurement axis, and coordinate values equal to zero on the others. The normalized reachable states are arbitrary convex combinations of these posteriors. They form a convex set with the shape of an octahedron (Fig. 2.11). This octahedron leaves out some states which would be possible from the point of view of logical consistency but can never be reached in practice. For example, the state with Cartesian coordinates $(1,1,1)$ describes the situation where we know for sure that the gem is located in box 2. Logically, this is perfectly possible. Yet, due to the knowledge constraints, there is no way that we can ever acquire this information experimentally. The normalized states which are logically possible form another convex set with the shape of a cube, whose eight corners are located at the points with coordinates $(\pm 1, \pm 1, \pm 1)$. This cube is larger than the octahedron.

As our last examples we consider the two continuous toy models, the circle model and the sphere model. In the circle model the convex cone is an actual three-dimensional cone (Fig. 2.12), whereas in the sphere model the convex cone is four-dimensional and a bit more difficult to visualize (Fig. 2.13). In both cases the pure states (whose

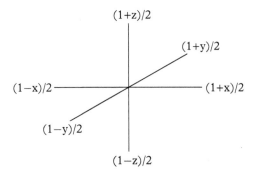

Fig. 2.10 *Parameters specifying a normalized state in the Spekkens toy model. The outcome probabilities of the three allowed measurements, assigned to the ends of the respective axes, are parametrized by Cartesian coordinates, x, y, and z. Each coordinate can take values between -1 and $+1$.*

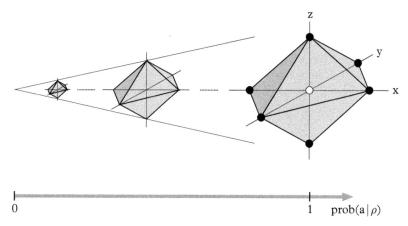

Fig. 2.11 *Convex cone for the Spekkens toy model. The convex cone consists of rescaled versions of the convex set of normalized reachable states, parametrized by the probability of the overarching proposition, a. This convex set has the shape of an octahedron. Pure states are located at its tips (black dots). In total there are six such pure states, corresponding to the points +1 and −1 on each coordinate axis. The state of total ignorance is located at the origin.*

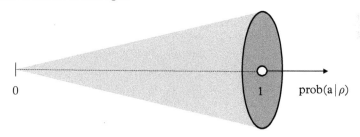

Fig. 2.12 *Convex cone for the circle model. The circular base of the cone (dark grey) comprises the convex set of all normalized reachable states. On its circumference (black) lie the pure states, at its origin (open circle) the state of total ignorance. Subnormalized states, where* $\mathrm{prob}(a|\rho) \leq 1$, *lie on or inside rescaled versions of the base (light grey). Together, all states form a three-dimensional cone.*

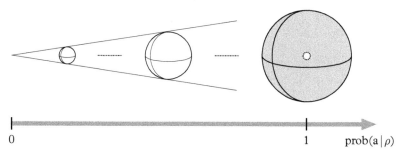

Fig. 2.13 *Convex cone for the sphere model. Pure states lie on the surface of a ball (right, shaded in grey); all other normalized states in the interior of this ball. At the origin (open circle) is the state of total ignorance. Subnormalized states lie on or inside rescaled versions of the ball (left and centre). Together, all states form an abstract cone of dimension four.*

precise definition and meaning shall be discussed in Section 2.3) form a continuum, either a circle or the surface of a sphere. All other normalized reachable states arise from convex combinations of these pure states. Thus, they constitute the interior of a circular disk or ball, respectively. In terms of the dimensionality, the sphere model comes closest to a continuum version of the Spekkens toy model; here the convex set of normalized reachable states—a ball—is three-dimensional, like its counterpart, the octahedron (Fig. 2.11).

2.3 Maximal Knowledge

If there are constraints on what can be learnt by measurement, it is no longer self-evident what it means for knowledge to be maximal. I consider two alternative ways of defining maximal knowledge under such circumstances, and how they are related. This leads to the basic notions of 'pure' and 'mixed' states. I derive a first constraint on the number of continuous parameters that specify a pure state.

In Section 2.2 we introduced the notion of total ignorance and the state which represents it, ι. Here we turn to the opposite extreme, namely a situation of 'maximal knowledge'. In our general framework, where it may happen that propositions cannot be tested jointly, the definition of maximal knowledge is less straightforward than the definition of total ignorance; it is not obvious how to recognize a state which represents maximal knowledge. In the absence of any knowledge constraints, this would be easy: it would be a state where all probabilities are either zero or one, and hence the truth values of all propositions are known with certainty. As soon as there are knowledge constraints, however, this simple definition no longer works. For instance, in the 3×3 problem and in the Spekkens toy model the rules of the game make it impossible to ever reach a state where all probabilities are either zero or one. So *within the constraints* of the game, what does it then mean to say that a candidate possesses maximal knowledge—in the sense of knowledge that she can actually access? In order to answer this question, it is worthwhile recalling the role of the state as the nexus between past data and future expectations, as illustrated in Fig. 1.17. On the one hand, the state synthesizes our knowledge about past interventions; on the other, it catalogues our expectations as to the outcomes of future measurements. This suggests two alternative definitions of maximal knowledge—one 'past-' or 'preparation-oriented', the other 'future-' or 'expectation-oriented':

1. We might say that we possess maximal knowledge when the state has been prepared as accurately as possible. Preparation takes place via a sequence of ascertainments, s, as defined in Eq. (2.3). In general, the preparation starts in some prior state, σ, which need not be the state of total ignorance. Rather, σ may encapsulate the results of preceding measurements and, possibly, personal biases on the part of the observer. The measurement sequence, s, achieves a complete specification of the posterior state if and only if it completely overwrites all these earlier results and biases; to wit, if it alone determines the posterior, regardless of the prior.

Mathematically, this means that $\sigma_{|s}$ is the same for all σ (provided the latter permits the sequence at all, $\text{prob}(s|\sigma) \neq 0$). Such exhaustive preparation leads to a posterior which, by this definition, represents maximal knowledge. Worded the other way round, a state, ρ, represents maximal knowledge if and only if there exists a sequence of ascertainments, s, such that $\rho = \sigma_{|s}$ *independently of* σ.

2. Alternatively, we might say that we possess maximal knowledge when our expectations about the outcomes of future measurements are as precise as possible. This is the case whenever we can be certain about the truth of a most accurate proposition; for example, we know for sure that a car will be revealed behind the left door, or that a photon will exit 'up' behind a given beam splitter. Formally, this means that a state, ρ, represents maximal knowledge if and only if there exists some most accurate proposition, e, which in this state is true with certainty, $\text{prob}(e|\rho) = 1$.

In the absence of knowledge constraints, these two characterizations of maximal knowledge are equivalent and amount to the straightforward definition given earlier. First of all, without knowledge constraints all propositions can be tested jointly. Thus, in a sequence of ascertainments, s, their order does not matter; the confirmed propositions, $\{x_i\}$, are all true simultaneously, amounting to a single combined proposition 'x_1 *and* x_2 *and* $\ldots$'. This proposition specifies the posterior uniquely, regardless of the prior, if and only if it is most accurate. So indeed, complete specification of the state (the first definition of maximal knowledge) then implies the truth of a most accurate proposition (the second definition of maximal knowledge), and vice versa. Furthermore, in the absence of knowledge constraints all most accurate propositions are mutually exclusive. Hence, if a most accurate proposition is true with certainty, all others must be false with certainty; and then, indeed, all probabilities are either zero or one.

As soon as we have knowledge constraints, the equivalence of the two proposed definitions can no longer be taken for granted. For example, in the 3×3 problem the sequence $s = (\neg_a e_{22}) \leftarrow e_{11}$ will lead to a unique posterior; yet afterwards there will be no most accurate proposition which is true with certainty. (You will investigate this in greater detail in Exercise (2.5).) Hence the posterior will meet the criterion of the first, but not of the second, definition of maximal knowledge. The converse, however, continues to hold: maximal knowledge according to the second definition still implies maximal knowledge according to the first. This may be seen as follows. All our toy models, as well as our experimental evidence about the quantum realm, share the feature that confirming the truth of a most accurate proposition supplants all information gathered previously. A case in point is the polarizing beam splitter depicted in Fig. 1.5. The beam splitter tests two mutually exclusive most accurate propositions, $\{\Uparrow_\theta, \Downarrow_\theta\}$, and, depending on the outcome, sends a photon one way or the other. For the photons in one particular exiting beam, the pertinent most accurate proposition is true with certainty. This alone determines the outcome probabilities of all subsequent measurements; whatever happened before the beam splitter no longer plays a role. Thus, the ascertainment of a most accurate proposition amounts to a complete specification of the state. It is as if there had been no prior history at all, and one had started straight from the state of total ignorance. In mathematical terms, this means that upon ascertaining a most accurate proposition, e, any prior, σ, is updated to the same posterior,

$$\sigma_{|e} = \iota_{|e} \ \forall \sigma. \tag{2.9}$$

This includes the special case where in the prior state the most accurate proposition is already known to be true with certainty: $\sigma = \rho$ with $\text{prob}(e|\rho) = 1$. By Eq. (1.30), this particular state will not change when the truth of e is reaffirmed, $\rho_{|e} = \rho$. Thus, by Eq. (2.9), it must itself coincide with $\iota_{|e}$ and with all other $\sigma_{|e}$. In effect,

$$\text{prob}(e|\rho) = 1 \quad \Rightarrow \quad \rho = \sigma_{|e} \ \forall \sigma. \tag{2.10}$$

On the right-hand side of this implication, the ascertainment of e constitutes a special case of an ascertainment of a sequence, s. Therefore, we may also read the implication as follows: the truth of a most accurate proposition implies the existence of a measurement sequence—namely, $s = e$—which fixes the state uniquely, to wit, $\rho = \sigma_{|s}$ regardless of the prior, σ. Indeed, the former corresponds to the second, and the latter to the first definition of maximal knowledge.

Notwithstanding the above caveats, the experimental evidence suggests that in the quantum realm the implication does in fact also work the other way round, like in a theory without knowledge constraints. For instance, the only way to produce a beam whose properties do not depend on any peculiarities of the source is to place a beam splitter (or a polarization filter) in between. Thus, total independence from the prior can only be achieved by confirming a most accurate proposition:

$$\exists s : \rho = \sigma_{|s} \ \forall \sigma \quad \Rightarrow \quad \exists e : \text{prob}(e|\rho) = 1. \tag{2.11}$$

The implication works as well in all the toy models which were designed to mimic the experimental evidence, to wit, the Spekkens toy model, the circle model, and the sphere model. Therefore, we shall henceforth assume that the two definitions of maximal knowledge are equivalent, and we shall exclude exotic models like the 3×3 problem where this is not the case.

There is a third, less operational but very useful way to characterize a state of maximal knowledge. This characterization, too, will work both with and without knowledge constraints. Let ρ be a state that represents maximal knowledge according to our earlier definitions. In this state some most accurate proposition, e, must be true with certainty. Now suppose that this state can be written as a convex combination of two other states,

$$\rho = t\rho_1 + (1-t)\rho_2, \quad \rho_1 \neq \rho_2, \quad 0 < t < 1.$$

In order to ensure $\text{prob}(e|\rho) = 1$, these other states would also have to assign probability one to the most accurate proposition e: $\text{prob}(e|\rho_1) = \text{prob}(e|\rho_2) = 1$. Yet by Eq. (2.10), there exists only one state in which e is true with certainty: $\iota_{|e}$. So actually, the two states must be identical, $\rho_1 = \rho_2 = \iota_{|e}$, contrary to our original assumption. We conclude that a state of maximal knowledge can never be written as a convex combination of other states. Conversely, consider a state, ρ, which cannot be written as a convex combination of other states. Like every reachable state, however, it can be represented in the form of Eq. (2.5). These two assertions are compatible only if all summands in the latter representation are in fact identical. Consequently, the state in question must itself be a posterior; there must exist some sequence, s, such that $\rho = \iota_{|s}$. Since the state of total ignorance lies in the interior (rather than on the boundary) of the convex set of normalized states, it can always be written as a convex combination of an arbitrary given state, σ, and some other state, $\bar{\sigma}$:

$$\forall \sigma \neq \iota\ \exists \bar{\sigma}, 0 < t < 1: \quad \iota = t\sigma + (1-t)\bar{\sigma};$$

this idea is illustrated in Fig. 2.14. By Eq. (1.38), its update, $\rho = \iota_{|s}$, then has the form

$$\rho \propto t\,\text{prob}(s|\sigma)\,\sigma_{|s} + (1-t)\,\text{prob}(s|\bar{\sigma})\,\bar{\sigma}_{|s}.$$

We assumed initially that ρ must not be a convex combination of other states. This is assured only if ρ, $\sigma_{|s}$, and $\bar{\sigma}_{|s}$ are in fact all identical. As σ could be chosen freely, this means that $\rho = \sigma_{|s}$ for arbitrary σ. Thus, ρ meets the criterion of our first definition of maximal knowledge. We conclude that a state which cannot be written as a convex combination of other states must be a state of maximal knowledge. In sum, there is a one-to-one correspondence between states of maximal knowledge and normalized states which cannot be written as a convex combination of other states. The latter are called *pure* states; all others are *mixed*. For a coin flip, the Monty Hall problem, the Spekkens toy model, and the circle and sphere models, the pure states are marked in Figs 2.8, 2.9, 2.11, 2.12, and 2.13, respectively. From now on, we shall use the terms 'state of maximal knowledge' and 'pure state' interchangeably.

Maximal knowledge is preserved under measurement updates; a state of maximal knowledge is always updated to a (possibly different) state of maximal knowledge. Indeed, according to our first definition (left-hand side of Eq. (2.11)), maximal

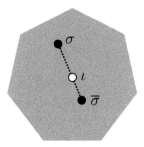

Fig. 2.14 *The state of total ignorance, ι, as a convex combination of two other states, σ and $\bar{\sigma}$.*

knowledge means that there exists a sequence, s, such that $\rho = \sigma_{|s}$ for all σ. Then for any update of this state, $\rho_{|x}$, there also exists a sequence—namely, $s' = x \leftarrow s$—such that $\rho_{|x} = \sigma_{|s'}$ for all σ; so the updated state represents maximal knowledge, too. The update from one state of maximal knowledge to another is mirrored by an update of most accurate propositions, in the following sense. According to the second definition (right-hand side of Eq. (2.11)), each state of maximal knowledge assigns probability one to some most accurate proposition. Hence, if the original state, ρ, assigns probability one to the most accurate proposition e, then its update, $\rho_{|x}$, must assign probability one to some most accurate proposition, too. We shall denote the latter by $e_{|x}$. Formally:

$$\exists e : \mathrm{prob}(e|\rho) = 1 \quad \Rightarrow \quad \exists e_{|x} : \mathrm{prob}(e_{|x}|\rho_{|x}) = 1. \tag{2.12}$$

In the absence of knowledge constraints, it is $e_{|x} = e$: if I already know that the most accurate proposition e: 'this die shows "5"' is true, then whatever else I choose to measure afterwards—however else I choose to look at the die—will not change the truth of that proposition. But in settings with knowledge constraints, it may well be that $e_{|x} \neq e$. A case in point is the update shown in Fig. 2.4 for the Spekkens toy model. Prior to measurement, the most accurate proposition e: 'boxes 3 and 4 are empty' is true with certainty. Upon ascertaining x: 'the two outer boxes are empty', this ceases to be the case. Rather, a different most accurate proposition becomes true with certainty. In this particular example, the tested proposition, x, is itself most accurate, so it is $e_{|x} = x$. More generally, $e_{|x}$ is a refinement of x,

$$e_{|x} \subseteq x. \tag{2.13}$$

Indeed, it is precisely the confirmation of x which has led from e to $e_{|x}$. Since the pertinent measurement must be reproducible, a second test of x—this time given the truth of $e_{|x}$—will reconfirm x with certainty. Thus, the truth of $e_{|x}$ implies the truth of x.

The results of this section allow us to draw some interesting conclusions about probability theories that exhibit a continuum of most accurate propositions. We start out by clarifying what we mean by a probability *theory*, as opposed to a probabilistic model tailored to some specific situation. The prime example of the former is classical probability theory. It is a theory which can be applied to a large variety of situations, all characterized by the absence of knowledge constraints. There is no limitation on the dimension, d, that the associated logical structures may have; it might be $d = 2$ for a coin flip, $d = 3$ for the Monty Hall problem, and so on. If two applications of classical probability theory have the same dimension, they are mathematically equivalent, in the sense that they feature the same logical structure and the same convex cone of states. For instance, two applications which both have dimension $d = 6$ are (i) a die roll and (ii) a combination of a coin flip ($d_A = 2$) with the Monty Hall problem ($d_B = 3$), resulting in a composite game of dimension $d_A d_B = 6$. While the pertinent propositions will use different words that reflect the specific setup of the game, the logical structures and convex cones will be exactly the same. By analogy, we demand of any alternative probability theory that it be *universal* in the following double sense:

1. it is applicable to problems of arbitrary dimension

2. in any application the logical structure and the convex cone depend on the dimension only.

By contrast, there are probabilistic models which are tailored to some specific situation of some specific dimension, yet which cannot be generalized to arbitrary dimensions. One case in point is the Spekkens toy model, which has dimension $d = 2$. It is not clear how this model might be extended to, say, $d = 3$ or $d = 7$. Therefore, in our terminology it shall be considered merely a 'model' but not a 'theory'.

In addition to the dimension of the logical structure, d (presumed to be finite), we define a new quantity, X, as the number of continuous parameters needed to specify a pure state. Thanks to the one-to-one correspondence between pure states and most accurate propositions, this coincides with the number of continuous parameters needed to specify a most accurate proposition. In any model or theory with a discrete set of most accurate propositions, this number equals zero. In particular, it equals zero in classical probability theory. It may be non-zero only in continuous models of the kind studied in Section 2.2; for example, in the circle model it is $X = 1$, whereas in the sphere model it is $X = 2$. In a probability theory, as opposed to merely a specific model, the dimension, d, may vary freely. Then the number of parameters needed to specify a pure state or most accurate proposition, respectively, is a function of this dimension, $X(d)$. We can infer some basic properties of this function from the following consistency argument. Rather than considering a most accurate proposition, e, as being directly a member of a logical structure of dimension d, we may view it as being part of a smaller substructure which, in turn, is embedded in the full logical structure. To be specific, we fix an arbitrary proposition, $x \subseteq a$, of dimension $k := d(x) \leq d$. As long as e neither excludes nor logically implies this proposition, $e \not\subseteq \neg_a x$ and $e \not\subseteq x$ (which is possible only in theories with knowledge constraints), there exist both $e_{|x}$ and $e_{|\neg_a x}$ (Eq. (2.12)). The former has two basic properties. First, by definition, it is

$$\mathrm{prob}(e_{|x} | x, e) = 1. \tag{2.14}$$

(Here we did not write out the state prior to ascertaining e because, by Eq. (2.9), that prior state is irrelevant.) Secondly, thanks to Eq. (2.13), there exists some y such that $x = e_{|x} \oplus y$. Using both properties in the product rule in the form of Eq. (1.34), with x replaced by $e_{|x}$ and ρ by $\rho_{|e}$, yields

$$\mathrm{prob}(e_{|x} | e) = \mathrm{prob}(x | e). \tag{2.15}$$

Thus we find

$$\mathrm{prob}(e_{|x} \oplus (\neg_a x) | e) = \mathrm{prob}(e_{|x} | e) + \mathrm{prob}(\neg_a x | e) = \mathrm{prob}(x | e) + \mathrm{prob}(\neg_a x | e) = 1.$$

In other words, whenever e has been confirmed as true, a subsequent test will reveal either $e_{|x}$ or $\neg_a x$ as true; the former logically implies the latter:

$$e \subseteq e_{|x} \oplus (\neg_a x). \tag{2.16}$$

Hence we may consider e as a member of a logical structure of the smaller dimension

$$d(e_{|x} \oplus (\neg_a x)) = d - k + 1.$$

Specifying e within this smaller structure requires only $X(d - k + 1)$ parameters. To this we must add the number of parameters required to specify the smaller structure itself. At fixed x, this amounts to specifying $e_{|x}$. Since the latter is a refinement of the fixed x, Eq. (2.13), its specification requires another $X(k)$ parameters. Regardless of k, the total number of parameters must be the same as in the direct approach:

$$X(d) = X(d - k + 1) + X(k) \quad \forall k = 1, \ldots, d.$$

This implies that the number of parameters grows linearly with the dimension,

$$X(d) = X(2)(d - 1). \tag{2.17}$$

This scaling must hold for every non-classical probability theory that purports to be universal in the sense defined above. An immediate and rather obvious consequence is that if in the smallest possible logical structure, $d = 2$, the set of pure states is discrete, $X(2) = 0$, then it must be so in all higher-dimensional structures, $X(d) = 0$. Conversely, if at $d = 2$ the pure states form a continuum, then they will do so in all dimensions. In particular, should it be possible to extend the continuous circle and sphere models to higher dimensions in a manner consistent with universality, the number of degrees of freedom of a pure state would have to scale according to $X_{\text{circ}}(d) = d - 1$ and $X_{\text{sph}}(d) = 2(d - 1)$, respectively.

2.4 The Operations AND and OR

> The classical logical operations AND and OR presuppose that all propositions can be tested jointly. I show how these operations can be generalized to situations with knowledge constraints, and I discuss their properties and operational meaning.

Up to this point we have carefully avoided extending the notion of binary AND and OR operations—which in the classical set picture corresponded to the straightforward intersection and union of sets, respectively—to situations with knowledge constraints. Classically, one way to define the proposition 'x and y' (denoted by $x \wedge y$) is as the least precise proposition implying both x and y,

$$z \subseteq x \wedge y \quad :\Leftrightarrow \quad z \subseteq x, \; z \subseteq y, \tag{2.18}$$

while 'x or y' (denoted by $x \vee y$) can be defined as the most precise proposition implied by both x and y,

$$x \vee y \subseteq z \quad :\Leftrightarrow \quad x \subseteq z, \ y \subseteq z. \tag{2.19}$$

In the following, we will convince ourselves that these definitions carry over to our more general framework with knowledge constraints. To this end, we must prove two results:

1. propositions with the properties required by Eqs (2.18) or (2.19), respectively, exist
2. given that such propositions exist, they are unique.

The latter part is easy. Suppose there are two propositions, $(x \wedge y)_1$ and $(x \wedge y)_2$, which both satisfy Eq. (2.18). Then $z \subseteq (x \wedge y)_1$ implies $z \subseteq x$ and $z \subseteq y$, which, in turn, implies $z \subseteq (x \wedge y)_2$. For $z = (x \wedge y)_1$, this entails $(x \wedge y)_1 \subseteq (x \wedge y)_2$. The same reasoning applies with the subscripts interchanged, yielding $(x \wedge y)_2 \subseteq (x \wedge y)_1$. Antisymmetry, Eq. (1.14), then stipulates that the two propositions must be identical: $(x \wedge y)_1 = (x \wedge y)_2$. Thus, indeed, $x \wedge y$ is unique. A similar argument shows that $x \vee y$ is unique as well.

The question of whether $x \wedge y$ and $x \vee y$ exist at all is far less trivial. Since we can no longer rely on the classical set picture, we must prove their existence by explicit construction. For this construction we will use some of the results obtained in Section 2.3. To begin with, the existence of a unique update for most accurate propositions, $e \to e_{|x}$ (Eq. (2.12)), for arbitrary e and x allows us to define the map

$$\theta_x : \emptyset \to \emptyset, \quad \theta_x : e \to \begin{cases} \emptyset : & e, x \text{ mutually exclusive} \\ e_{|x} : & \text{otherwise} \end{cases} \tag{2.20}$$

for any given x. It maps the absurd proposition to itself, and an arbitrary most accurate proposition to either the absurd or another most accurate proposition. The map from a most accurate to the absurd proposition occurs when e and x contradict each other, so that, given the truth of e, it is impossible to ever confirm x; in this case an update, $e_{|x}$, does not exist. Since, by Eq. (2.13),

$$\theta_x(e) \subseteq x, \tag{2.21}$$

and the map is idempotent,

$$\theta_x \circ \theta_x = \theta_x, \tag{2.22}$$

it may be regarded as a 'projection' onto x. In a first step we use this projection to construct the OR operator for pairs consisting of an arbitrary proposition, x, and a most accurate proposition, e:

$$x \vee e = x \oplus \theta_{\neg_a x}(e). \tag{2.23}$$

If the two propositions can be tested jointly, this construction agrees with the classical definition. Indeed, in that case there are just two possibilities: either $e \subseteq x$, in which case $x \vee e = x$; or e contradicts x, in which case $x \vee e = x \oplus e$. Yet the construction also works when the two propositions cannot be tested jointly. In order to see this, we must prove that it satisfies Eq. (2.19). For the implication from left to right we need $x \subseteq x \vee e$ and $e \subseteq x \vee e$. The former is obvious from the construction, while the latter follows from Eq. (2.16), with x replaced by $\neg_a x$. For the reverse implication, from right to left, we can argue as follows. By assumption, the proposition z is logically implied by x, and thus can be tested jointly with x and also with $\neg_a x$. Hence Bayes' rule, Eq. (1.39), applies:

$$\text{prob}(z|\neg_a x, e) = \frac{\text{prob}(\neg_a x|z, e)\,\text{prob}(z|e)}{\text{prob}(\neg_a x|e)}.$$

Furthermore, z is logically implied by e. This means, on the one hand, that the second factor in the numerator is equal to one, $\text{prob}(z|e) = 1$; and on the other, by Eq. (1.30), that the first factor in the numerator equals the denominator. Thus we obtain

$$\text{prob}(z|\theta_{\neg_a x}(e)) = \text{prob}(z|\neg_a x, e) = 1,$$

and hence $\theta_{\neg_a x}(e) \subseteq z$. Together with $x \subseteq z$ this indeed yields $x \vee e \subseteq z$.

Having constructed the OR operator for pairs involving a most accurate proposition, we can now iterate the procedure to arrive at a construction for arbitrary pairs:

$$x \vee y = (\dots((x \vee e_1) \vee e_2)\dots) \vee e_n, \quad y = \bigoplus_{i=1}^{n} e_i. \tag{2.24}$$

Here the $\{e_i\}$ constitute a partition of y into most accurate refinements. In the presence of knowledge constraints, this partition might not be unique; nevertheless, the resultant $x \vee y$ is. All we have to show is, once again, that $x \vee y$ satisfies Eq. (2.19). The argument consists of two parts. First, we know from our earlier construction that $e_1 \subseteq x \vee e_1$, $e_2 \subseteq (x \vee e_1) \vee e_2$, and so on, so that $e_i \subseteq x \vee y$ for all i. This ensures that, indeed, $y \subseteq x \vee y$. (The other implication, $x \subseteq x \vee y$, is obvious.) Secondly, if $y \subseteq z$, then it is also $e_i \subseteq z$ for all i. Together with $x \subseteq z$, this entails $x \vee e_1 \subseteq z$, $(x \vee e_1) \vee e_2 \subseteq z$, and so on, such that, indeed, $x \vee y \subseteq z$. In other words, the above construction does indeed define the unique OR operator for arbitrary propositions; $x \vee y$ always exists. This OR coincides with the earlier defined 'either … or' if and only if the propositions contradict each other:

$$x \vee y = x \oplus y \quad \Leftrightarrow \quad x, y \text{ mutually exclusive.} \tag{2.25}$$

Equipped with the OR operator, we can use twofold complementation ('a but not …') to construct the AND operator:

$$x \wedge y = \neg_a((\neg_a x) \vee (\neg_a y)). \tag{2.26}$$

Here we are assuming that all propositions involved are refinements of the overarching proposition, a, given as true. (If not, a should be broadened until they are.) The properties of complementation, Eqs (1.19) and (1.20), ensure that the above construction of the AND operator satisfies Eq. (2.18); you will verify this in Exercise (2.7). The construction makes the binary operators automatically satisfy *De Morgan's laws*: upon complementation, AND and OR are swapped,

$$\neg_a(x \wedge y) = (\neg_a x) \vee (\neg_a y), \quad \neg_a(x \vee y) = (\neg_a x) \wedge (\neg_a y). \tag{2.27}$$

As you will show in Exercise (2.7), both operators are commutative,

$$x \wedge y = y \wedge x, \quad x \vee y = y \vee x, \tag{2.28}$$

as well as associative,

$$x \wedge (y \wedge z) = (x \wedge y) \wedge z, \quad x \vee (y \vee z) = (x \vee y) \vee z. \tag{2.29}$$

However, in contrast to the classical set picture, the distributivity property,

$$x \wedge (y \vee z) = (x \wedge y) \vee (x \wedge z), \tag{2.30}$$

does in general *not* hold, unless all propositions involved can be tested jointly. You will investigate a concrete counterexample in Exercise (2.7). Finally, the dimensions of $x \vee y$ and $x \wedge y$ satisfy

$$d(x \vee y) = d(x) + d(y) - d(x \wedge y), \tag{2.31}$$

which generalizes the previous sum rule, Eq. (1.23), and coincides with the classical sum rule discussed in Exercise (1.1). You will prove this more general formula in Exercise (2.7).

The AND operator can be given an operational meaning, even in the presence of knowledge constraints. Thanks to Eqs (2.18) and (1.30), it is

$$\text{prob}(x \leftarrow y \leftarrow x \leftarrow \ldots | x \wedge y, \rho) = 1$$

for alternating sequences of arbitrary length and for arbitrary ρ. Conversely, by twofold application of the product rule, Eq. (1.32), together with Eq. (1.33), we find

$$\text{prob}(x \wedge y | x \leftarrow y \leftarrow x \leftarrow \ldots, \rho) = \frac{\text{prob}(x \wedge y | y \leftarrow x \leftarrow \ldots, \rho)}{\text{prob}(x | y \leftarrow x \leftarrow \ldots, \rho)}$$

$$= \frac{\text{prob}(x \wedge y | x \leftarrow \ldots, \rho)}{\text{prob}(x | y \leftarrow x \leftarrow \ldots, \rho) \, \text{prob}(y | x \leftarrow \ldots, \rho)}.$$

In the limit of an infinitely long sequence, the numerator in the second line equals the probability on the left-hand side. Then both probabilities in the denominator must be equal to one; in particular,

$$\text{prob}(y | x \leftarrow y \leftarrow x \leftarrow \ldots, \rho) = 1.$$

At the same time, it is obviously

$$\text{prob}(x | x \leftarrow y \leftarrow x \leftarrow \ldots, \rho) = 1.$$

When both x and y are true with certainty, then so is $x \wedge y$:

$$\text{prob}(x \wedge y | x \leftarrow y \leftarrow x \leftarrow \ldots, \rho) = 1.$$

Thus, not only does the truth of $x \wedge y$ ensure that the system will pass any sequence of filters testing for x or y, but also, conversely, passing an *infinite* sequence of alternating filters that test for x and y, respectively, entails that $x \wedge y$ is true. The two assertions are equivalent; symbolically:

$$x \wedge y = x \leftarrow y \leftarrow x \leftarrow \ldots \qquad (2.32)$$

In classical probability theory, the order of measurements does not matter, so the sequence on the right-hand side may be reduced to just two measurements in arbitrary order, $x \leftarrow y$ or $y \leftarrow x$; but in more general settings with knowledge constraints, it is crucial to have the infinite sequence. Where we are dealing with a quantum particle, the proposition $x \wedge y$ may be tested by sending it through an infinite sequence of filters, testing x and y in an alternating fashion. The proposition $x \wedge y$ is true if and only if the particle passes all these filters. In a game of chance, we must play an infinite number of rounds, testing alternating propositions. This measurement procedure is clearly an idealization insofar as any real experimental setup will contain only a finite number of filters or allow only a finite number of rounds; but by increasing this number, we can reach any desired precision. For the OR operator, there is no equally straightforward experimental procedure. However, with the help of De Morgan's laws, a test of $x \vee y$ may always be converted into a test involving the AND operator.

We conclude with a few general remarks:

1. It is important to bear in mind that all the results in this section are valid only under the assumptions that we made in Section 2.3, in particular with regard to the equivalence of the two definitions of maximal knowledge. So while these results apply to the Spekkens toy model, the circle and sphere models, and—most importantly—quantum theory, they do not necessarily hold for more exotic toy models such as the 3×3 problem. You will investigate this further in Exercise (2.7).

2. All derivations presumed that we are dealing with propositions of finite dimension. While most results carry over to the infinite-dimensional case, the justification of this limit entails mathematical intricacies which are beyond the scope of this book. It is also not relevant for us, as the quantum systems used for information processing are finite-dimensional.

3. The existence of the binary AND and OR operators endows the set of propositions with the mathematical structure of a 'lattice'. (More precisely, taking its other properties into account also, it is an 'orthomodular lattice'.) We chose to deduce this lattice structure from earlier assumptions that we had made about logic and probabilities. Alternatively, we might have posited the lattice structure from the outset; such a requirement could be motivated by, say, the operational definition of the AND operator. Taking the lattice structure as a derived concept, as we did, is in line with various axiomatic approaches to quantum theory, known in the literature as 'operational statistics', the theory of 'test spaces', 'orthoalgebras', or 'general probabilistic theories'. By contrast, the approach which takes the lattice structure as a primitive is known as 'quantum logic'.

2.5 Composition

Multiple systems may be combined into a single composite system. I explain what this means for the logical structure, its dimension, and the number of parameters that specify a state. I introduce the notions of composite and reduced states, product states, statistical independence, and correlations.

So far our discussions have focused on individual systems or individual games of chance. We considered propositions and associated probabilities that pertained to, say, polarization measurements on a single photon, flips of a single coin, or the opening of doors or boxes in a single game. By contrast, in this section we shall turn to situations where there are *multiple* systems being experimented upon, or multiple games played, in parallel. We start with the simplest scenario of just two systems, A and B; for instance, measuring the polarization of a photon (A) and the magnetic moment of an electron (B), or flipping a coin (A) and rolling a die (B). Whenever propositions refer to distinct systems, we shall denote the logical AND by $\otimes$ rather than $\wedge$:

$$x \otimes y \equiv x^{\mathrm{A}} \wedge y^{\mathrm{B}}. \tag{2.33}$$

This notation signals that the two propositions pertain to distinct systems, while doing away with the superscripts. It is understood that the first proposition pertains to the first system, A, and the second proposition pertains to the second system, B. Thus, the order of the propositions is important: it is $x \otimes y \neq y \otimes x$. Since the systems are distinct, experiments performed on A will not restrict the experiments which may be performed on B in any way, and vice versa; propositions which pertain to different systems may be tested jointly. Therefore, when in Eq. (2.30) one replaces the operator $\wedge$ with $\otimes$ and, in addition, $\vee$ with $\oplus$ (Eq. (2.25)), the three propositions x, y, and z are guaranteed to be testable jointly: x and y, as well as x and z, because they pertain to distinct systems; and y and z because they are mutually exclusive. Under these circumstances the law of distributivity is guaranteed to hold:

$$x \otimes (y \oplus z) = (x \otimes y) \oplus (x \otimes z). \tag{2.34}$$

If the logical structures pertaining to two individual subsystems, or *constituents*, have respective dimensions d_A and d_B, then the dimension of the composite is given by their product,

$$d_{AB} = d_A d_B. \tag{2.35}$$

To derive this product rule, we consider the respective overarching propositions pertaining to A and B, a and b, with $d(a) = d_A$ and $d(b) = d_B$. Let $\{e_i\}$ and $\{f_j\}$ be most accurate partitions of these overarching propositions,

$$a = \bigoplus_i^{d_A} e_i, \quad b = \bigoplus_j^{d_B} f_j.$$

Then, thanks to distributivity, Eq. (2.34), it is

$$a \otimes b = \left(\bigoplus_i e_i\right) \otimes \left(\bigoplus_j f_j\right) = \bigoplus_{ij} (e_i \otimes f_j).$$

The left-hand side, $a \otimes b$, constitutes the overarching proposition pertaining to the composite; for instance, 'there is a coin *and* there is a die'. As the right-hand side shows, this overarching proposition may be decomposed into the $d_A d_B$ joint propositions $\{e_i \otimes f_j\}$. The latter are most accurate propositions about the composite; for example, 'the coin shows "Heads" *and* the die shows "2"' or 'the photon passes the horizontal polarization filter *and* the atom is deflected upwards in the Stern–Gerlach apparatus'. So indeed, the maximum number of mutually exclusive most accurate propositions about the composite equals the product $d_A d_B$.

Another number of interest, for both individual and composite systems, is the number of real parameters, S, needed to characterize an arbitrary state, ρ. In general, it is sufficient to know the probabilities of just a limited set of propositions, from which one can then infer the probabilities of all others; such a subset of propositions is called *informationally complete*. The number of state parameters equals the minimum number of propositions in such an informationally complete set. As long as the probability of the overarching proposition, $\mathrm{prob}(a|\rho)$, is not normalized to one but allowed to take arbitrary values, this number of parameters is at least as large as the dimension of the logical structure,

$$S \geq d. \tag{2.36}$$

Indeed, if in the Monty Hall problem we allow for the possibility that there might be no car at all, and hence $\mathrm{prob}(a|\rho) \leq 1$ rather than $\mathrm{prob}(a|\rho) = 1$, then we need a total of $S = d = 3$ probabilities—say, those of l, m, and r—to specify an arbitrary state. Given the probabilities of this minimal informationally complete set, all other probabilities can be calculated with the help of the sum rule. In the 3×3 problem, on the other hand, we need $S = 5$ probabilities: for instance, those of $\{e_{ij}|i,j = 1,2\}$ and of the overarching proposition, a (which, again, is not required to be true with certainty). In this case it is $S > d$. Similarly, in the Spekkens toy model an arbitrary subnormalized state is specified by $S = 4$ parameters (Fig. 2.11), which is also strictly larger than the dimension, $d = 2$. These examples follow a general pattern which you will derive in Exercise (2.10): it is $S = d$ whenever all propositions can be tested jointly, like in the Monty Hall problem, whereas it is $S > d$ whenever there are knowledge constraints, like in the 3×3 problem and the Spekkens toy model.

Given two systems, A and B, measurements may be performed solely on A, solely on B, or jointly on both A and B. The respective expectations as to their outcomes are encoded in the states ρ_{A}, ρ_{B}, and ρ_{AB}. The states pertaining to A and B individually, ρ_{A} and ρ_{B}, are called *reduced states*, whereas ρ_{AB} is termed the *composite state*. The reduced states are characterized by S_{A} and S_{B} parameters, respectively, and the composite state, by S_{AB} parameters. In the absence of knowledge constraints, it is $S = d$, so the product rule for the dimension, Eq. (2.35), immediately implies a product rule for the number of state parameters:

$$S_{\mathrm{AB}} = S_{\mathrm{A}} S_{\mathrm{B}}. \tag{2.37}$$

As soon as there are knowledge constraints, this product rule need no longer hold. We illustrate this with the following game. Several (possibly biased) coins are flipped simultaneously. A subsequent measurement consists not in counting the number of Heads and Tails separately, but in counting the maximum number of identical sides. So when two coins are flipped, there are just two possible outcomes: '1' (corresponding to one Heads, one Tails) or '2' (corresponding to either two Heads or two Tails). We denote the associated most accurate propositions by $\{e_1, e_2\}$. Let A and B be two such games with two coins each. Pertaining to A or B individually, a state is specified by $S_{\mathrm{A}} = S_{\mathrm{B}} = 2$ parameters, namely the two probabilities $\{\mathrm{prob}(e_i|\rho_{\mathrm{A}})\}$ or $\{\mathrm{prob}(e_j|\rho_{\mathrm{B}})\}$, respectively. One may also play a combined game, AB. For this combined game there exist two different versions:

1. In the first version, two types of measurement are allowed: *either* the outcomes are ascertained for A and B individually, yielding the possible combined outcomes $(1,1)$, $(1,2)$, $(2,1)$, or $(2,2)$; *or* the maximum number of identical sides is counted when all four coins are taken together, with possible outcomes '2', '3', or '4'. Between the outcomes of these two measurements there exist some logical relationships; for instance, $(1,1)$ implies '2' and, conversely, '4' implies $(2,2)$. However, not in every case does the outcome of one type of measurement fully determine the outcome of the other.

2. In an alternative version, only one type of measurement is allowed, namely ascertaining the outcomes for A and B individually, with possible combined outcomes $(1,1)$, $(1,2)$, $(2,1)$, or $(2,2)$. In addition, there is a hard-wired constraint on the joint probabilities: by fiat, they must be invariant under a swap of A and B,

$$\text{prob}(e_i \otimes e_j | \rho_{AB}) = \text{prob}(e_j \otimes e_i | \rho_{AB}) \; \forall i, j.$$

This entails that of the $S_A S_B = 4$ joint probabilities, $\{\text{prob}(e_i \otimes e_j | \rho_{AB})\}$, only three are independent.

In both versions of the combined game, the product rule, Eq. (2.37), is violated. In the first version it is impossible to calculate the probability of '4', $\text{prob}(\text{'4'}|\rho_{AB})$, solely from the joint probabilities $\{\text{prob}(e_i \otimes e_j | \rho_{AB})\}$; to wit, from probabilities and correlations pertaining to the constituent games. This makes '4' a genuinely *global proposition* about the composite. Whenever there are such global propositions, joint probabilities alone no longer suffice to specify a composite state. It is then $S_{AB} > S_A S_B$. Indeed, in this version of the game the composite state is specified not by $S_A S_B = 4$ but by $S_{AB} = 5$ parameters; the extra parameter needed is precisely the probability of the global proposition, '4'. In the second version of the game, the violation of the product rule has a different cause. There, the imposed symmetry constitutes a hard-wired constraint on the allowed joint probabilities. Such a constraint reduces the number of parameters needed to specify the composite state, so that $S_{AB} < S_A S_B$. Indeed, in our modified game the composite state is now specified not by $S_A S_B = 4$ but by only $S_{AB} = 3$ parameters. You will investigate further aspects of these peculiar games in Exercise (2.11).

In our counterexample the product rule was violated because there were global propositions or hard-wired constraints, respectively. In fact, as we will show in the next paragraph, global propositions and hard-wired constraints are the only two possible causes of such a violation; when these are excluded, the product rule continues to hold. In other words, the product rule remains valid whenever the following two assumptions are satisfied:

1. There are no global propositions. In other words, knowledge of all joint probabilities, $\text{prob}(x \otimes y | \rho_{AB})$ for arbitrary x and y, amounts to knowledge of the composite state.

2. There are no hard-wired constraints on the joint probabilities. This means that if $\{x_i\}_{i=1}^{S_A}$ and $\{y_j\}_{j=1}^{S_B}$ are minimal informationally complete sets pertaining to the systems A and B, respectively, their joint probabilities, $\{\text{prob}(x_i \otimes y_j | \rho_{AB})\}$, are independent. None of them can be expressed as a function of the others.

We would expect that these assumptions are met in any probability theory which, like classical probability theory, is operational and universal. The absence of hard-wired constraints is necessary to ensure that constituents are truly distinct systems, in the sense that they may be prepared and coupled to each other in arbitrary ways. And, as we will discover in Sections 2.7 and 2.8, the absence of global propositions is necessary for our ability to measure and prepare states, and thus to endow the notion of 'state' with operational meaning. For these reasons, we shall henceforth presume the validity of the product rule, Eq. (2.37), and leave aside pathological models like our peculiar game.

In the following, we briefly outline why the above two assumptions do, indeed, ensure the product rule. We assume that we are given the $S_A S_B$ joint probabilities $\{\mathrm{prob}(x_i \otimes y_j | \rho_{AB})\}$, with $\{x_i\}_{i=1}^{S_A}$ and $\{y_j\}_{j=1}^{S_B}$ being the respective minimal informationally complete sets. By our second assumption, these are all independent, so the number of composite state parameters cannot be smaller than $S_A S_B$. What we have yet to demonstrate is that the number of composite state parameters cannot be larger than $S_A S_B$ either. To this end, we must prove that the given probabilities of the $\{x_i \otimes y_j\}$ suffice to determine the full composite state—which, thanks to our first assumption, is the case if and only if they determine the probabilities of arbitrary joint propositions $x \otimes y$. The latter can be shown as follows. Without loss of generality, we choose the minimal informationally complete sets pertaining to A and B in such a way that its first d_A or d_B members, respectively, constitute a partition of the overarching proposition:

$$a = \bigoplus_{i=1}^{d_A} x_i, \quad b = \bigoplus_{j=1}^{d_B} y_j.$$

Using distributivity, Eq. (2.34), in a more general form,

$$x \otimes \left(\bigoplus_j y_j \right) = \bigoplus_j (x \otimes y_j), \tag{2.38}$$

as well as the sum rule, Eq. (1.29), we can then calculate the S_A probabilities

$$\mathrm{prob}(x_i \otimes b | \rho_{AB}) = \sum_{j=1}^{d_B} \mathrm{prob}(x_i \otimes y_j | \rho_{AB}),$$

and likewise the S_B probabilities $\{\mathrm{prob}(a \otimes y_j | \rho_{AB})\}$. Next, the classical variant of Eq. (2.32)—that is, with the infinite sequence replaced by just $x \leftarrow y$—and Eq. (1.33) allow us to rewrite $x_i \otimes y_j$ in the form

$$x_i \otimes y_j = x_i \leftarrow y_j = x_i \leftarrow a \leftarrow y_i = x_i \leftarrow (a \otimes y_i).$$

Plugging the right-hand side into the product rule for probabilities, Eq. (1.32), yields the conditional probabilities

$$\mathrm{prob}(x_i | a \otimes y_j, \rho_{AB}) = \frac{\mathrm{prob}(x_i \otimes y_j | \rho_{AB})}{\mathrm{prob}(a \otimes y_j | \rho_{AB})},$$

and likewise their counterparts with swapped arguments, $\{\text{prob}(y_j|x_i \otimes b, \rho_{AB})\}$. Since the sets $\{x_i\}$ and $\{y_j\}$ are informationally complete, these probabilities in fact determine the conditional probabilities $\text{prob}(x|a \otimes y_j, \rho_{AB})$ and $\text{prob}(y|x_i \otimes b, \rho_{AB})$ for *arbitrary x* and *y*. A further application of the product rule then leads to

$$\text{prob}(x \otimes y_j|\rho_{AB}) = \text{prob}(x|a \otimes y_j, \rho_{AB})\,\text{prob}(a \otimes y_j|\rho_{AB})$$

for arbitrary x, as well as (via distributivity and the sum rule) to $\text{prob}(x \otimes b|\rho_{AB})$. In the same fashion we obtain $\text{prob}(x_i \otimes y|\rho_{AB})$ and $\text{prob}(a \otimes y|\rho_{AB})$ for arbitrary y. These probabilities, in turn, yield

$$\text{prob}(x_i|a \otimes y, \rho_{AB}) = \frac{\text{prob}(x_i \otimes y|\rho_{AB})}{\text{prob}(a \otimes y|\rho_{AB})}$$

for arbitrary y, and likewise $\text{prob}(y_j|x \otimes b, \rho_{AB})$ for arbitrary x. Once again, informational completeness of $\{x_i\}$ and $\{y_j\}$ implies that the given probabilities in fact determine $\text{prob}(x|a \otimes y, \rho_{AB})$ and $\text{prob}(y|x \otimes b, \rho_{AB})$ for arbitrary x and y. Finally, one last application of the product rule yields

$$\text{prob}(x \otimes y|\rho_{AB}) = \text{prob}(y|x \otimes b, \rho_{AB})\,\text{prob}(x \otimes b|\rho_{AB})$$

for arbitrary x and y. Thus, indeed, the probabilities of the $S_A S_B$ joint propositions $\{x_i \otimes y_j\}$ determine the probabilities of arbitrary joint propositions $x \otimes y$.

When a probability theory is universal, in the sense defined in Section 2.3, the number of parameters needed to specify an arbitrary state must be a function of the dimension of the logical structure only, $S(d)$. The respective product rules for the dimension and for the number of parameters, Eqs (2.35) and (2.37), then imply that the number of parameters must obey a power law,

$$S(d) = d^n, \quad n = 1, 2, \dots \tag{2.39}$$

Classical probability theory corresponds to the case $n = 1$. In addition, there is a constraint which links $S(d)$ to the number of parameters needed to specify a *pure* state, $X(d)$. The normalized states form a convex set, which is some manifold of dimension $(S(d) - 1)$. As examples, we have encountered a line (coin flip, Fig. 2.8), a triangle (Monty Hall problem, Fig. 2.9), an octahedron (Spekkens toy model, Fig. 2.11), a disk (circle model, Fig. 2.12), and a ball (sphere model, Fig. 2.13). In all cases the boundary of this manifold contains, and sometimes even coincides with, the set of pure states. Therefore, the number of continuous parameters needed to specify a pure state cannot exceed the dimension of that boundary,

$$X(d) \leq S(d) - 2. \tag{2.40}$$

It implies, in particular, that

$$X(2) \geq 1 \quad \Rightarrow \quad S(2) \geq 3 \quad \Rightarrow \quad n \geq 2.$$

In other words, in any probability theory featuring a continuum of pure states—for instance, in a hypothetical theory which extends the circle or sphere model to arbitrary dimension—the number of state parameters must grow at least quadratically with the dimension; faster than in classical probability theory. Conversely, any theory where the number of state parameters grows only linearly with the dimension is necessarily classical, with a discrete set of pure states.

Composite states are classified according to whether or not they exhibit correlations. To make this notion more precise, we consider first the situation where we know the reduced states of both constituents, ρ_A and ρ_B, and, moreover, we are told that all joint probabilities factorize; that is to say, the probability of any joint proposition, $x \otimes y$, equals the product of the probabilities of x and y, calculated with the respective reduced states. Provided there are no global propositions, this factorization property uniquely specifies the composite state, ρ_{AB}. The latter is then called a *product state* and written as $\rho_{AB} = \rho_A \otimes \rho_B$. Formally,

$$\rho_{AB} = \rho_A \otimes \rho_B \quad :\Leftrightarrow \quad \mathrm{prob}(x \otimes y | \rho_{AB}) = \mathrm{prob}(x|\rho_A) \cdot \mathrm{prob}(y|\rho_B) \; \forall x, y. \quad (2.41)$$

In such a product state the outcomes of measurements on A and the outcomes of measurements on B are *statistically independent*. As long as the composite system is in a product state, any measurement on one of the constituents triggers an update of the reduced state of that particular constituent only; the reduced state of the other constituent remains unaffected:

$$[\rho_A \otimes \rho_B]|_{x^A} = \rho_{A|x^A} \otimes \rho_B. \quad (2.42)$$

For instance, if we roll two dice independently and learn that the first die shows an even number, this will affect our expectations as to the first die only. By contrast, if there is at least one joint proposition whose probability does not factorize, then one says that the composite state exhibits *correlations*. In this case there exists at least one pair of measurements on A and B whose outcomes are statistically correlated. Then a measurement on A might reveal information about B, too; and the ensuing state update may also alter the reduced state of B. For example, the two dice might be perfectly synchronized in such a way that they always show the same number of spots. If we then learn that the first die shows an even number, we know that the second die must show an even number, too; and hence the ensuing state update must affect the reduced state of the second die as well.

2.6   Exchangeable Assemblies

Some experiments require multiple copies of a system that have all been prepared identically. I examine this notion of 'identical preparation' more closely and, to this end, introduce the concept of exchangeability. I show that exchangeable states have a particular form, known as the de Finetti representation.

We are now ready to move from two to many systems. However, we shall limit ourselves to one particular case: all systems or games are of the same kind—say, rolling multiple dice—and, moreover, all systems have been prepared, or all games set up, in exactly the same way. So, for example, all dice were produced in the same factory, with exactly the same imperfections; all atoms in a beam entering a Stern–Gerlach apparatus originated from the same source; or all photons hitting a polarization filter were emitted by the same laser. Thus, we are dealing with multiple copies of a system, each with the same prior history. The number of copies, L, is presumed to be finite, but other than that is allowed to vary freely; we are always able to eliminate or add copies with the same history. An assembly of this kind may be thought of as having been drawn randomly from a fictitious infinite sequence of systems (the 'source') in which the ordering is irrelevant (Fig. 2.15). These general characteristics impose two constraints on the composite state of the assembly, $\rho^{(L)}$:

1. The state is invariant under permutations of the constituents. Thus, it must be

$$\text{prob}\left(\bigotimes_{K=1}^{L} x_{i(K)} \middle| \rho^{(L)}\right) = \text{prob}\left(\bigotimes_{K=1}^{L} x_{i(\pi^{-1}(K))} \middle| \rho^{(L)}\right) \tag{2.43}$$

   for an arbitrary set of propositions, $\{x_i\}$, and an arbitrary permutation of the constituents, π. On the left-hand side the proposition $x_{i(K)}$ pertains to member K of the assembly, whereas on the right-hand side this proposition pertains to the permuted member $\pi(K)$.

2. It must be possible to enlarge an assembly of L copies to an assembly of $L+1$ copies, of which the former is then a subassembly. Like $\rho^{(L)}$, the state of this enlarged assembly, $\rho^{(L+1)}$, must be invariant under permutations; and

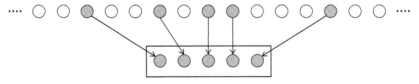

Fig. 2.15 *Exchangeable assembly. One obtains a finite exchangeable assembly of L systems by randomly drawing L systems from a fictitious infinite symmetric sequence.*

to all propositions which refer to the original assembly it must assign the same probabilities as $\rho^{(L)}$:

$$\text{prob}\left(\bigotimes_{K=1}^{L} x_{i(K)} \middle| \rho^{(L)}\right) = \text{prob}\left(\bigotimes_{K=1}^{L} x_{i(K)} \otimes a \middle| \rho^{(L+1)}\right). \tag{2.44}$$

Here a is the overarching proposition 'this constituent exists', presumed to be true, referring to the added, $(L+1)$th copy.

Whenever a composite state meets these requirements, the assembly is called *exchangeable*. (Often an exchangeable assembly also goes by the name of an 'exchangeable sequence'. However, I prefer the term 'assembly', to avoid the impression that its members have to be ordered, spatially or temporally, in any particular way.) In this section we shall elaborate on the mathematical description of exchangeable assemblies. In Sections 2.7 and 2.8 we will see that they play a crucial role in the operational definition of states.

Let an exchangeable assembly comprise L copies, and let $\{x_i\}_{i=1}^{S}$ be a minimal informationally complete set of propositions pertaining to an individual constituent. We test on the first L_1 members of the assembly the proposition x_1, on the next L_2 members the proposition x_2, and so on, with $\sum_i L_i = L$. For instance, we might roll $L = 100$ identically manufactured dice and then test on each of the first $L_1 = 15$ dice the proposition x_1: 'the die shows "1"', on each of the next $L_2 = 20$ dice the proposition x_2: 'the die shows "2"', and so on. The probability that all these tests yield a positive result can be calculated in two ways: either in the state of this assembly, $\rho^{(L)}$, or in the state of an assembly which has been enlarged to N $(N > L)$ copies, $\rho^{(N)}$. In both cases the probability must be the same. By analogy with Eq. (2.44), iterated $(N-L)$ times, this amounts to the condition

$$\text{prob}\left(\bigotimes_{i=1}^{S} x_i^{\otimes L_i} \middle| \rho^{(L)}\right) = \text{prob}\left(\bigotimes_{i=1}^{S} x_i^{\otimes L_i} \otimes a^{\otimes(N-L)} \middle| \rho^{(N)}\right).$$

Here we used the shorthand

$$x^{\otimes K} := \underbrace{x \otimes \ldots \otimes x}_{K}. \tag{2.45}$$

We may partition the enlarged assembly into S subassemblies of sizes $\{N_i\}_{i=1}^{S}$, where $\sum_i N_i = N$ and $N_i > L_i$ for all i. The first subassembly comprises the first L_1 members of the original assembly, on which the proposition x_1 is tested, plus the first $(N_1 - L_1)$ of the added copies. Likewise for all other subassemblies (Fig. 2.16). On the first $(N_1 - L_1)$ copies that we added to the assembly, we may partition the overarching proposition, a, into x_1 and its complement, $\neg_a x_1$; on the next $(N_2 - L_2)$ copies, into x_2 and $\neg_a x_2$; and so on. Thus we obtain

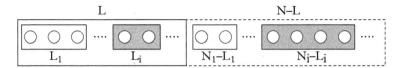

Fig. 2.16 *Partitioning the enlarged assembly. The enlarged assembly has N ($N > L$) constituents. It is partitioned into S subassemblies of sizes $\{N_i\}_{i=1}^{S}$. The ith subassembly (grey) comprises L_i members of the original assembly and $(N_i - L_i)$ of the added copies, in the order shown.*

$$\text{prob}\left(\bigotimes_{i=1}^{S} x_i^{\otimes L_i}\middle|\rho^{(L)}\right) = \text{prob}\left(\bigotimes_{i=1}^{S} x_i^{\otimes L_i} \otimes \bigotimes_{j=1}^{S} (x_j \oplus \neg_a x_j)^{\otimes(N_j - L_j)}\middle|\rho^{(N)}\right).$$

By distributivity, Eq. (2.38), the combined proposition featuring on the right-hand side may be written as a partition,

$$\bigotimes_{i=1}^{S} x_i^{\otimes L_i} \otimes \bigotimes_{j=1}^{S} (x_j \oplus \neg_a x_j)^{\otimes(N_j - L_j)} = \bigoplus \underbrace{(\ldots \otimes \ldots \ldots \otimes \ldots)}_{N},$$

into more refined alternatives, each of which is a conjunction of N single-constituent propositions. In these conjunctions the proposition x_i features K_i times and its complement, $\neg_a x_i$, $(N_i - K_i)$ times, where K_i may vary between L_i and N_i. For given $\{K_i\}$, there are

$$m_{\{K_i\}} = \prod_{j=1}^{S} \binom{N_j - L_j}{K_j - L_j}$$

conjunctions in the partition which differ only in the ordering of the propositions. Thanks to exchangeability, this ordering is of no import for the probability. Therefore, by the sum rule, Eq. (1.29), it is

$$\text{prob}\left(\bigotimes_{i=1}^{S} x_i^{\otimes L_i}\middle|\rho^{(L)}\right) = \sum_{\{K_i\}} m_{\{K_i\}}\text{prob}\left(\bigotimes_{i=1}^{S} x_i^{\otimes K_i} \otimes (\neg_a x_i)^{\otimes(N_i - K_i)}\middle|\rho^{(N)}\right).$$

The probability on the right-hand side pertains to one particular ordering of the propositions $\{x_i\}$ and $\{\neg_a x_i\}$. Yet again, this particular ordering does not matter. The probability may just as well be expressed as the probability that in N_1 tests of the proposition x_1, N_2 tests of the proposition x_2, and so on, the proposition x_1 is revealed as true K_1 times and false $(N_1 - K_1)$ times, x_2 is revealed as true K_2 times and false $(N_2 - K_2)$ times, and so on, *regardless of the order*, divided by the number of possible orderings of the test results:

$$\text{prob}\left(\bigotimes_{i=1}^{S} x_i^{\otimes K_i} \otimes (\neg_a x_i)^{\otimes (N_i - K_i)} \middle| \rho^{(N)}\right) = \frac{\text{prob}(\{K_i\} | \{N_i\}, \rho^{(N)})}{\prod_{j=1}^{S} \binom{N_j}{K_j}}.$$

Thus we arrive at

$$\text{prob}\left(\bigotimes_{i=1}^{S} x_i^{\otimes L_i} \middle| \rho^{(L)}\right) = \sum_{\{K_i\}} \text{prob}\left(\{K_i\} \middle| \{N_i\}, \rho^{(N)}\right) \prod_{j=1}^{S} \frac{\binom{N_j - L_j}{K_j - L_j}}{\binom{N_j}{K_j}}.$$

Written in this form, the right-hand side no longer makes reference to any specific ordering of the propositions or constituents.

The size of the enlarged assembly, N, can be chosen freely. So can the sizes of the S subassemblies, $\{N_i\}_{i=1}^{S}$, into which it is partitioned. In particular, we may consider the limit where all these numbers become very large, $N_i \to \infty$. In this limit the sum over $\{K_i\}$ is dominated by values of K_i which are very large, too. We may then approximate the binomial coefficients with the help of Stirling's formula,

$$N! \sim N^N e^{-N} \sqrt{2\pi N}, \tag{2.46}$$

to obtain the asymptotic behaviour

$$\frac{\binom{N_j - L_j}{K_j - L_j}}{\binom{N_j}{K_j}} \sim \left(\frac{K_i}{N_i}\right)^{L_i}.$$

Since the propositions $\{x_i\}$ constitute a minimal informationally complete set, there exists a unique single-constituent state, ρ, which satisfies the conditions

$$\text{prob}(x_i | \rho) = \frac{K_i}{N_i} \quad \forall i = 1, \ldots, S. \tag{2.47}$$

When expressed in terms of this state, the asymptotic behaviour of the combinatorial factors takes the simple form

$$\prod_{j=1}^{S} \frac{\binom{N_j - L_j}{K_j - L_j}}{\binom{N_j}{K_j}} \sim \prod_{j=1}^{S} \left(\frac{K_i}{N_i}\right)^{L_i} = \text{prob}\left(\bigotimes_{i=1}^{S} x_i^{\otimes L_i} \middle| \rho^{\otimes L}\right),$$

where, by analogy with Eq. (2.45), we defined

$$\rho^{\otimes L} := \underbrace{\rho \otimes \ldots \otimes \rho}_{L}. \tag{2.48}$$

This leaves us with

$$\mathrm{prob}\left(\bigotimes_{i=1}^{S} x_i^{\otimes L_i}\,\Big|\,\rho^{(L)}\right) = \lim_{N_i \to \infty} \sum_{\{K_i\}} \mathrm{prob}\left(\{K_i\}\,\Big|\,\{N_i\},\rho^{(N)}\right) \mathrm{prob}\left(\bigotimes_{i=1}^{S} x_i^{\otimes L_i}\,\Big|\,\rho^{\otimes L}\right).$$

By virtue of Eq. (2.47), there is associated with each set $\{K_i\}$, at given $\{N_i\}$, a distinct single-constituent state, ρ. Consequently, we may just as well sum over these associated ρ, rather than $\{K_i\}$, and replace the probability distribution for $\{K_i\}$ with a probability distribution for ρ. Moreover, in the limit $N_i \to \infty$, these single-constituent states associated with the $\{K_i\}$ lie dense in the manifold of normalized single-constituent states; they form practically a continuum. We may therefore replace the sum over associated states with an integral over the entire manifold of normalized single-constituent states,

$$\mathrm{prob}\left(\bigotimes_{i=1}^{S} x_i^{\otimes L_i}\,\Big|\,\rho^{(L)}\right) = \int d\rho\,\mathrm{pdf}(\rho)\,\mathrm{prob}\left(\bigotimes_{i=1}^{S} x_i^{\otimes L_i}\,\Big|\,\rho^{\otimes L}\right), \tag{2.49}$$

with some—as yet undetermined—probability density function, $\mathrm{pdf}(\rho)$.

The integral representation given in the last equation holds for the probabilities of joint propositions which are formed from members of the informationally complete set, $\{x_i\}$. For a composite system with just two constituents, we learnt in Section 2.5 that these joint probabilities in fact determine the probabilities of *all* joint propositions. This finding extends to composite systems with any number of constituents. Therefore, the integral representation continues to hold when we replace the conjunction of x_i's with a conjunction of *arbitrary* single-constituent propositions; in short, it applies to all joint probabilities. As long as there are no global propositions, the entirety of joint probabilities, in turn, completely specifies the composite state, $\rho^{(L)}$. Thus, in the absence of global propositions, the integral representation also applies to the state itself,

$$\rho^{(L)} = \int d\rho\,\mathrm{pdf}(\rho)\,\rho^{\otimes L}. \tag{2.50}$$

This representation of the state of an exchangeable assembly is known as the *de Finetti representation*. Loosely, the de Finetti representation may be interpreted as follows. In an exchangeable assembly, all members have been prepared identically—yet in which state is generally not known. If one did know that the state of each constituent was ρ, the composite state of the assembly would simply be the product state, $\rho^{\otimes L}$. Alas, the single-constituent state is *not* known, so the composite state must be a mixture of such product states, weighted with some probability density function, $\mathrm{pdf}(\rho)$. The latter describes a kind of 'meta-probability', in the sense that it does not refer directly to testable propositions. Rather, it assigns probabilities to states (which, in turn, assign probabilities to propositions).

To make these ideas more concrete, we return to our initial example of $L = 100$ identically manufactured dice. We suppose that there are three factories of varying quality in which dice are produced. The first factory produces perfect dice which, when rolled, obey a uniform probability distribution,

$$\{\mathrm{prob}(x_i|\rho_1)\} = \{\tfrac{1}{6},\tfrac{1}{6},\tfrac{1}{6},\tfrac{1}{6},\tfrac{1}{6},\tfrac{1}{6}\}.$$

Here x_i denotes the proposition 'the die shows i spots' and ρ_1, the state—that is, catalogue of probabilities—of a die produced in the first factory. The other two factories produce dice with some systematic error, and hence non-uniform probability distributions

$$\{\mathrm{prob}(x_i|\rho_2)\} = \{0.15, 0.15, 0.16, 0.17, 0.18, 0.19\}$$

and

$$\{\mathrm{prob}(x_i|\rho_3)\} = \{0.17, 0.16, 0.17, 0.17, 0.18, 0.15\},$$

respectively. Of all dice currently in circulation, one half were produced in factory 1, one third were produced in factory 2, and one sixth in factory 3. About our assembly of $L = 100$ dice we only know that they come from the same factory, but we do not know which. Accordingly, we assign to our assembly the composite state

$$\rho^{(100)} = \frac{1}{2}\rho_1^{\otimes 100} + \frac{1}{3}\rho_2^{\otimes 100} + \frac{1}{6}\rho_3^{\otimes 100},$$

which has precisely the form of a de Finetti representation. This simple example features a sum rather than an integral because, from the outset, we limited the single-constituent state to only three possibilities.

In our derivation of the general formula we took pains not to require that propositions can be tested jointly. This means the de Finetti representation holds not only in classical probability but also in theories with knowledge constraints, provided there are no global propositions.

2.7 Measuring a State

A state, understood as a catalogue of probabilities, can be measured by producing many identical copies of the system at hand and performing measurements on these copies. I describe the pertinent experimental procedure, which also works in the presence of knowledge constraints. I show how the state estimate changes, and its accuracy improves, as measurement data accumulate.

Up until now, our theoretical framework has met the standard of operationalism, as laid out in Section 2.1, only in part. While we have been careful to admit solely propositions that can actually be tested, and assigned probabilities to such testable propositions only, we have yet to give an operational meaning to the probabilities themselves. This means spelling out an experimental procedure for measuring them.

Since the entirety of probabilities constitutes the state, that is tantamount to a procedure for measuring the state. In this section we shall describe such a procedure. It will be general enough to work for theories both with and without knowledge constraints. Subsequently, in the light of that procedure, we will have to revisit the preliminary discussion of Section 2.2 about the preparation of states; this will be done in Section 2.8.

It is obvious that the state, understood as a catalogue of probabilities, cannot be determined solely on the basis of a single measurement on a single system. Indeed, the probability of a coin showing Heads cannot be measured by flipping it once; the probability of a Kinder Surprise® egg containing a little dinosaur cannot be measured by opening one such egg; and the probability of a photon from a given source passing through some polarization filter cannot be determined by sending an individual photon through this filter. Rather, the measurement of the state must involve measurements on multiple copies of the system or game of chance, respectively, which have all been prepared, or set up, in exactly the same way. In other words, measurements must be performed on multiple constituents of an exchangeable assembly. In the case of coin flips, one may realize such copies either by flipping the same coin multiple times or by flipping multiple coins in parallel, provided these coins all come from the same mint where they were manufactured with exactly the same (if any) systematic imperfections. In the case of chocolate eggs or photons, on the other hand, it is not possible to reuse the same system for sequential measurements. In these cases there remains only the second possibility: the copies in the exchangeable assembly must be distinct systems—chocolate eggs or photons—with identical prior history; that is, originating from the same chocolate factory or photon source, respectively.

The task at hand can now be formulated in more precise terms. *Given an exchangeable assembly*, we seek an experimental procedure which allows us—at least in principle—to determine the state of one of its constituents. So, for example, given a sufficiently large number of surprise eggs from the same factory, we seek a procedure to ascertain the probabilities of the various possible contents of an egg; or, given a sufficiently large number of photons from the same source, we seek a procedure to ascertain the probabilities with which an individual photon from this source passes through various polarization filters. The pertinent procedure will involve taking samples from the assembly, each comprising several copies, and performing measurements on them; say, opening 1,000 surprise eggs, or sending 1,000 photons through a particular polarization filter. The data gleaned from these samples will then inform our expectations as to the outcomes of further measurements on other constituents, or other samples, of the same assembly (Fig. 2.17). If only the number of samples and their sizes are large enough, and the measurements chosen properly, these data should be sufficient to determine the state of a constituent to arbitrary precision. By sending sufficiently many photons from the same source through sufficiently many different polarization filters and counting how many of them pass, we expect to learn the probabilities with which another photon from this source will pass through any polarization filter—and hence its state.

Let us translate these qualitative ideas into mathematics. We assume that the theoretical model which describes the physical system at hand or the pertinent game of chance

Fig. 2.17 *The role of exchangeable assemblies in empirical science. A sample is taken from an exchangeable assembly and subjected to measurement. The results inform expectations as to the outcomes of measurements on other, possibly larger samples.*

does not feature global propositions. Thus we exclude exotic cases like the peculiar coin flip game described (in its first version) in Section 2.5. Under this assumption, the state of the exchangeable assembly has a de Finetti representation, Eq. (2.50). Our initial ignorance about the state of a constituent is reflected in the probability density function, $\text{pdf}(\rho)$. To learn more about this state, we take a sample of size M ($M < L$) from the exchangeable assembly and subject it to some measurement, yielding data, D. Afterwards, the remaining $(L - M)$ constituents (that is, the original assembly minus the sample) still form an exchangeable assembly. Its state also has a de Finetti representation; yet this de Finetti representation features a probability density function which is generally different from the original one. In order to describe how the probability density function changes, we assume that the sample comprises the first M members of the assembly. If the assembly were in a product state, $\rho^{\otimes L}$, then, by Eq. (2.42), measuring D would trigger an update of this state to

$$(\rho^{\otimes L})_{|D} = (\rho^{\otimes M})_{|D} \otimes \rho^{\otimes (L-M)}.$$

In reality, the state of the assembly is not a product state but a mixture of such product states, Eq. (2.50). According to Eq. (1.38), this mixture is updated to

$$(\rho^{(L)})_{|D} \propto \int d\rho\, \text{pdf}(\rho)\, \text{prob}(D|\rho^{\otimes M})\, (\rho^{\otimes M})_{|D} \otimes \rho^{\otimes (L-M)}. \qquad (2.51)$$

Indeed, the remaining $(L - M)$ members, which were not subject to measurement, still form an exchangeable sequence, whose state has a de Finetti representation. This de Finetti representation features a new probability density function,

$$\text{pdf}(\rho|D) \propto \text{prob}(D|\rho^{\otimes M})\, \text{pdf}(\rho). \qquad (2.52)$$

Here the constant of proportionality does not depend on ρ; its value is such that the new density function is still normalized. This update rule for the probability density function has the structure of Bayes' rule, Eq. (1.39), with the left-hand side corresponding to the posterior and the two factors on the right-hand side corresponding to likelihood function and prior, respectively. Unlike the original Bayes' rule, however, which applied to ordinary probabilities assigned to testable propositions, the present rule applies to the 'meta-probabilities' assigned to single-constituent states. Loosely speaking, it is an extension of Bayes' rule to the meta-level. It encapsulates, in a quantitative fashion, the process of *learning* about the state from sample data.

To give a concrete example, we consider coins produced by a particular mint known to have quality issues. All coins produced by this mint show the same systematic imperfection, which makes the probabilities of Heads and Tails in a coin flip experiment differ from one half. The actual probability distribution is not known, however, and needs to be measured. For a coin flip experiment, the convex set of normalized states is depicted in Fig. 2.8 (thick straight line). A normalized state is specified by a single parameter. As this parameter, we choose the probability of Heads, p. For the sake of argument, we expect a priori that while the state will likely differ from the distribution associated with a perfect coin, $p = 1/2$ (represented in the figure by the open circle), the deviation from this perfect distribution will not be too large, either. We model this initial bias with a Gaussian,

$$\text{pdf}(p) \propto \exp\left[-\frac{(p-\frac{1}{2})^2}{2\varsigma^2}\right], \tag{2.53}$$

of some non-zero width, $\varsigma > 0$. In order to learn more about p, we take a sample of M coins from the mint and flip them. (Given that they are all produced in an identical fashion, this is equivalent to flipping a single coin M times.) We find that they show Heads K times and Tails $(M - K)$ times. The likelihood of this frequency is given by a binomial distribution,

$$\text{prob}(\{K, M-K\}|\{p, 1-p\}^{\otimes M}) = \binom{M}{K} p^K (1-p)^{M-K}, \tag{2.54}$$

which, for large M and K, may be approximated by another Gaussian,

$$\text{prob}(\{K, M-K\}|\{p, 1-p\}^{\otimes M}) \propto \exp\left[-\frac{(p-\frac{K}{M})^2}{2\tau^2}\right]. \tag{2.55}$$

It is peaked at $K/M = p$ and has variance

$$\tau^2 = \frac{p(1-p)}{M}.$$

Rather than as a function of the frequencies, given p, this Gaussian may also be regarded as a function of p, given the frequencies. Viewed as the latter, it is peaked at $p = K/M$ and has variance

$$\tau^2 = \frac{K(M-K)}{M^3}.\tag{2.56}$$

By virtue of the update rule, Eq. (2.52), the posterior is then the product of the two Gaussians. It is once again a Gaussian,

$$\mathrm{pdf}(p|\{K, M-K\}) \propto \exp\left[-\frac{(p-\nu)^2}{2\omega^2}\right],\tag{2.57}$$

whose centre,

$$\nu = \frac{\varsigma^2}{\varsigma^2 + \tau^2}\frac{K}{M} + \frac{\tau^2}{\varsigma^2 + \tau^2}\frac{1}{2},\tag{2.58}$$

interpolates between the measured relative frequency, K/M, and the centre of the prior, $1/2$. (In statistics, this is known as 'Bayesian interpolation'.) Its variance is given by

$$\omega^2 = \frac{\varsigma^2\tau^2}{\varsigma^2 + \tau^2}.\tag{2.59}$$

In the limit of very large samples, $M \to \infty$, the likelihood function approaches a δ-function, $\tau \to 0$, and so does the posterior, $\omega \to 0$. In the interpolation formula for its centre, Eq. (2.58), the residual influence of the prior fades away; the location of the centre is entirely dominated by the experimental data, $\nu \to K/M$. Thus we have ascertained the state: $p = K/M$. This is in fact the first instance where a probability is identified with an empirical relative frequency. There was no need to posit this connection from the outset; it arose naturally out of a careful consideration of exchangeable assemblies.

The various steps of our argument in the coin flip example may be generalized (Fig. 2.18):

1. Initially, an exchangeable assembly is characterized by some prior probability density function, $\mathrm{pdf}(\rho)$, on the convex set of normalized single-constituent states. If, based on earlier experience or theoretical grounds, one expects the members of the assembly to be in a state close to σ, then this prior will be peaked around σ. It has a non-zero width, reflecting the finite degree of confidence in this initial bias.

2. Investigation of a sample of size M yields data, D. Associated with this data is a likelihood function, $\mathrm{prob}(D|\rho^{\otimes M})$, which is typically peaked around some other state that might be close, but is usually not equal, to σ. The likelihood function,

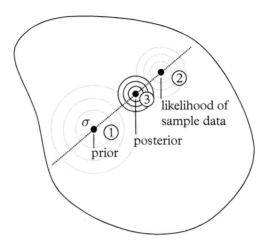

Fig. 2.18 *Learning from sample data.*

too, has a non-zero width, reflecting the finite size of the sample and possibly other forms of noise.

3. According to Bayes' rule on the meta-level, multiplying the prior with the likelihood function yields the posterior, $\mathrm{pdf}(\rho|D)$. The latter is typically narrower than the prior, reflecting the growing confidence in the state estimate as experimental data accumulate. The centre of the posterior has shifted from the original bias, σ, to a new state interpolating between σ and the centre of the likelihood function.

This procedure can be iterated, with the posterior taking the place of the prior in the next iteration. Especially on systems more complex than a coin flip, one might perform different measurements on different samples until, eventually, all aspects of the system have been investigated. By collecting a sufficient amount of experimental data, the posterior can be made as narrow as desired. In this fashion, the state of a constituent of the assembly can be determined to any required precision.

The measurement procedure described above does not just yield the state of an individual constituent; it actually yields the state of the entire assembly. As the posterior approaches a δ-function, $\mathrm{pdf}(\rho|D) \to \delta(\rho - \rho_0)$, peaked at the measured single-constituent state, ρ_0, the state of the remaining exchangeable assembly approaches the corresponding product state:

$$\rho^{(L-M)} \to \rho_0^{\otimes(L-M)}. \tag{2.60}$$

This is significant for two reasons. First, being thus measurable, it also endows the composite state of the assembly with an operational meaning. Secondly, and more importantly, it underpins one of the central tenets of the scientific method. We generally assume that we may study samples in a laboratory and then extrapolate our findings to alike systems elsewhere in the universe, even if these are much larger. For instance, it should be possible to study the properties of hydrogen in a laboratory and then extrapolate the findings to the enormous clouds of hydrogen in remote galaxies; or to study hot, dense matter in a high-energy particle collider and then extrapolate the findings to the interior of stars. Such courageous extrapolation rests on the premise that it would make no difference if, say, a hydrogen atom in the laboratory and a hydrogen atom in the galactic cloud swapped places; in other words, that the laboratory sample, being composed of the same kind of constituents and having been prepared under conditions (temperature, pressure, etc.) which mirror those of the remote system, and the remote system itself are in fact part of a single exchangeable assembly. Moreover, there is the premise that the insights gleaned from the laboratory sample suffice to predict the properties of the remote system *completely*. The remote system's larger size does not entail new features which cannot possibly be investigated in the laboratory. Translated into the language of exchangeable assemblies, this means that measurements on a sample must allow one to determine the state not just of an individual constituent but also of other, possibly larger samples—and thus, indeed, of the assembly as a whole. This idea is alluded to graphically in Fig. 2.17.

In a nutshell, the procedure for measuring a state comprises three elements: embedding into an exchangeable assembly; sampling; and Bayesian update. It is here that, for the first time, probabilities are linked to measured relative frequencies. Via this procedure we learn not just the state of an individual constituent but also the state of the assembly into which it is embedded. The latter is crucial for our ability to extrapolate measurement results to other, possibly much larger systems, and thus, ultimately, our ability to study the world in a laboratory. Even though we illustrated the state-measurement procedure with a classical coin flip example, it is by no means limited to classical probability theory; it works just as well in more general settings with knowledge constraints. Indeed, nowhere in the derivation of the update rule, Eq. (2.52), did we have to assume that all propositions can be tested jointly. Its derivation does, however, rely on the de Finetti representation for exchangeable assemblies, Eq. (2.50). Therefore, like the latter, it hinges on the assumption that there are no global propositions. Our ability to comprehend the world around us through the study of laboratory samples presumes the absence of global propositions.

2.8 Preparing a State

In an operational theory it ought to be possible to take an exchangeable assembly with unknown prior history and mould it into a definite product state, by means of measurements and the selective discarding of constituents. I investigate the pertinent experimental procedures and the ensuing constraints on the set of states. I argue that these constraints leave only one alternative to classical probability theory that is at the same time consistent, universal, and fully operational. This will turn out to be quantum theory.

Throughout this chapter we have emphasized that the concept of 'state' has a place in an operational theory if and only if we can spell out the experimental procedures for preparing and measuring it. Specifically, we have imposed two requirements so far:

1. Every state must be reachable from the state of total ignorance, via a combination of measurement and mixing (Eq. (2.5)).
2. Every state must be measurable, by manufacturing an exchangeable assembly, sampling, and Bayesian update. The latter presupposes, in particular, that there are no global propositions.

These two requirements are in fact linked, and must be strengthened further, by the following consideration. As for measuring a state, we laid out the pertinent procedure in Section 2.7. It involves preparing a large number of identical copies, from which samples are then taken and subjected to measurement. Thus, strictly speaking, a state ρ_0 can be the outcome of a state measurement only if a large number of copies were previously prepared in that same state; preparing just a single copy is not enough. In other words, the measurability of ρ_0 presupposes that it is possible to prepare not just ρ_0 itself but also arbitrary product states, $\rho_0^{\otimes N}$, for any N. This constitutes an additional, non-trivial assumption which will eventually lead to a further constraint on the number of state parameters. In this section we shall inspect this assumption and its ramifications more closely.

A product state is a special case of an exchangeable state. One should be able to prepare it by starting from an exchangeable assembly (in some unspecified initial state) and then subjecting this assembly to a series of suitable experimental procedures, maintaining at all times its invariance under permutations. These procedures should be designed to narrow the probability density function in the de Finetti representation until it approaches a δ-function, $\mathrm{pdf}(\rho) \to \delta(\rho - \rho_0)$, peaked at the desired state, ρ_0. In particular, we consider two types of procedures: (i) measurement and (ii) discarding constituents. For instance, the raw material might consist in photons originating from some light source, assuming they form an exchangeable assembly. The photons are then sent through a polarization filter. The filter serves a dual purpose: it measures the polarization in a certain direction; and, depending on the outcome of this measurement, it may discard (that is, absorb) a photon. Behind the filter emerges a beam of photons which still constitutes an exchangeable assembly. Compared to the original assembly, it contains fewer photons, and these are now all polarized in the same direction. Another example might be an assembly of perfect dice which are all rolled in parallel and automatically discarded (in a manner hidden from the observer) if they show an odd number. The net result is a smaller assembly of dice showing even numbers only. Thus, indeed, the suitable combination of measurement and selective discarding leads to a narrowing of the range of possible single-constituent states. (By contrast, the other intervention considered in Section 2.2—mixing—would be counterproductive. In the present context it would correspond to merging two different exchangeable assemblies, with variable relative weights. Yet such merging would only increase the uncertainty about the single-constituent state and lead to a broadening, rather than narrowing, of the pertinent probability density function.) Hereafter, we shall investigate which product states, $\rho_0^{\otimes N}$, can be prepared in this way. Only those states may be considered truly operational.

We begin by scrutinizing the effect of measurement on the exchangeable assembly. In order to allow for the subsequent selective discarding of constituents, this assembly must comprise a sufficiently large number of constituents, M; certainly more than the assembly we seek to prepare, $M > N$. Any measurement performed on this initial assembly will involve testing propositions about individual constituents, in a manner that maintains the permutation symmetry of the assembly. The latter is ensured if the measurement merely counts the frequency with which certain propositions are found to be true rather than attributing these results to specific constituents. For instance, in the earlier die roll example, a comparison of the sizes of initial and final assembly will reveal the frequencies with which the dice showed odd and even numbers, respectively; but the setup is such that it does not permit an attribution to individual dice. In mathematical terms, let $\{x_i\}$ denote the set of propositions being tested on individual constituents. They partition the overarching proposition, $\bigoplus_i x_i = a$. The measurement on the assembly then consists in testing these propositions on all constituents and counting the respective frequencies with which they are found to be true; the order of the constituents is irrelevant. A specific outcome of absolute frequencies, $\{M_i\}$, with $\sum_i M_i = M$, corresponds to the truth of the proposition

$$\{M_i\} := \bigoplus_\pi \pi \left[\bigotimes_i x_i^{\otimes M_i} \right], \tag{2.61}$$

where π denotes all possible inequivalent permutations, or 'combinations'. (To avoid a proliferation of symbols, I use the same notation for the proposition as for the associated set of numbers.) If, prior to measurement, the assembly is already in a product state, $\rho^{\otimes M}$, the posterior reads

$$(\rho^{\otimes M})_{|\{M_i\}} = \frac{1}{\#\{\pi\}} \sum_\pi \pi \left[\bigotimes_i \rho_{|x_i}^{\otimes M_i} \right], \tag{2.62}$$

where

$$\#\{\pi\} = \binom{M}{M_1, M_2, \ldots} = \frac{M!}{M_1! M_2! \cdots}$$

counts the number of combinations. You will derive this posterior in Exercise (2.8). If, on the other hand, the prior state has the more general de Finetti form, Eq. (2.50) (with $L = M$), the posterior reads

$$(\rho^{(M)})_{|\{M_i\}} \propto \int d\rho \, \mathrm{pdf}(\rho | \{M_i\}) \, (\rho^{\otimes M})_{|\{M_i\}}, \tag{2.63}$$

with

$$\text{pdf}(\rho|\{M_i\}) \propto \text{pdf}(\rho)\,\text{prob}(\{M_i\}|\rho^{\otimes M}). \tag{2.64}$$

These formulae are special cases of Eqs (2.51) and (2.52), for $L = M$ and $D = \{M_i\}$.

After the measurement, a filter may selectively discard constituents—such as, in our earlier example, dice showing an odd number. Discarding constituents reduces the size of the assembly from M to N. Constituents are discarded based on the test results for the propositions $\{x_i\}$. Of the constituents for which a certain proposition, x_j, was found to be true, the filter discards some fraction, reducing their number from M_j to N_j; and it does so for all j. If the state after measurement, but prior to filtering, was $(\rho^{\otimes M})_{|\{M_i\}}$ (Eq. (2.62)), then discarding the constituents changes that to $(\rho^{\otimes N})_{|\{N_i\}}$. If, on the other hand, the post-measurement state had the more general form of Eq. (2.63), then the filter yields an assembly in the state

$$\rho^{(N)} \propto \int d\rho\,\text{pdf}(\rho|\{M_i\})\,(\rho^{\otimes N})_{|\{N_i\}}. \tag{2.65}$$

In the limit where the filtered assembly is still very large, $N \to \infty$, the last term in the integrand approaches a product state,

$$(\rho^{\otimes N})_{|\{N_i\}} \to (\rho_{|\{f_i\}})^{\otimes N}, \quad \rho_{|\{f_i\}} := \sum_i f_i \rho_{|x_i}, \quad f_i := \frac{N_i}{N}. \tag{2.66}$$

This can be seen as follows. Multiplying out the N-fold product yields

$$(\rho_{|\{f_i\}})^{\otimes N} = \sum_{\{N_i'\}} \left(\prod_k f_k^{N_k'} \right) \sum_\pi \pi \left[\bigotimes_i \rho_{|x_i}^{\otimes N_i'} \right] = \sum_{\{N_i'\}} \text{prob}(\{N_i'\}|\{f_i\})\,(\rho^{\otimes N})_{|\{N_i'\}},$$

where the summation runs over all possible $\{N_i'\}$, $\sum_i N_i' = N$, and the probability refers to the multinomial distribution,

$$\text{prob}(\{N_i'\}|\{f_i\}) = \binom{N}{N_1', N_2', \ldots} \prod_k f_k^{N_k'}.$$

As $N \to \infty$, this multinomial distribution becomes narrowly peaked at $N_i' = f_i N = N_i$ for all i. Thus, indeed, $(\rho_{|\{f_i\}})^{\otimes N}$ approaches $(\rho^{\otimes N})_{|\{N_i\}}$, and vice versa. In this limit the state of the filtered assembly acquires the form

$$\rho^{(N)} \propto \int d\rho\,\text{pdf}(\rho|\{M_i\})\,(\rho_{|\{f_i\}})^{\otimes N}. \tag{2.67}$$

This state is still exchangeable, as we anticipated. It possesses a de Finetti representation, Eq. (2.50), with $L = N$ and modified probability density function

$$\text{pdf}(\rho|\text{filtered}) = \int d\rho'\, \delta(\rho - \rho'_{|\{f_i\}})\, \text{pdf}(\rho'|\{M_i\}). \qquad (2.68)$$

The questions we posed at the outset can now be phrased in more specific terms: is it possible to engineer the experimental procedures in such a way that the filtered assembly will be in a product state, $\rho^{(N)} = \rho_0^{\otimes N}$? And are there any constraints on the states which can be thus produced? For the filtered assembly to be in a product state, the probability density function must approach a δ-function, $\text{pdf}(\rho|\text{filtered}) \to \delta(\rho - \rho_0)$. This can only happen if $\rho'_{|\{f_i\}} = \rho_0$ for all ρ' which contribute to the integral, $\text{pdf}(\rho'|\{M_i\}) \neq 0$. As the map from ρ to $\rho_{|\{f_i\}}$, Eq. (2.66), is idempotent,

$$\left(\rho_{|\{f_i\}}\right)_{|\{f_i\}} = \rho_{|\{f_i\}},$$

this condition implies a constraint on ρ_0:

$$\rho_0 = \rho_{0|\{f_i\}}. \qquad (2.69)$$

In other words, with our two basic procedures—measurement and selective disposal of constituents—we can only prepare states which satisfy this additional constraint. Therefore, only states of this kind should feature in our operational theory.

The additional constraint on operational states has consequences for the number of state parameters, $S(d)$. In Eq. (2.69) both states are normalized. Thus, the state on the left-hand side is specified by $(S(d) - 1)$ parameters. This must equal the number of parameters needed to specify the state on the right-hand side. According to Eq. (2.66), specifying the latter amounts to specifying the relative frequencies, $\{f_i\}$, the measurement setup, $\{x_i\}$, and the respective posteriors, $\{\rho_{0|x_i}\}$, regarded as states in theories of respective dimensions $\{d(x_i)\}$. For simplicity, we shall limit ourselves to the special case where all the tested propositions are most accurate: $d(x_i) = 1$ for all i. (You will study the more general case in Exercise (2.15).) Then the posteriors, $\{\rho_{0|x_i}\}$, are in fact uniquely determined by the most accurate propositions and require no extra parameters. As for the most accurate propositions themselves, they must partition the overarching proposition, so there are d of them, and they are mutually exclusive. For the first proposition, e_1, one needs $X(d)$ parameters. The second proposition, e_2, must contradict e_1, and hence be some refinement of the complement of e_1, $e_2 \subseteq \neg_a e_1$. This complement has dimension $(d-1)$, so in order to specify e_2, one needs only $X(d-1)$ parameters. In an analogous fashion we can argue that the specification of e_3 requires $X(d-2)$ parameters, and so on. Summing up and using Eq. (2.17), one obtains a total of

$$\sum_{k=1}^{d} X(k) = X(2)\frac{d(d-1)}{2} \qquad (2.70)$$

parameters. To this we must add the number of continuous parameters needed to specify the d relative frequencies. Since we assumed the limit $N \to \infty$, the allowed frequencies do indeed form a continuum, so their specification requires d continuous parameters. Due to normalization, however, only $(d-1)$ of these are independent. Comparing the total number of parameters on both sides of Eq. (2.69) then yields an extra constraint on the number of state parameters:

$$S(d) = X(2)\frac{d(d-1)}{2} + d. \tag{2.71}$$

This constraint is consistent with, and further strengthens, the relationship between $S(d)$ and $X(d)$ which we found earlier, Eq. (2.40). It implies that the number of state parameters grows linearly with the dimension, $S(d) = d$, if and only if the set of most accurate propositions is discrete, $X(2) = 0$. This is the case of classical probability theory. Otherwise, for $X(2) \geq 1$, the number of state parameters can only grow quadratically with the dimension. This means that in the power law, Eq. (2.39), we must then have $n = 2$. We are thus left with just two possibilities:

$S(d)$	$X(d)$	case	
d	0	classical	(2.72)
d^2	$2(d-1)$	quantum	

The first case corresponds to classical probability theory. The second case represents the sole non-classical alternative which is logically consistent, universal, and fully compliant with all our operational requirements. This is quantum theory, to which we will turn in Chapter 3.

Chapter Summary

- An operational theory admits only concepts that are associated with a concrete experimental procedure. In particular, it admits only propositions which can be tested and only states which can be both prepared and measured.

- Some quantum phenomena may be mimicked by toy models with artificially imposed knowledge constraints. Examples are the Spekkens toy model, the circle model, and the sphere model.

- Every state must be reachable from the state of total ignorance, via a combination of measurement and mixing. In the presence of knowledge constraints there exist states which would be logically consistent yet cannot be reached in this manner and, therefore, must remain outside an operational theory. The reachable states form a convex set.

- Subnormalized states are states where the overarching proposition need not have probability one. Reachable subnormalized states form a convex cone.
- Maximal knowledge can be characterized in two ways: (i) as the result of a preparation so accurate that it completely overwrites all prior results and biases; or (ii) as certainty about the truth of some most accurate proposition.
- States of maximal knowledge cannot be written as a convex combination of other states. They are also called pure states; all others are mixed.
- Upon measurement, a pure state is updated to another pure state. In settings with knowledge constraints, prior and posterior states might differ.
- The classical AND and OR operations can be extended to situations with knowledge constraints, thus endowing the set of propositions with the structure of a lattice. They are commutative and associative but in general no longer distributive.
- When multiple systems are combined into one composite system, both the dimension of the logical structure and the number of parameters that specify a subnormalized state satisfy a product rule.
- If all joint probabilities factorize, to wit, the outcomes of measurements on different constituents are statistically independent, then the composite system is in a product state. Otherwise, the state is said to exhibit correlations.
- Multiple copies of a system, each prepared independently and in the same (but possibly unknown) fashion, form an exchangeable assembly. Its state has the de Finetti form.
- A state cannot be determined on the basis of a single measurement on a single system. Rather, measuring a state involves (i) manufacturing an exchangeable assembly; (ii) taking a sample and subjecting it to measurement; and (iii) updating, via Bayes' rule, the probability density function in the de Finetti representation. The last two steps are iterated, possibly with varying types of measurements, until the probability density function is narrowly peaked. The measured state is given by the location of the peak.
- When all members of an exchangeable assembly are subjected to the same measurement, some are discarded based on the outcome, and the remainder are randomly reshuffled, the result is again an exchangeable assembly. In this way one can prepare assemblies in any desired product state.
- A probability theory is universal if it is applicable to problems of arbitrary dimension, and if the logical structure and convex cone of reachable states depend on that dimension only.
- The requirements of logical consistency, universality, and operational meaning leave only one alternative to classical probability theory. This will be quantum theory.

Further Reading

The operational approach to physical theories resonated strongly with some of the early protagonists of quantum theory. It is still instructive today to read, for instance, the pertinent writings of Niels Bohr, collected in Bohr (1987). The convex cone of reachable states plays a key role in a number of analyses of the probabilistic structure of quantum theory, for example by Mackey (1963), Kraus (1983), Varadarajan (1985), or Gudder (1988). That propositions about the outcomes of quantum measurements exhibit the mathematical structure of a lattice was first recognized by Birkhoff and v. Neumann (1936). You can learn more about lattice theory in general from Birkhoff (1967). The lattice structure represents a generalized 'quantum logic' that includes, from the start, the binary operators AND and OR. This framework of quantum logic yields many interesting results, which you will find summarized in books by Jauch (1968) and Piron (1976), as well as in the remarkable thesis of Piron (1964). Exchangeable assemblies and the de Finetti representation in classical probability theory are discussed in the book by de Finetti (1990), their extension to quantum theory and generalized probability theories, in articles by Caves *et al.* (2002), Renner (2007), and Barrett and Leifer (2009). That in quantum theory, too, the state update for an exchangeable assembly is governed by a version of Bayes' rule, Eq. (2.52), was first noted by Schack *et al.* (2001). As an introduction to inference and learning in general, I recommend the book by MacKay (2003); more details about Bayesian interpolation in particular can be found in MacKay (1992). Measuring, or at least estimating, a state on the basis of multiple measurements on identically prepared copies is a research topic of its own, known as 'state tomography'. For an overview of that field you may consult D'Ariano *et al.* (2004) and Banaszek *et al.* (2013). Finally, the idea that basic desiderata of logical consistency, operational meaning, and universality essentially single out quantum theory as the sole alternative to classical probability theory is behind an area of research known as 'reconstruction' of quantum theory. Some of the references mentioned earlier fall into that category. For an overview of further, more recent activities I suggest the collection of essays in Chiribella and Spekkens (2016).

EXERCISES

2.1. Specker's parable

Specker (1960) tells the following parable of the overprotective seer (translated from the original German, and with clarifications added in square brackets, by Liang *et al.* (2011)):

> At the Assyrian School of Prophets in Arba'ilu in the time of King Asarhaddon, there taught a seer from Nineva. He was [a] distinguished representative of his faculty (eclipses of the sun and moon) and aside from the heavenly bodies, his interest was almost exclusively in his daughter. His

teaching success was limited, the subject proved to be dry, and required
a previous knowledge of mathematics which was scarcely available. If he
did not find the student interest which he desired in class, he did find it
elsewhere in overwhelming measure. His daughter had hardly reached a
marriageable age when he was flooded with requests for her hand from
students and young graduates. And though he did not believe that he would
always have her by his side, she was in any case still too young and her
suitors in no way worthy. In order that they might convince themselves of
their worthiness, he promised her to the one who could solve a "prediction
problem" which he set [as follows].

The suitor was taken before a table on which three little boxes stood in a
row and was asked to say which boxes contained a gem and which did not
[with the total number of gems being indeterminate]. But no matter how
many tried, the task seemed impossible. In accordance with his prediction,
each of the suitors was requested by the father to open two boxes which he
had marked as both empty or both full. But it always turned out that one
contained a gem and the other one did not, and furthermore the stone was
sometimes in the first box [opened] and sometimes in the second. But how
should it be possible, given three boxes, neither to mark two as empty nor
two as full?

The daughter would have remained single until her father's death had she
not followed the advice of a prophet's son and [when a suitor she fancied had
made his prediction] quickly opened two boxes, one of which was marked
full and the other empty [and the suitor's prediction was found, in this
case, to be correct]. Following the weak protest of her father that he had
wanted two other boxes opened, she tried to open the third. But this proved
impossible whereupon the father grudgingly admitted that the prediction
was correct [or at least could not be proven incorrect].

Draw the Hasse diagram representing all *testable* propositions (that is, only those
which may be tested according to the father's rules) and their logical relationships.
Does it satisfy all the properties discussed in the main text?

2.2. Totally mixed state

Argue that in a probability theory which is universal and compatible with the
requirements laid out in Chapter 2 the state of total ignorance can be written in
the form

$$\iota = \frac{1}{d(a)} \sum_{\oplus_i e_i = a} \iota_{|e_i}$$

for any most accurate partition, $\{e_i\}$, of the overarching proposition, a. Thus, the
state of total ignorance may be thought of as the result of mixing—in the sense of
convex combination—all the pure states associated with this partition, with equal
weight. This explains why the state of total ignorance is also called the 'totally mixed'
state.

2.3. Reachable states in classical probability

Verify that in classical probability theory the convex set of reachable states (Section 2.2) coincides with the convex set of logically consistent states (Section 1.3).

2.4. Sequence of measurements

(a) Reproducibility demands that for any two propositions, x and y, which individually pertain to outcomes of reproducible measurements, we have

$$\text{prob}(x|x,\rho) = \text{prob}(y|y,\rho) = 1 \quad \forall \rho.$$

However, this does not guarantee that the sequence $x \leftarrow y$ will be equally reproducible; there may be situations where

$$\text{prob}(x \leftarrow y|x \leftarrow y,\rho) \neq 1.$$

Thus, a sequence of reproducible measurements can in general *not* be regarded as one combined reproducible measurement. Give an example.

(b) Show that the following four statements are equivalent:
   i. The sequence $x \leftarrow y$ is reproducible,

$$\text{prob}(x \leftarrow y|x \leftarrow y,\rho) = 1 \quad \forall \rho.$$

   ii. The order of ascertainments is irrelevant,

$$\text{prob}(x \leftarrow y|\rho) = \text{prob}(y \leftarrow x|\rho) \quad \forall \rho,$$

   or, in short, $x \leftarrow y = y \leftarrow x$.
   iii. The two propositions, x and y, can be tested jointly.
   iv. The sequence of ascertainments amounts to the joint proposition 'x and y',

$$x \leftarrow y = x \wedge y.$$

Therefore, a sequence of reproducible measurements may be regarded as one combined reproducible measurement if and only if the propositions involved are jointly testable.

(c) Show

$$x \leftarrow (x \oplus y) = (x \oplus y) \leftarrow x = x$$

and

$$x \leftarrow y = \emptyset \quad \Leftrightarrow \quad x \subseteq \neg_a y.$$

(d) Verify that the update rule for mixtures, Eq. (1.38), applies just as well after ascertaining an entire sequence of propositions, s, rather than just an individual proposition, y. This holds regardless of whether the sequence is reproducible.

2.5. Maximal knowledge

(a) Verify that in the 3×3 problem the two alternative definitions of maximal knowledge are not equivalent. Find specific situations where you have maximal knowledge according to one definition but not the other.

(b) In the main text we used the assumption that the first criterion for maximal knowledge is as strong as the second, Eq. (2.11), to derive the preservation of maximal knowledge under measurement updates, Eq. (2.12). Show that, conversely, assuming the preservation of maximal knowledge under measurement updates implies Eq. (2.11). In other words, the two requirements are equivalent.

(c) For the 3×3 problem describe a situation where a measurement update does *not* preserve maximal knowledge, in the sense that Eq. (2.12) is violated.

(d) Without using Eq. (2.11) or Eq. (2.12), and hence also for exotic models like the 3×3 problem, show that there cannot exist a state in which two different most accurate propositions are true at the same time. That is to say,

$$\text{prob}(e|\rho) = \text{prob}(f|\rho) = 1 \quad \Rightarrow \quad e = f.$$

Remember that without making said assumptions, you are not allowed to argue with the logical AND operation.

2.6. Card game

Kirkpatrick (2003) devised the following card game. There is a deck of 30 cards featuring different faces and suits (Table 2.1). Out of this deck one takes an—initially random—subdeck. On the subdeck one may perform one of three measurements: one may measure the variable 'face', 'suit', or 'colour' (red if Diamonds or Hearts,

Table 2.1 *Deck of cards used in Kirkpatrick's card game. It features three different suits (Diamonds, Hearts, Spades) and three different faces (King, Queen, Jack). The deck contains ten cards of each suit and ten cards of each face. In total, there are 30 cards.*

	Diamonds	Hearts	Spades
King	5	4	1
Queen	1	5	4
Jack	4	1	5

black if Spades). Such a measurement proceeds in three steps:

1. Shuffle the subdeck.
2. Report the value of the variable (face, suit, or colour) of the *top* card of the subdeck.
3. Construct a new subdeck consisting of all cards for which the variable has the reported value. For instance, if we just measured 'face' and found 'Queen', the new subdeck consists of all ten Queens.

Assignments:

(a) Draw the Hasse diagram representing all *testable* propositions and their logical relationships.
(b) Starting from an initially random subdeck, then performing successive measurements of 'face' and 'suit', what is the probability of finding first 'Queen' and then 'Spades'? If this is followed by another measurement of 'face', what is the probability of finding 'Queen' again?
(c) Do the logical structure and probabilities of this game satisfy all the desiderata laid out in the main text?
(d) Which of the quantum phenomena presented in Section 1.1 are also exhibited (qualitatively) by this game? Which are not?
(e) Are the two alternative definitions of maximal knowledge equivalent? What if only the variables 'face' and 'suit' may be measured, but not 'colour'?

2.7. **Binary operators** $\wedge, \vee$

(a) Verify that the AND ($\wedge$) operator constructed via Eq. (2.26) satisfies Eq. (2.18).
(b) Show that the binary operators $\wedge$ and $\vee$ are commutative.
(c) Are the binary operators $\wedge$ and $\vee$ well defined for the 3×3 problem? If so, what is $e_{11} \vee e_{22}$? If not, why?
(d) Show that in general the binary operators $\wedge$ and $\vee$ do not satisfy the distributivity property, Eq. (2.30). Give a concrete counterexample based on one of the toy models where the binary operators are defined. What does the experimental evidence say about distributivity in the quantum realm?
(e) Prove the generalized sum rule for the dimension, Eq. (2.31).

2.8. **Measurement update**

Prove the following assertions:

(a) For any $e \subseteq x$, it is

$$\mathrm{prob}(e|\rho) = 0 \quad \Leftrightarrow \quad \mathrm{prob}(e|\rho_{|x}) = 0.$$

For $e \not\subseteq x$, on the other hand, the two equations are not equivalent. Give a concrete counterexample.

(b) For any ρ, it is

$$\neg_a x = \bigvee\nolimits_{\mathrm{prob}(e|\rho_{|x})=0} \theta_{\neg_a x}(e).$$

2.9. Narrowing the state

Given a state, ρ, we may define its *width* as the minimum dimension required for a proposition to be true with certainty:

$$d(\rho) := \min_{\text{prob}(x|\rho)=1} d(x).$$

In the classical set picture this coincides with the number of most accurate propositions that have a non-vanishing probability,

$$d_{\text{class}}(\rho) = \#\{e| \text{prob}(e|\rho) \neq 0\},$$

and hence with the size of the support (as measured by the number of elements) in sample space. In the absence of knowledge constraints, it is clear that a measurement can only lead to a narrowing, or at most a preservation, of this support:

$$d(\rho_{|x}) \leq d(\rho) \; \forall \rho, x.$$

For instance, before rolling a die, the pertinent state has width $d(\rho) = 6$. Then, hidden from the observer's view, the die is rolled and the truth of the proposition x: 'it is an odd number' is announced. This triggers an update to a new state, $\rho \to \rho_{|x}$, which has a smaller width, $d(\rho_{|x}) = 3$. The narrowing of the state reflects our learning as we process experimental data. As soon as the width is down to one, we have achieved maximal knowledge.

(a) Show that in the 3×3 problem there exist measurements which violate the above inequality. In other words, there are measurements which lead to a broadening of the state.
(b) By contrast, prove that in all models where the two definitions of maximal knowledge are equivalent the above inequality continues to hold, even if there are knowledge constraints. (Project)

2.10. State parameters

Prove that the number of state parameters is at least as large as the dimension of the logical structure, Eq. (2.36). Explain why it is $S = d$ if all propositions can be tested jointly, whereas it is $S > d$ in the presence of knowledge constraints.

2.11. Constrained coin flip games

Consider the peculiar coin flip game described in Section 2.5 after Eq. (2.37).

(a) Draw a Hasse diagram of the logical structure for both the single game and the combined game (in both versions).
(b) Determine the dimensions of the respective logical structures. Do they satisfy the product rule for the dimension, Eq. (2.35)?

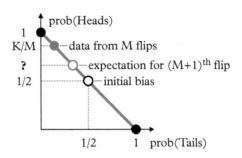

Fig. 2.19 *Flipping an imperfect coin. A coin is possibly imperfect, but the precise nature of its imperfection is not known. It is flipped M times, yielding Heads K times. The expectation for the $(M+1)th$ flip is shaped both by these data and by the prior expectation of a not-too-unfair coin.*

2.12. Flipping an imperfect coin

A mint produces coins of varying quality. Their centre of mass may shift randomly so that, when one particular coin is tossed, the probability of Heads, p, is not necessarily equal to one half. The probability density function of p is given by the Gaussian in Eq. (2.53), with width $\varsigma = 1/100$. You flip one particular coin $M = 100$ times and obtain Heads $K = 90$ times. Which probability would you assign to Heads in the 101st flip? The problem is illustrated schematically in Fig. 2.19. How confident are you about that assignment?

2.13. Unknown beam splitter

Consider the experimental setup depicted in Fig. 1.7, but with the following twist. The parameter setting of the first splitter (in the figure: θ) is not known; it could be either θ_1^+ or θ_1^-. Consequently, it is not certain in which state the particles exit the first splitter; a priori, both settings are equally likely. The parameter setting of the second splitter (in the figure: $\theta + \Delta\theta$), on the other hand, is known to be θ_2. The conditional probability that a particle will exit the second device 'up', given that it left the first device 'up', depends on the unknown setting of the first splitter. The two possibilities differ by a factor

$$r := \frac{\text{prob}(\Uparrow_{\theta_2} \mid \Uparrow_{\theta_1^-})}{\text{prob}(\Uparrow_{\theta_2} \mid \Uparrow_{\theta_1^+})}.$$

On the screen you count M hits at the upper position, and none at the lower.

(a) Taking your experimental observation into account, calculate the posterior probabilities of the two parameter settings as functions of r and M.
(b) Given $r = 0.96$, what is the minimum value of M so that the posterior probability of θ_1^+ exceeds 99%?

2.14. State preparation

Consider the preparation procedure outlined in Section 2.8. Through measurement and filtering, it yields an exchangeable assembly whose state is given by Eq. (2.67).

(a) Derive Eq. (2.62).

(b) There may be situations where the measurement counts, $\{M_i\}$, are not known. Discuss how this changes the formulae for the post-measurement and post-filter states, Eqs (2.63), (2.65), and (2.67), respectively.

(c) Consider a die roll experiment where $M = 10,000$ dice are rolled in parallel. Then there is a filter which randomly selects $N_1 = 800$ dice showing an even number and $N_2 = 200$ dice showing an odd number. The dice are known to be perfect. Calculate the resultant post-filter state.

(d) Consider the same experiment, yet involving dice that have unknown and possibly severe imperfections. This is described mathematically by a prior distribution on the convex set of single-constituent states, $\text{pdf}(\rho)$, which is broad and centred around the state of total ignorance. Calculate the resultant post-filter state.

(e) In the latter scenario involving imperfect dice, you subsequently learn that the initial measurement counts were $M_1 = 6,000$ times 'even' and $M_2 = 4,000$ times 'odd'. Does that change the post-filter state, and if so, how?

You may consider the (possibly unknown) $\{M_i\}$ and $\{N_i\}$ to be large enough to use approximations.

2.15. Parameter counting

Repeat the argument leading up to Eq. (2.71) in the general case where the dimensions of the propositions tested, $k_i := d(x_i)$, are arbitrary.

(a) Show that the number of parameters needed to specify the measurement setup, $\{x_i\}$, is given by

$$X(\{k_i\}) := X(2)\sum_{i<j} k_i k_j.$$

(b) Derive the constraint

$$S(d) = X(2)\sum_{i<j} k_i k_j + \sum_i S(k_i).$$

Show that this constraint is satisfied by both cases of Eq. (2.72).

2.16. Sneak preview of entanglement

Consider a system composed of two constituents, A and B, with respective dimensions d_A and d_B ($d_{A,B} \geq 2$).

(a) Show that if the reduced states of the constituents, ρ_A and ρ_B, are pure, then the composite state is a product state, $\rho_{AB} = \rho_A \otimes \rho_B$, and this product state is also pure.

(b) How many continuous parameters do you need to specify such a pure product state in either the classical or the 'quantum' case (Eq. (2.72))? Compare this to the number of parameters needed to specify an *arbitrary* pure state of the composite system.

The mismatch in the quantum case suggests that there exist pure composite states which are not product states. Such states describe a situation where one has maximal knowledge about the whole but not about the parts. They are called 'entangled' states and will play a prominent role in quantum theory.

3
Probability in Hilbert Space

3.1 Linear Algebra in a Nutshell

Linear algebra is the branch of mathematics that deals with vector spaces and operations on those vector spaces. I assume that you have taken a first course in linear algebra and are familiar with the basics; for this reason I provide only a concise summary. Its main purpose is to ensure that we share a common terminology, and to collect all key definitions in a single place for future reference. In addition, I introduce a few concepts which might not have been covered in an introductory course but will prove useful for the formulation of quantum theory, such as the notion of 'kets' and 'bras'.

A *linear space*, or *vector space*, is a set of elements, called *vectors*. In the context of quantum theory a vector is commonly indicated by the *ket* symbol, $|\cdot\rangle$. Vectors may be added and multiplied by numbers; the result is again a vector. That is to say, if $|u\rangle$ and $|v\rangle$ are vectors, and α and β arbitrary numbers, then $\alpha|u\rangle + \beta|v\rangle$ is again a vector. Like the addition of numbers, the addition of vectors is commutative and associative:

$$|u\rangle + |v\rangle = |v\rangle + |u\rangle, \quad (|u\rangle + |v\rangle) + |w\rangle = |u\rangle + (|v\rangle + |w\rangle). \tag{3.1}$$

Moreover, there exists a unique zero vector, 0 (commonly denoted without the ket), such that

$$|u\rangle + 0 = |u\rangle \ \forall \, |u\rangle; \tag{3.2}$$

and for each vector $|u\rangle$ there exists a unique vector $-|u\rangle$ such that

$$|u\rangle + (-|u\rangle) = 0. \tag{3.3}$$

Multiplying a vector by a number is compatible with the ordinary multiplication of numbers, in the sense that

$$\alpha(\beta|u\rangle) = (\alpha\beta)|u\rangle, \tag{3.4}$$

Quantum Theory: An Information Processing Approach. Jochen Rau, Oxford University Press (2021). © Jochen Rau.
DOI: 10.1093/oso/9780192896308.003.0003

and is distributive both ways:

$$\alpha(|u\rangle + |v\rangle) = \alpha|u\rangle + \alpha|v\rangle, \quad (\alpha + \beta)|u\rangle = \alpha|u\rangle + \beta|u\rangle. \tag{3.5}$$

Multiplying a vector by one yields the identical vector, and multiplying it by zero yields the zero vector:

$$1|u\rangle = |u\rangle, \quad 0|u\rangle = 0. \tag{3.6}$$

Depending on whether the numbers by which one may multiply are real or complex, the vector space is called a real or a complex vector space. (Strictly speaking, more exotic possibilities also exist, such as quaternionic vector spaces.) A familiar example of a real vector space is ordinary three-dimensional space. Complex vector spaces, on the other hand, will become relevant for the description of quantum systems; they shall therefore be the focus of this introduction.

A set of vectors, $\{|v_i\rangle\}$, is said to be *linearly independent* if it is impossible to write a member of this set as a linear combination of the others; that is, if the equation

$$\sum_i \alpha_i |v_i\rangle = 0$$

can hold only when $\alpha_i = 0$ for all i. Otherwise, we say that the vectors are *linearly dependent*. There is an upper limit to the number of vectors which can be linearly independent. This upper limit is the *dimension* of the vector space, d. In principle, it can be infinite, but for our purposes it will be sufficient to consider the finite-dimensional case only. A set of d linearly independent vectors, $\{|i\rangle, i = 1, \ldots, d\}$, constitutes a *basis* in the vector space. Every element of the vector space has a unique representation as a linear combination of these basis vectors,

$$|u\rangle = \sum_{i=1}^{d} u_i |i\rangle. \tag{3.7}$$

The pertinent coefficients, $\{u_i\}$, are called the *components* of $|u\rangle$ in that particular basis.

A vector space is called an *inner product space* or *Hilbert space* if, in addition to the linear structure described above, it is endowed with an inner, or 'scalar', product. (Technically, the definition of a Hilbert space involves some further requirements, but these make no difference in the finite-dimensional case considered here.) This scalar product assigns to each pair of vectors, $|u\rangle$ and $|v\rangle$, a complex number denoted by $(|u\rangle, |v\rangle)$ (or a real number, if the vector space is real). By definition, a scalar product is:

1. 'conjugate symmetric', which means that interchanging the order of the vectors entails complex conjugation,

$$(|v\rangle, |u\rangle) = (|u\rangle, |v\rangle)^*; \tag{3.8}$$

2. linear in its second argument,

$$(|u\rangle, \alpha|v\rangle + \beta|w\rangle) = \alpha(|u\rangle, |v\rangle) + \beta(|u\rangle, |w\rangle). \tag{3.9}$$

Due to conjugate symmetry, this implies conjugate linearity, or 'antilinearity', in the first argument,

$$(\alpha|u\rangle + \beta|v\rangle, |w\rangle) = \alpha^*(|u\rangle, |w\rangle) + \beta^*(|v\rangle, |w\rangle); \tag{3.10}$$

3. positive definite,

$$(|u\rangle, |u\rangle) > 0 \quad \forall |u\rangle \neq 0. \tag{3.11}$$

If the scalar product of two vectors is zero, the vectors are said to be *orthogonal*. With the help of the scalar product one defines the *norm*, or *length*, of a vector as

$$\| |u\rangle \| := \sqrt{(|u\rangle, |u\rangle)}. \tag{3.12}$$

This norm is positive definite,

$$\| |u\rangle \| > 0 \quad \forall |u\rangle \neq 0, \tag{3.13}$$

and absolutely homogeneous,

$$\| \alpha|u\rangle \| = |\alpha| \, \| |u\rangle \|, \tag{3.14}$$

and satisfies the *triangle inequality*,

$$\| |u\rangle + |v\rangle \| \leq \| |u\rangle \| + \| |v\rangle \|. \tag{3.15}$$

You will prove the latter in Exercise (3.1). Vectors with norm one, $\| |u\rangle \| = 1$, are termed *unit vectors*. A basis where all basis vectors are mutually orthogonal is called an *orthogonal basis*. If, in addition, they are all unit vectors, the basis is called *orthonormal*.

A map from the vector space to the complex numbers (or real numbers if the vector space is real) which is linear,

$$f(\alpha|v\rangle + \beta|w\rangle) = \alpha f(|v\rangle) + \beta f(|w\rangle),$$

is called a *linear functional* on the vector space. The linear functionals themselves form a vector space: if f and g are linear functionals, and α and β arbitrary numbers, then their linear combination

$$(\alpha f + \beta g)(|v\rangle) := \alpha f(|v\rangle) + \beta g(|v\rangle)$$

is again a linear functional. One says that this vector space of linear functionals is *dual* to the original vector space. In a Hilbert space every vector, $|u\rangle$, defines a unique linear functional, f_u, via

$$f_u(|v\rangle) := (|u\rangle, |v\rangle).$$

According to the *Riesz representation theorem*, the converse is also true: associated with each linear functional, f, is a unique vector, $|u_f\rangle$, such that

$$f(|v\rangle) = (|u_f\rangle, |v\rangle)$$

for all $|v\rangle$. Thus, the correspondence between vectors and linear functionals is one to one. It is customary to indicate the linear functional associated with a vector $|u\rangle$ by the mirror image of the ket symbol: the *bra* symbol, $\langle u|$. The original vector space is then referred to as the 'ket-space', and its dual as the 'bra-space'. Their members are called 'ket vectors' (or just 'kets') and 'bra vectors' (or just 'bras'), respectively. This notation and terminology, introduced by Paul Dirac, allows one to write the scalar product very suggestively as a 'braket',

$$\langle u|v\rangle \equiv (|u\rangle, |v\rangle). \tag{3.16}$$

Due to the conjugate symmetry of the scalar product, the correspondence between ket and bra vectors is not linear but antilinear. If $|u\rangle$ and $|v\rangle$ correspond to $\langle u|$ and $\langle v|$, respectively, then the linear combination $\alpha |u\rangle + \beta |v\rangle$ corresponds to $\alpha^* \langle u| + \beta^* \langle v|$.

A *subspace* of the Hilbert space is a subset of vectors which themselves form a vector space. If the subspace has dimension one, it is also termed a *ray*. A rather peculiar subspace is the *zero space*, which consists of the zero vector only. A subspace is *included* in another subspace if all members of the former are also members of the latter. Two subspaces are *orthogonal* if every member of one is orthogonal to every member of the other. The *intersection* of two subspaces is the set of all vectors that are contained in both. It is again a subspace, namely the largest possible subspace to be included in the two original subspaces. Given any set of vectors, $\{v_i\}$, their linear combinations, $\{\sum_i \alpha_i v_i\}$, constitute a subspace of the Hilbert space; this is the subspace *spanned* by the $\{v_i\}$. Likewise, two subspaces span a third, larger subspace which contains all linear combinations of vectors from the two original subspaces. The resulting subspace is the smallest possible subspace to include both original subspaces, and is referred to as the *sum* of the two subspaces. If two subspaces are orthogonal and their sum equals the entire Hilbert space, they are *orthogonal complements* of each other. More generally, if several mutually orthogonal subspaces add up to the full Hilbert space or some subspace thereof, they are said to constitute an *orthogonal decomposition* of the Hilbert space or of that subspace, respectively.

To make these concepts more concrete, we shall illustrate them in the case of ordinary (real) three-dimensional space. The three Cartesian axes, the diagonal in the x–y plane, and the x–y plane itself are all examples of subspaces of three-dimensional space. The

Cartesian axes and the diagonal are one-dimensional and hence rays; the x–y plane is two-dimensional. Whereas the x axis, the y axis, and the diagonal are included in the x–y plane, the z axis is not. The Cartesian axes are mutually orthogonal, as are the z axis and the diagonal, as well as the z axis and the x–y plane. By contrast, the x axis and the diagonal are not orthogonal. Any two different rays have as their intersection the zero space. By contrast, the intersection of the x–y plane with the diagonal gives the diagonal, and with the x–z plane, the x axis. Each of the following three pairs of rays—the x axis and the y axis, the x axis and the diagonal, and the y axis and the diagonal—spans the x–y plane. That is to say, in each pair the members add up to the x–y plane. The x–y plane has as its orthogonal complement the z axis, and vice versa; the two constitute an orthogonal decomposition of the full three-dimensional space. The x–y plane can be decomposed further into smaller orthogonal subspaces, for example, the x and y axes. Together, all three Cartesian axes constitute another orthogonal decomposition of the full three-dimensional space. Since it consists of rays only, this decomposition cannot be refined any further.

An *operator* on a vector space maps vectors to vectors: if $\hat{A}$ is an operator and $|u\rangle$ is a vector, then $\hat{A}|u\rangle$ is another vector. The operator is uniquely defined by its action on every vector in the space. In order to distinguish operators from vectors and ordinary numbers, they shall be denoted by letters with a hat. The operator is *linear* if it satisfies

$$\hat{A}(\alpha|u\rangle + \beta|v\rangle) = \alpha\hat{A}|u\rangle + \beta\hat{A}|v\rangle. \tag{3.17}$$

Two examples in real three-dimensional space are the operator which effects a rotation by some prescribed angle about the x axis and the operator which stretches all vectors by a factor of two in the y direction. Operators can be added and multiplied by numbers. The results are again operators, defined by

$$(\hat{A} + \hat{B})|u\rangle := \hat{A}|u\rangle + \hat{B}|u\rangle, \quad (\lambda\hat{A})|u\rangle := \lambda(\hat{A}|u\rangle), \tag{3.18}$$

respectively. Sum and multiplication by a number satisfy Eqs (3.1) to (3.6), with the vectors replaced by operators. Thus, operators on a vector space themselves form a (different) vector space. The linear operators constitute a subspace of this vector space. Operators can also be multiplied with each other. The product of two operators, $\hat{A}\hat{B}$, is defined by

$$(\hat{A}\hat{B})|u\rangle := \hat{A}(\hat{B}|u\rangle). \tag{3.19}$$

This multiplication is associative, $\hat{A}(\hat{B}\hat{C}) = (\hat{A}\hat{B})\hat{C}$, but in general *not* commutative. The extent of non-commutativity is described by the *commutator* of two operators,

$$[\hat{A}, \hat{B}] := \hat{A}\hat{B} - \hat{B}\hat{A}. \tag{3.20}$$

Whenever it vanishes for two particular operators, one says that these operators *commute*.

Given a basis in the vector space, $\{|i\rangle\}$, every vector may be represented in the form of Eq. (3.7); this applies to both $|u\rangle$ and

$$\hat{A}|u\rangle =: |v\rangle = \sum_i v_i |i\rangle.$$

If $\hat{A}$ is linear, the respective components are related by a matrix equation,

$$v_i = \sum_k A_{ik} u_k, \tag{3.21}$$

where the $\{A_{ik}\}$ are the *matrix elements* of $\hat{A}$ in the given basis, defined via

$$\hat{A}|k\rangle = \sum_i A_{ik} |i\rangle. \tag{3.22}$$

These matrix elements specify the linear operator uniquely. In a d-dimensional vector space they are the entries of a $d \times d$ matrix, which we denote by a boldface letter, $\mathbf{A}$. Indeed, there is a one-to-one correspondence between linear operators on a d-dimensional vector space and $d \times d$ matrices. This correspondence preserves the structure of the space of linear operators: if $\hat{A}$ and $\hat{B}$ correspond to $\mathbf{A}$ and $\mathbf{B}$, respectively, then $\lambda \hat{A}$ corresponds to $\lambda \mathbf{A}$, $\hat{A} + \hat{B}$ to $\mathbf{A} + \mathbf{B}$, and $\hat{A}\hat{B}$ to the matrix product $\mathbf{A}\mathbf{B}$. Still, we must be careful to distinguish the abstract linear operator, $\hat{A}$, which is basis-independent, from its matrix representation, $\mathbf{A}$, which pertains to a particular basis. If the vector space is endowed with a scalar product and the basis is orthonormal, $\langle i|k \rangle = \delta_{ik}$, the components of a vector in that basis are given by the scalar product

$$u_i = \langle i|u \rangle, \tag{3.23}$$

and the matrix elements of a linear operator, by

$$A_{ik} = \langle i|\hat{A}|k \rangle. \tag{3.24}$$

In such an orthonormal basis the sum of the diagonal elements is termed the *trace* of the linear operator,

$$\mathrm{tr}(\hat{A}) := \sum_i \langle i|\hat{A}|i \rangle. \tag{3.25}$$

As you will verify in Exercise (3.3), it is in fact independent of the chosen orthonormal basis and hence is a property of the abstract operator. From any ket, $|v\rangle$, and any bra, $\langle w|$, one can construct a particular linear operator called their *dyadic product* and denoted by $|v\rangle\langle w|$. As the notation suggests, its action on an arbitrary vector, $|u\rangle$, is given by

$$(|v\rangle\langle w|)\,|u\rangle := |v\rangle\,\langle w|u\rangle = \langle w|u\rangle\,|v\rangle\,. \tag{3.26}$$

In an orthonormal basis every linear operator may be written as a linear combination of dyads,

$$\hat{A} = \sum_{ik} A_{ik}\,|i\rangle\langle k|\,, \tag{3.27}$$

where the A_{ik} are its matrix elements, Eq. (3.24).

Each linear operator, $\hat{A}$, that acts on a Hilbert space possesses a unique *adjoint*, $\hat{A}^{\dagger}$, defined via

$$(\hat{A}^{\dagger}\,|u\rangle\,,|v\rangle) := (|u\rangle\,,\hat{A}\,|v\rangle) \quad \forall\,|u\rangle\,,|v\rangle\,. \tag{3.28}$$

In bra–ket notation this is equivalent to

$$\langle u|\hat{A}^{\dagger}|v\rangle = \langle v|\hat{A}|u\rangle^{*} \quad \forall\,|u\rangle\,,|v\rangle\,. \tag{3.29}$$

In particular, taking the adjoint of a dyad swaps bra and ket:

$$(|v\rangle\langle w|)^{\dagger} = |w\rangle\langle v|\,. \tag{3.30}$$

The properties of the scalar product imply the following properties of the adjoint:

$$(\hat{A}^{\dagger})^{\dagger} = \hat{A}, \tag{3.31}$$

$$(\lambda\hat{A})^{\dagger} = \lambda^{*}\hat{A}^{\dagger}, \quad (\hat{A}+\hat{B})^{\dagger} = \hat{A}^{\dagger}+\hat{B}^{\dagger}, \tag{3.32}$$

$$(\hat{A}\hat{B})^{\dagger} = \hat{B}^{\dagger}\hat{A}^{\dagger}. \tag{3.33}$$

A linear operator which is its own adjoint, $\hat{A} = \hat{A}^{\dagger}$, is called *self-adjoint* or *Hermitian*. (Technically, the definitions of these two terms differ slightly, but they are equivalent in the finite-dimensional Hilbert spaces considered here.) In an orthonormal basis such an operator is represented by a Hermitian matrix, for which $A_{ik} = A_{ki}^{*}$. The Hermitian operators on a Hilbert space themselves form a real linear space: if $\hat{A}$ and $\hat{B}$ are Hermitian, then so is any linear combination $\alpha\hat{A} + \beta\hat{B}$, provided α and β are real. If the Hermitian operators act on a complex Hilbert space of dimension d, this real linear space has dimension d^{2}.

Besides Hermitian operators, two other important types of linear operators are unitary operators and projectors. To begin with, the *unit operator*, $\hat{I}$, is the operator which maps each vector to itself, $\hat{I}\,|v\rangle = |v\rangle$; it is represented in any basis by the unit matrix. A linear operator, $\hat{U}$, is *unitary* (or, in a real Hilbert space, *orthogonal*) if it preserves all scalar products,

$$(\hat{U}|v\rangle, \hat{U}|w\rangle) = (|v\rangle, |w\rangle) \quad \forall |v\rangle, |w\rangle. \tag{3.34}$$

This is tantamount to the condition

$$\hat{U}^\dagger \hat{U} = \hat{U}\hat{U}^\dagger = \hat{I}. \tag{3.35}$$

In ordinary three-dimensional space the operators which satisfy this condition are those which preserve all lengths and angles. Geometrically, these correspond to rotations and reflections.

A projection operator, or *projector*, projects any vector onto some subspace of the Hilbert space. Such a projection is illustrated—in real three-dimensional space—in Fig. 3.1. The projection is orthogonal in the sense that

$$((|v\rangle - \hat{P}|v\rangle), \hat{P}|v\rangle) = 0 \quad \forall |v\rangle.$$

Since this condition holds for arbitrary vectors, $|v\rangle$, it amounts to the operator equation $\hat{P} = \hat{P}^\dagger \hat{P}$. The latter, in turn, implies that a projector must be both Hermitian and *idempotent*,

$$\hat{P}^\dagger = \hat{P}, \quad \hat{P}^2 = \hat{P}. \tag{3.36}$$

Idempotence reflects the fact that once a vector has been projected onto the subspace, another application of the same projector will have no further effect. One can always choose an orthonormal basis where the first k basis vectors span the subspace onto which the projector projects, whereas the remaining basis vectors are orthogonal to that subspace. In Fig. 3.1 this is the basis comprising $|u_1\rangle$, $|u_2\rangle$, and the unnamed vector orthogonal to the subspace, assuming these all have unit norm and are mutually orthogonal. In such a basis the matrix representing the projector is diagonal,

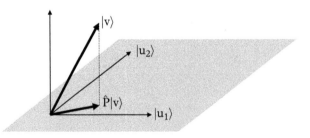

Fig. 3.1 *Projection onto a subspace. The projection operator, $\hat{P}$, projects an arbitrary vector, $|v\rangle$, onto the subspace spanned by the vectors $\{|u_i\rangle\}$.*

$\mathbf{P} = \mathrm{diag}(1,\ldots,1,0,\ldots)$, with the first k entries being one and all others being zero. Its trace equals k, to wit, the dimension of the subspace:

$$\mathrm{tr}(\hat{P}) = \dim(\text{subspace}). \tag{3.37}$$

Where $\hat{P}$ projects onto a ray, spanned by the unit vector $|u\rangle$, its application to an arbitrary vector must yield a vector in that ray, to wit, some multiple of $|u\rangle$. The proportionality factor is given by the scalar product of the two vectors:

$$\hat{P}|v\rangle = \langle u|v\rangle\,|u\rangle.$$

Hence this particular projector is a dyad,

$$\hat{P}_u = |u\rangle\langle u|. \tag{3.38}$$

More generally, if $\hat{P}$ projects onto a subspace spanned by the orthonormal basis $\{|u_i\rangle, i = 1,\ldots,k\}$, the projector takes the form

$$\hat{P} = \sum_{i=1}^{k} |u_i\rangle\langle u_i|. \tag{3.39}$$

You will verify the latter in Exercise (3.4).

When a linear operator maps a vector to a multiple of itself,

$$\hat{A}|v\rangle = a|v\rangle, \tag{3.40}$$

that particular vector is termed an *eigenvector* of the operator, and the scalar a the associated *eigenvalue*. If there exist several eigenvectors associated with the same eigenvalue, then any linear combination of these will again be an eigenvector associated with said eigenvalue. Thus, the eigenvectors associated with a given eigenvalue form a subspace of the Hilbert space, called an *eigenspace*. Where this eigenspace is one-dimensional, the eigenvalue is said to be *non-degenerate*; otherwise, it is *degenerate*, with the dimension of the eigenspace being its *degree of degeneracy*, or *multiplicity*. The set of all eigenvalues is called the *spectrum* of the linear operator. There is a powerful general result, known as the *spectral theorem*, about the spectrum and the eigenspaces of a Hermitian operator on a complex Hilbert space: a Hermitian operator can always be written in the form

$$\hat{A} = \sum_i a_i \hat{P}_i, \quad a_i \in \mathbb{R}, \quad \hat{P}_i \hat{P}_k = \delta_{ik} \hat{P}_i, \tag{3.41}$$

called its *eigen-* or *spectral decomposition*, where the $\{a_i\}$ are its eigenvalues and the $\{\hat{P}_i\}$ are the projectors onto the associated eigenspaces. The eigenvalues are all real, and the eigenspaces constitute an orthogonal decomposition of the Hilbert space. The latter implies, in particular, that there exists an orthonormal basis consisting solely of eigenvectors of $\hat{A}$; such a basis is an *eigenbasis*. In an eigenbasis the Hermitian operator is represented by a diagonal matrix. The entries on the diagonal are the eigenvalues, with each eigenvalue appearing one or multiple times according to its multiplicity. From the spectral decomposition we also learn that the trace of a Hermitian operator equals the sum of its eigenvalues multiplied by their respective multiplicities,

$$\mathrm{tr}(\hat{A}) = \sum_i a_i \, \mathrm{tr}(\hat{P}_i). \tag{3.42}$$

An arbitrary power of the operator, $\hat{A}^n$, has the same eigenspaces as the original operator, yet with associated eigenvalues $\{(a_i)^n\}$. Consequently, it may be written in the form

$$\hat{A}^n = \sum_i (a_i)^n \hat{P}_i.$$

More generally, an arbitrary function of a Hermitian operator is given by

$$f(\hat{A}) = \sum_i f(a_i) \hat{P}_i. \tag{3.43}$$

A Hermitian operator is said to be *positive definite*, $\hat{A} > 0$, if the scalar product $\langle u|\hat{A}|u\rangle$ is strictly positive for any non-zero vector $|u\rangle$. This is tantamount to all eigenvalues being strictly positive:

$$\hat{A} > 0 \quad :\Leftrightarrow \quad \langle u|\hat{A}|u\rangle > 0 \; \forall \, |u\rangle \neq 0 \quad \Leftrightarrow \quad a_i > 0 \; \forall i. \tag{3.44}$$

When strict positivity (> 0) is relaxed to also allow the value zero (≥ 0), the operator is called *positive semi-definite*.

3.2   Propositions and States

> I establish the connection between the general considerations of Chapters 1 and 2 and complex Hilbert space. I discuss how propositions and states are represented in a complex Hilbert space, and I explain the basic rules for calculating probabilities and updating the state after a measurement. I verify that this Hilbert space representation has all the properties of an operational probability theory.

In addition to the mathematical properties discussed in Section 3.1, a Hilbert space possesses a feature which will be of particular interest to us: it exhibits a *logical structure*, with all the characteristics required of an operational theory. How the various elements of this logical structure are represented in Hilbert space, and how this compares to the classical set picture, is summarized in Table 3.1. First of all, the overarching proposition corresponds to the Hilbert space as a whole. If, say, the overarching proposition asserts

Table 3.1 *Generic logical structure in operational probability and its respective representations in classical theory (without knowledge constraints) and in Hilbert space.*

Operational concept, symbol		Classical	Hilbert space
overarching proposition	a	sample space	Hilbert space
testable proposition	$x, y, \ldots$	subset thereof	subspace thereof
logical implication	$x \subseteq y$	set inclusion	subspace inclusion
absurd proposition	$\emptyset$	empty set	zero space
most accurate proposition	e, f	singleton	ray
negation	$\neg_a x$	set complement	orthogonal complement
mutually exclusive		disjoint	orthogonal
may be tested jointly		(always)	have joint orthogonal basis
logical AND	$x \wedge y$	intersection	intersection
logical OR	$x \vee y$	union	sum
partition	$y = \bigoplus_i x_i$	set partition	orthogonal decomposition
dimension	$d(x)$	cardinality	subspace dimension

that 'there is a polarized photon' (of which only the polarization degrees of freedom interest us), then this is reflected in the choice of Hilbert space with the appropriate dimension. All other testable propositions are represented by linear subspaces of this Hilbert space. A special case is the absurd proposition, which corresponds to the zero space. A proposition logically implies another proposition, $x \subseteq y$, if and only if the subspace associated with x is included in the subspace associated with y. The smallest subspaces (apart from the zero space) are one-dimensional; they are the rays. These represent propositions which are most accurate. A proposition and its negation, x and $\neg_a x$, correspond to subspaces which are mutually orthogonal, and which together span the entire Hilbert space. In other words, the subspace associated with the negation, $\neg_a x$, is the orthogonal complement of the original subspace. More generally, two propositions are mutually exclusive, $x \subseteq \neg_a y$, if and only if the associated subspaces are orthogonal. In contrast to classical logic, it is no longer assured that any two propositions may be tested jointly. This is the case only if the associated subspaces share a common orthogonal basis.

Propositions may be combined with a logical AND or a logical OR. The logical AND corresponds to the intersection of the pertinent subspaces, as illustrated schematically in Fig. 3.2. The logical OR, on the other hand, corresponds to the sum of the two subspaces. The sum of two subspaces is again a subspace of Hilbert space, namely that subspace which contains all possible sums of vectors from the original subspaces. The partition of a proposition into mutually exclusive alternatives, $y = \bigoplus_i x_i$, corresponds to

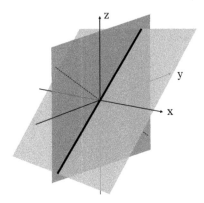

Fig. 3.2 *Intersection of subspaces, illustrated schematically in a real three-dimensional vector space. Two different two-dimensional subspaces (planes) intersect in a one-dimensional subspace (thick line). Provided the two planes share a joint orthogonal basis (here: the x axis, the thick line, and the dashed line), the propositions represented by the two planes are jointly testable. The intersection corresponds to their combination with a logical* AND.

the decomposition of a subspace into even smaller, mutually orthogonal subspaces. For instance, in Fig. 3.2 each plane can be decomposed into two orthogonal one-dimensional subspaces. In one case these might be the thick line and the x axis, and in the other case, the thick line and the dashed line, respectively. Finally, the dimension of a proposition denotes the number of most accurate propositions into which it can be partitioned; that is, in Hilbert space, the number of mutually orthogonal one-dimensional subspaces into which a given subspace can be decomposed. This means that the dimension of a proposition coincides with the dimension of the associated subspace.

The logical structure thus defined satisfies all the basic properties spelt out in Section 1.2. Indeed, subspace inclusion is transitive, reflexive, and antisymmetric, in accordance with Eqs (1.12) to (1.14). The zero space is the unique subspace contained in all others (Eq. (1.15)). The sum of mutually orthogonal subspaces is commutative, Eq. (1.16), and associative, Eq. (1.17), and it yields one of the original subspaces if and only if the other is the zero space (Eq. (1.18)). If a subspace is included in another subspace, then, conversely, the orthogonal complement of the latter is included in the orthogonal complement of the former (Eq. (1.19)); and taking the orthogonal complement of the orthogonal complement returns the original subspace (Eq. (1.20)). Lastly, the subspace dimension obeys the sum rule, Eq. (1.23).

What is more, the logical structure in Hilbert space allows one to define the binary operators ∧ (intersection) and ∨ (sum) for *arbitrary* pairs of subspaces, and these operators exhibit exactly the properties discussed in Section 2.4. Indeed, the intersection of two subspaces is the largest subspace which is included in both (Eq. (2.18)); and their sum is the smallest subspace which includes both (Eq. (2.19)). The orthogonal complement of an intersection equals the sum of the original orthogonal complements, in accordance with De Morgan's law, Eq. (2.27). Intersection and sum are commutative,

Eq. (2.28), and associative, Eq. (2.29). Moreover, provided the subspaces involved share a common orthogonal basis—that is, the associated propositions can be tested jointly—they also satisfy the distributivity property, Eq. (2.30). (You will prove the latter in Exercise (3.5).) For arbitrary subspaces, however, distributivity does not hold, as can be seen in a simple counterexample: consider three distinct rays which lie in one plane (and hence cannot all be mutually orthogonal). Adding two rays yields the plane, and subsequent intersection with the third ray yields this third ray; whereas first intersecting the two rays, individually, with the third ray yields the zero space in both cases, and subsequent addition still returns the zero space—rather than the third ray. Finally, the subspace dimension also obeys the more general version of the sum rule, Eq. (2.31).

There is a one-to-one correspondence between a subspace of the Hilbert space and the projection operator which projects onto that subspace. Therefore, rather than by the subspace itself, a proposition may also be represented by the associated projector. The various elements of the logical structure then translate into definitions and properties which pertain to projection operators; these are summarized in Table 3.2. You will verify in Exercise (3.6) that the various expressions in terms of projectors do indeed mirror the subspace properties listed in Table 3.1. The table includes one particular feature which merits some extra remarks. A most accurate proposition is represented in Hilbert space by a ray. A ray, in turn, is spanned by a single vector, $|\psi\rangle$; this can be any (non-zero) vector which lies in that ray. Without loss of generality, it may be taken to be a unit vector, $\langle\psi|\psi\rangle = 1$. Then the projector onto the ray coincides with the dyadic product $|\psi\rangle\langle\psi|$ (Eq. (3.38)). This opens an alternative way of representing a most accurate

Table 3.2 *Representation of the logical structure in terms of projection operators.*

Operational concept, symbol		Projector representation
overarching proposition	a	$\hat{I}$ (unit operator)
testable proposition	$x, y, \ldots$	$\hat{P}, \hat{Q}, \ldots$ (projectors)
logical implication	$x \subseteq y$	$\hat{P}\hat{Q} = \hat{Q}\hat{P} = \hat{P}$
absurd proposition	$\emptyset$	0
most accurate proposition	e, f	$|\psi\rangle\langle\psi|$ (unit vector, $\langle\psi|\psi\rangle = 1$)
negation	$\neg_a x$	$\hat{I} - \hat{P}$
mutually exclusive		$\hat{P}\hat{Q} = \hat{Q}\hat{P} = 0$
may be tested jointly		$[\hat{P}, \hat{Q}] = 0$
logical AND	$x \wedge y$	if jointly testable: $\hat{P}\hat{Q}$
logical OR	$x \vee y$	if jointly testable: $\hat{P} + \hat{Q} - \hat{P}\hat{Q}$
partition	$y = \bigoplus_i x_i$	$\hat{Q} = \sum_i \hat{P}_i, \quad \hat{P}_i\hat{P}_k = \delta_{ik}\hat{P}_i$
dimension	$d(x)$	$\mathrm{tr}(\hat{P})$

proposition—namely, by this unit vector $|\psi\rangle$. In contrast to the ray and the associated projector, however, this unit vector is not unique. We will argue later in this section that in quantum theory the Hilbert space must be over the *complex* numbers. Consequently, the same ray may be spanned by various unit vectors which differ from each other by a phase factor, $|\psi'\rangle = \exp(i\phi)|\psi\rangle$. These unit vectors all represent the same most accurate proposition.

In line with our general definition, Eq. (1.24), a *state*, ρ, assigns to each testable proposition, x, a probability between zero and one. When propositions are represented by projection operators, this becomes a map from the set of projectors to the unit interval, $[0,1]$. This map must satisfy the pertinent representation of the sum rule, Eq. (1.29). In the projector representation, the partition on the left-hand side of the sum rule becomes a sum of mutually exclusive projectors (Table 3.2). Thus, at least when applied to such a sum of mutually exclusive projectors, the map must be linear. In fact, according to *Gleason's theorem* (which I will not prove here), the map must be linear throughout and, more precisely, must have the form

$$\text{prob}(x|\rho) = \text{tr}(\hat{\rho}\hat{P}_x), \qquad (3.45)$$

where $\hat{P}_x$ is the projector representing the proposition x and $\hat{\rho}$ is a linear operator called the *statistical operator*. This statistical operator must be:

1. Hermitian,

$$\hat{\rho}^\dagger = \hat{\rho}, \qquad (3.46)$$

   in order to ensure that all probabilities are real
2. positive semi-definite,

$$\hat{\rho} \geq 0, \qquad (3.47)$$

   to ensure that all probabilities are non-negative
3. subnormalized or normalized, respectively,

$$\text{tr}(\hat{\rho}) \leq 1, \qquad (3.48)$$

   to reflect the corresponding condition on the probability of the overarching proposition, $\text{prob}(a|\rho)$. In many instances the state is assumed to be normalized. Then the trace of the statistical operator is exactly equal to one.

This general result of Gleason applies to arbitrary Hilbert spaces (real, complex, or quaternionic) of dimension at least three; and in Section 3.5 we will give an argument as to why it is reasonable to assume that it holds in a Hilbert space of dimension two as well.

In a state of *maximal knowledge* there is some most accurate proposition, e, which is true with certainty. In Hilbert space, this amounts to a constraint on the statistical operator:

$$\exists\,|\psi\rangle,\,\langle\psi|\psi\rangle = 1: \quad \mathrm{tr}(\hat{\rho}\,|\psi\rangle\langle\psi|) = \langle\psi|\hat{\rho}|\psi\rangle = 1.$$

Here the unit vector, $|\psi\rangle$, represents the most accurate proposition which is true with certainty. Being Hermitian, the statistical operator has a spectral decomposition, Eq. (3.41); and since it is, moreover, positive semi-definite and has a trace bounded by one, its eigenvalues, $\{\rho_k\}$, are all non-negative and add up to at most one:

$$\hat{\rho} = \sum_k \rho_k \hat{P}_k, \quad \rho_k \geq 0, \quad \sum_k \rho_k \,\mathrm{tr}(\hat{P}_k) \leq 1. \tag{3.49}$$

In terms of these eigenvalues and the associated projectors, the condition for maximal knowledge becomes

$$\sum_k \langle\psi|\hat{P}_k|\psi\rangle\,\rho_k = 1.$$

The left-hand side constitutes a convex combination of the eigenvalues. It is equal to one if and only if one of the eigenvalues is equal to one and all others are zero. Then the statistical operator itself becomes a projector, and, as such, idempotent: $\hat{\rho}^2 = \hat{\rho}$. Due to the upper bound on the trace, Eq. (3.48), the subspace onto which this maximal-knowledge state projects must be one-dimensional; that is, a ray. It is precisely the ray which represents the true most accurate proposition:

$$\hat{\rho} = |\psi\rangle\langle\psi|.$$

Since a projector onto a ray cannot be written as a linear combination of other projectors, a maximal-knowledge state is pure. The converse is also true: whenever a statistical operator does *not* describe maximal knowledge, it is a linear combination of different projectors and thus mixed. This one-to-one correspondence between maximal knowledge and purity is in line with our general discussion in Section 2.3. So altogether, we have the following correspondences:

$$\text{max. knowledge} \Leftrightarrow \hat{\rho}\,\text{pure} \Leftrightarrow \hat{\rho}^2 = \hat{\rho} \Leftrightarrow \exists\,|\psi\rangle,\,\langle\psi|\psi\rangle = 1: \hat{\rho} = |\psi\rangle\langle\psi|. \tag{3.50}$$

A state of maximal knowledge may be represented just as well by the unit vector, $|\psi\rangle$. However, in contrast to the statistical operator, this unit vector is not unique; it is defined only up to an undetermined phase factor. When expressed in terms of the unit vector, Gleason's theorem, Eq. (3.45), turns into the *Born rule*,

$$\mathrm{prob}(x|\psi) = \langle\psi|\hat{P}_x|\psi\rangle. \tag{3.51}$$

If the probability pertains to a most accurate proposition, f, the projector can be written as a dyad, $\hat{P}_f = |\chi\rangle\langle\chi|$. In this special case the Born rule takes the simple form

$$\text{prob}(f|\psi) = |\langle\chi|\psi\rangle|^2. \tag{3.52}$$

Upon *measurement*, a statistical operator must be updated. We consider first the special case of pure states. As in our general framework, Eq. (2.12), measurement preserves maximal knowledge; once we have maximal knowledge about a system, subsequent measurements will not destroy that. Therefore, upon ascertaining the truth of some proposition, x, a pure state, $|\psi\rangle$, must be updated to another pure state, $|\psi_{|x}\rangle$. This updated pure state must satisfy two conditions. First, by Eq. (2.13), it must lie in the subspace associated with x. If x is represented by the projector $\hat{P}_x$, this amounts to the constraint

$$\hat{P}_x|\psi_{|x}\rangle = |\psi_{|x}\rangle.$$

Secondly, by Eq. (2.16), the pre-measurement state, $|\psi\rangle$, must lie in the subspace which is spanned by: (i) the orthogonal complement of the previously considered subspace, represented by the projector $(\hat{I} - \hat{P}_x)$; and (ii) the ray spanned by the post-measurement state, represented by $|\psi_{|x}\rangle\langle\psi_{|x}|$. This translates into a further constraint,

$$\left(\hat{I} - \hat{P}_x + |\psi_{|x}\rangle\langle\psi_{|x}|\right)|\psi\rangle = |\psi\rangle.$$

The latter implies that the post-measurement state results from an orthogonal projection of the pre-measurement state onto the subspace associated with x:

$$|\psi_{|x}\rangle \propto \hat{P}_x|\psi\rangle. \tag{3.53}$$

The proportionality factor is such that the post-measurement state continues to be normalized. In terms of projectors, this update rule reads:

$$|\psi_{|x}\rangle\langle\psi_{|x}| = \frac{\hat{P}_x|\psi\rangle\langle\psi|\hat{P}_x}{\text{tr}(|\psi\rangle\langle\psi|\hat{P}_x)}. \tag{3.54}$$

Where the confirmed proposition is a most accurate one, f, the associated subspace is a ray spanned by some unit vector, $|\chi\rangle$, and the pertinent projector is a dyad, $\hat{P}_f = |\chi\rangle\langle\chi|$. Then the posterior pure state coincides with $|\chi\rangle$,

$$|\psi_{|f}\rangle \propto |\chi\rangle, \tag{3.55}$$

modulo an irrelevant phase factor. In particular, the posterior is then independent of the prior, $|\psi\rangle$.

Next, we turn to the update of mixed states, which are necessarily represented by a statistical operator. In the spectral decomposition of the latter, Eq. (3.49), the projectors, $\{\hat{P}_k\}$, might pertain to subspaces of varying dimension, depending on the multiplicity of the associated eigenvalues, $\{\rho_k\}$. However, a projector onto a higher-dimensional sub-space can always be decomposed into mutually orthogonal one-dimensional projectors,

$$\hat{P}_k = \sum_{i=1}^{\text{tr}\,\hat{P}_k} |k,i\rangle\langle k,i|, \quad \langle k,i|k,j\rangle = \delta_{ij}.$$

Thus, a statistical operator can always be expressed as a mixture of pure states,

$$\hat{\rho} = \sum_{k}\sum_{i=1}^{\text{tr}\,\hat{P}_k} \rho_k\,|k,i\rangle\langle k,i|.$$

Upon ascertaining the truth of x, this mixture must be updated according to our general rule, Eq. (1.38) (with y replaced by x). Making use of our result for pure states, Eq. (3.54), this yields the updated statistical operator

$$\hat{\rho}_{|x} \propto \sum_{k}\sum_{i=1}^{\text{tr}\,\hat{P}_k} \rho_k\,\langle k,i|\hat{P}_x|k,i\rangle\,\frac{\hat{P}_x|k,i\rangle\langle k,i|\hat{P}_x}{\langle k,i|\hat{P}_x|k,i\rangle} \propto \hat{P}_x\hat{\rho}\hat{P}_x.$$

With the proper normalization factor this becomes *Lüders' rule*,

$$\hat{\rho}_{|x} = \frac{\hat{P}_x\hat{\rho}\hat{P}_x}{\text{tr}(\hat{\rho}\hat{P}_x)}. \tag{3.56}$$

It mirrors the update rule for pure states, Eq. (3.54), with the statistical operator, $\hat{\rho}$, taking the place of the projector, $|\psi\rangle\langle\psi|$. Where the proposition that was confirmed is a most accurate one, f, the associated projector is a dyad, $\hat{P}_f = |\chi\rangle\langle\chi|$. The post-measurement state then simplifies to

$$\hat{\rho}_{|f} = \frac{|\chi\rangle\langle\chi|\hat{\rho}|\chi\rangle\langle\chi|}{\langle\chi|\hat{\rho}|\chi\rangle} = |\chi\rangle\langle\chi|;$$

that is,

$$\hat{\rho}_{|f} = \hat{P}_f. \tag{3.57}$$

Thus, independently of the prior state, $\hat{\rho}$, the post-measurement state is the pure state associated with the confirmed most accurate proposition, f. This conforms with our earlier assertion that the ascertainment of a most accurate proposition yields a unique posterior, superseding all prior history (Eq. (2.9)).

Finally, we want to give a reason for our earlier assertion that the Hilbert space ought to be over the complex numbers. To this end, we count the parameters in the theory. A pure state is described by a unit vector, modulo an irrelevant phase factor. In a d-dimensional complex Hilbert space this vector has d complex components, amounting to $2d$ real parameters; from this, however, one must subtract one parameter for the normalization condition and another parameter for the irrelevant phase. Hence, in effect, a pure state is characterized by

$$X(d) = 2(d-1)$$

parameters. This is in line with our general result, Eq. (2.17). If the Hilbert space were real or quaternionic, a pure state would be specified by $(d-1)$ or $4(d-1)$ parameters, respectively, which would also be consistent with the general result. An arbitrary state, on the other hand, is described by a statistical operator. Given an orthonormal basis in the complex Hilbert space, the latter is represented by a Hermitian $d \times d$ matrix. Provided we also allow for subnormalized states, there is no sharp constraint on its trace. Such a matrix is specified by d^2 real parameters, so we have

$$S(d) = d^2,$$

in agreement with both the power law, Eq. (2.39), and the extra constraint in Eq. (2.71). Thus, probability theory in complex Hilbert space realizes precisely the second possibility in Eq. (2.72): the sole universal, consistent, and operational alternative to the classical case. By contrast, a Hilbert space over the real or quaternionic numbers would *not* conform with all our constraints on the number of parameters. In these cases we would have $S(d) = d(d+1)/2$ (the number of parameters which specify a real symmetric $d \times d$ matrix) or $S(d) = d(2d-1)$, respectively, both in violation of the power law and hence in violation of a necessary condition for endowing states with an operational meaning.

We conclude that in terms of being universal, logically consistent, and operational, the probability theory in complex (but not real or quaternionic) Hilbert space complies fully with all requirements that we laid out in Chapters 1 and 2. In fact, it appears to be the only reasonable alternative to classical theory. In Sections 3.3 and 3.5 we shall convince ourselves that this alternative probability theory can indeed explain all the disturbing experimental evidence which had been the starting point of our whole enterprise. It is complex Hilbert space, therefore, which proves to be the appropriate mathematical framework for quantum theory.

3.3 Two-Level System

The simplest quantum system is a two-level system, which is described in a Hilbert space of dimension two. Eventually, this will turn out to be the elementary building block of quantum information processing: the 'qubit'. I analyse the two-level system in detail. In particular, I show that the Hilbert space framework explains all the experimental evidence about binary measurements (except correlations). I also consider a further quantum effect, the so-called quantum Zeno effect.

The simplest possible quantum system is a *two-level system*. This is a system where every measurement is binary, to wit, has only two possible outcomes. A two-level system already exhibits many of the characteristic, and at times perplexing, features of quantum theory. It is used to model a wide variety of real physical systems, encapsulating their key quantum properties, while at the same time being easy to describe mathematically. Furthermore, it constitutes the basic information carrier—the *qubit*—in a quantum computer. Experimentally, a two-level system can be realized in many ways: for instance, by a spin-1/2 particle, which, along any spatial axis, can be in one of two possible spin states, 'up' or 'down'; by a photon, which can be in one of two possible polarization states, such as vertically or horizontally polarized; or by an atom in some appropriate experimental setup where only transitions between two selected energy levels are allowed. The experimental evidence which we discussed in Section 1.1 all pertained to such two-level systems.

A two-level system is described in a two-dimensional Hilbert space. This Hilbert space can be endowed with an orthonormal basis. Inspired by the use of two-level systems in quantum computing, the standard basis, or 'computational basis', of a two-level system is commonly denoted by $\{|0\rangle, |1\rangle\}$. Every pure state is represented by a unit vector, $|\psi\rangle$. It can always be written as a linear combination, or *superposition*, of these two basis states,

$$|\psi\rangle = \alpha_0 |0\rangle + \alpha_1 |1\rangle, \quad |\alpha_0|^2 + |\alpha_1|^2 = 1. \tag{3.58}$$

Conversely, every such superposition of basis states yields an admissible pure state; this is known as the *superposition principle*. Since the unit vector is defined only up to an irrelevant phase factor (also termed a 'global' phase factor), we are free to multiply it by a suitable phase factor so as to render the first coefficient, α_0, real and non-negative. Then the coefficients can be parametrized by two angles, θ and φ:

$$|\psi\rangle = \cos\left(\frac{\theta}{2}\right)|0\rangle + e^{i\varphi}\sin\left(\frac{\theta}{2}\right)|1\rangle, \quad \theta \in [0, \pi], \quad \varphi \in [0, 2\pi). \tag{3.59}$$

The range of θ is determined by the requirement that both $\cos(\theta/2)$ and $\sin(\theta/2)$ lie between zero and one. The second angle, φ, is called the 'relative' phase, which, in contrast to the global phase, *is* relevant. This parametrization in terms of two angles allows for a convenient pictorial representation of the states of a two level-system. The two angles may be regarded as the spherical coordinates of a point on the surface of a unit sphere, called the *Bloch sphere* (Fig. 3.3). (In quantum optics this is also known as the 'Poincaré sphere'.) There is thus a one-to-one correspondence between pure states of a two-level system and points on the surface of the Bloch sphere. Pairs of orthogonal states are represented by antipodes on the sphere. In particular, the orthogonal standard basis

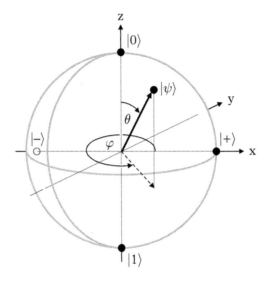

Fig. 3.3 *Bloch sphere.*

states, $|0\rangle$ and $|1\rangle$, correspond to the north and south poles of the sphere, respectively. There is another frequently used pair of orthogonal basis states,

$$|\pm\rangle := \frac{1}{\sqrt{2}}(|0\rangle \pm |1\rangle),\tag{3.60}$$

which lie on the equator ($\theta = \pi/2$) of the Bloch sphere at $\varphi = 0$ and $\varphi = \pi$, respectively.

An arbitrary, possibly mixed state of a two-level system must be described by a statistical operator, $\hat{\rho}$. This statistical operator is an element of the real linear space of Hermitian operators. The latter has dimension four and thus can be endowed with a basis comprising four elements. One particular choice for these elements is the following: the unit operator, $\hat{I}$ (also denoted by $\hat{\sigma}_0$), with matrix representation

$$\mathbf{I} = \begin{pmatrix} 1 & 0 \\ 0 & 1 \end{pmatrix}\tag{3.61}$$

in any basis; and the three *Pauli operators*, $\hat{X}$, $\hat{Y}$, and $\hat{Z}$ (also denoted by $\hat{\sigma}_i$, $i = 1,2,3$), with respective matrix representations

$$\mathbf{X} = \begin{pmatrix} 0 & 1 \\ 1 & 0 \end{pmatrix}, \quad \mathbf{Y} = \begin{pmatrix} 0 & -i \\ i & 0 \end{pmatrix}, \quad \mathbf{Z} = \begin{pmatrix} 1 & 0 \\ 0 & -1 \end{pmatrix}\tag{3.62}$$

in the standard basis $\{|0\rangle, |1\rangle\}$. All three Pauli operators are Hermitian, as they should be, and their squares equal the unit operator,

$$\hat{X}^2 = \hat{Y}^2 = \hat{Z}^2 = \hat{I}. \tag{3.63}$$

In addition, they satisfy the relations

$$\hat{X}\hat{Y} = -\hat{Y}\hat{X} = i\hat{Z}, \quad \hat{Y}\hat{Z} = -\hat{Z}\hat{Y} = i\hat{X}, \quad \hat{Z}\hat{X} = -\hat{X}\hat{Z} = i\hat{Y}. \tag{3.64}$$

All Pauli operators have the eigenvalues $+1$ and -1, and hence zero trace. Expressed as a linear combination of these four basis elements, the statistical operator reads

$$\hat{\rho} = \frac{1}{2}\left[(\mathrm{tr}\,\hat{\rho})\hat{I} + \vec{r}\cdot\hat{\vec{\sigma}}\right]. \tag{3.65}$$

Here $\hat{\vec{\sigma}}$ is short for a vector with operator-valued components $(\hat{X}, \hat{Y}, \hat{Z})$, and $\vec{r}$ denotes the vector of the respective coefficients,

$$\vec{r} := (x, y, z), \quad x = \mathrm{tr}(\hat{\rho}\hat{X}), \quad y = \mathrm{tr}(\hat{\rho}\hat{Y}), \quad z = \mathrm{tr}(\hat{\rho}\hat{Z}). \tag{3.66}$$

You will prove the latter equalities in Exercise (3.10). Moreover, in this exercise you will verify that

$$|\vec{r}|^2 \leq 1, \tag{3.67}$$

with equality if and only if the state is pure.

If we limit ourselves to normalized states, $\mathrm{tr}(\hat{\rho}) = 1$, there is a one-to-one correspondence between states and the set of allowed coefficient vectors, $\{\vec{r}\}$; to wit, the set of real three-dimensional vectors of length less than or equal to one. The latter may be visualized as a unit *ball*. In its interior lie the mixed states, whereas on its boundary lie the pure states. The boundary of this ball is, of course, identical to the Bloch sphere which we discussed earlier—except that points on it are now described by the three components of a unit vector rather than by two angles. In Exercise (3.10) you will show that the relationship between the components of the unit vector and the angles θ, φ is just like the relationship between Cartesian and spherical coordinates. Thus, the components of $\vec{r}$ coincide with the Cartesian coordinates shown in Fig. 3.3; the coefficient vector may be regarded as the 'position vector' of a state on or inside the Bloch sphere.

Since pure states correspond one-to-one to most accurate propositions, the Bloch sphere may be regarded just as well as the continuum of most accurate propositions. Antipodes then represent most accurate propositions which are mutually exclusive. When they pertain to the two possible outcomes of a binary measurement—say, the spin is 'up' or 'down' along some particular spatial axis—the measurement itself may be visualized as the diameter connecting these antipodes. This picture is precisely the sphere model which we introduced in Section 2.1 (see Fig. 2.5, right). With the Hilbert space framework in hand, we can now turn this qualitative model into a quantitative model. Specifically, we can now calculate the conditional probability that a measurement will reveal a most accurate proposition, f, as true, given that previously some other proposition, e, was found to be true (Fig. 3.4). Due to the rotational symmetry of the sphere, this conditional probability should depend only on the angle, $\Delta\theta$, between e and f. Without loss of generality, we may assume that the propositions f and $\neg_a f$ correspond to the standard basis states $|0\rangle$ and $|1\rangle$, respectively. The proposition e corresponds to a pure state in the general form of Eq. (3.59), with $\theta = \Delta\theta$ and unspecified φ. Then by the Born rule, Eq. (3.52), the conditional probability reads

$$\text{prob}(f|e) = |\langle 0|\psi\rangle|^2 = \cos^2\left(\frac{\Delta\theta}{2}\right). \tag{3.68}$$

As the angle increases from zero to π—meaning that e moves away from f and towards its negation, $\neg_a f$—the conditional probability decreases monotonically from one to zero. The functional dependence is such that for arbitrary choices of e and f, the conditional probabilities of f and of its negation, $\neg_a f$, always add up to one,

$$\text{prob}(f|e) + \text{prob}(\neg_a f|e) = \cos^2\left(\frac{\Delta\theta}{2}\right) + \cos^2\left(\frac{\pi - \Delta\theta}{2}\right) = 1,$$

in compliance with the law of total probability. The behaviour of the conditional probability, as shown in Fig. 3.4, agrees perfectly with the experimental observation (Fig. 1.8). In particular, it satisfies precisely the properties listed in Eqs (1.1) to (1.6). Some special values for the conditional probability are listed in Table 3.3.

Once the measurement along the north–south axis has been performed, the state must be updated. If the measurement revealed the proposition f as true, then, according to

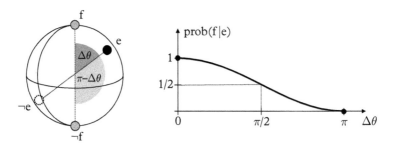

Fig. 3.4 *Conditional probability on the Bloch sphere. A most accurate proposition, e, has been ascertained as true (black dot), and its complement, $\neg_a e$, as false (open circle). The resultant pure state, $\iota_{|e}$, assigns to another most accurate proposition, f, a probability of somewhere between zero and one; and likewise to its complement, $\neg_a f$ (grey dots). Between the directions associated with e and f there is an angle, $\Delta\theta$. The probability of f given the truth of e, $\text{prob}(f|e)$, is a function of this angle (right).*

Table 3.3 *Special values for the conditional probability.*

$\Delta\theta$	$\cos^2\left(\frac{\Delta\theta}{2}\right)$	numerical
0	1	1
$\frac{\pi}{8}$	$\frac{2+\sqrt{2+\sqrt{2}}}{4}$	0.96
$\frac{\pi}{6}$	$\frac{(1+\sqrt{3})^2}{8}$	0.93
$\frac{\pi}{4}$	$\frac{2+\sqrt{2}}{4}$	0.85
$\frac{\pi}{3}$	$\frac{3}{4}$	0.75
$\frac{\pi}{2}$	$\frac{1}{2}$	0.5
π	0	0

Eq. (3.57), the state is updated from the prior (in which e was true) to the unique posterior in which f is true; that is to say, from $\hat{\rho} = \hat{P}_e = |\psi\rangle\langle\psi|$ to $\hat{\rho}_{|f} = \hat{P}_f = |0\rangle\langle 0|$, or, in terms of unit vectors, from $|\psi\rangle$ to $|0\rangle$. This is precisely the update rule of the sphere model as depicted in Fig. 2.6 (bottom right). This update rule, in combination with the formula for the conditional probability, Eq. (3.68), can explain most of the experimental evidence that we presented in Section 1.1. As an example, we consider the experiment shown in Fig. 1.10. The first splitter tests the most accurate proposition f: 'at parameter setting $\theta = 0$, the particle exits in the upper beam' and its opposite, $\neg_a f$: '... in the lower beam'. On the Bloch sphere these two propositions correspond to the north and south poles, respectively. The subsequent blocking lets pass only those particles for which f was found to be true; these are then all in the pure state corresponding to the north pole, $|0\rangle$. The second splitter tests e and $\neg_a e$, corresponding to 'up' and 'down' at parameter setting $\theta = \pi/2$. On the Bloch sphere these propositions are again antipodes, now situated on the equator, at respective longitudes φ and $\varphi + \pi$. The angle φ is not specified and will not affect the final result. The probability that e is revealed as true, given that previously f was found to be true, is given by $\text{prob}(e|f) = \cos^2(\pi/4) = 1/2$. Thus, the subsequent blocking cuts the beam intensity in half, letting pass only those particles for which e was found to be true. According to the update rule, the latter are now in the pure state corresponding to e; that is,

$$|\psi\rangle = \frac{1}{\sqrt{2}}\left(|0\rangle + e^{i\varphi}|1\rangle\right),$$

with unspecified φ. Finally, the third splitter tests once again f and $\neg_a f$, corresponding to 'up' and 'down' at the original parameter setting, $\theta = 0$. In the same logic as before, yet with e and f interchanged, the probability of both outcomes equals $\text{prob}(f|e) = \text{prob}(\neg_a f|e) = \cos^2(\pi/4) = 1/2$. So indeed, one will detect two beams of equal intensity. According to the update rule, these two final beams contain particles in the pure states $|0\rangle$ (upper beam) and $|1\rangle$ (lower beam), respectively.

Also among the evidence which may be understood in the Hilbert space framework is the phenomenon of lossless steering (Fig. 1.13). Our earlier explanation hinged on the assumption that the conditional probability, prob($\Uparrow_{\Delta\theta}$ | $\Uparrow$), has a Taylor expansion around $\Delta\theta = 0$, and that the subleading term in this expansion is of second order (Eq. (1.10)). This assumption can now be justified. Indeed, Eq. (3.68) may be expanded around $\Delta\theta = 0$, yielding a quadratic dependence for small angles:

$$\text{prob}(f|e) \approx 1 - \frac{(\Delta\theta)^2}{4}. \tag{3.69}$$

There exists another quantum phenomenon which is the mirror image of lossless steering, called the *quantum Zeno effect*. It is named after Zeno of Elea (fifth-century BC), a pre-Socratic Greek philosopher best known for his paradoxes that aim to show the impossibility of motion. In one of these paradoxes he considers a flying arrow. At each moment the arrow resides in a specific location, where—Zeno argues—it is at rest, like in a snapshot image. Yet if the arrow is at rest in every single instant, how can it possibly move? Inspired by this paradox, one considers a quantum system that, if left alone, should be evolving in time—that is, 'moving'. Of this quantum system one takes frequent 'snapshots' in the form of measurements. Could it be that these snapshots render any motion impossible? The simplest example is a two-level system which is initially in the pure state $|0\rangle$ and then evolves according to

$$|\psi(t)\rangle = \cos\left(\frac{t}{T} \cdot \frac{\pi}{2}\right)|0\rangle + \sin\left(\frac{t}{T} \cdot \frac{\pi}{2}\right)|1\rangle, \quad 0 \le t \le T.$$

This time evolution corresponds to a uniform motion on the Bloch sphere from the north to the south pole (Fig. 3.5). Now the system is subjected to repeated measurements, all along the north–south axis. Basically, one keeps asking: 'Are you still at the north pole—or already at the south pole?' The first such measurement occurs at time T/N, the second at $2T/N$, and so on, until the last measurement at time T; in total, it is repeated N times. According to Eq. (3.68), the first measurement yields the result 'north' with probability $\cos^2(\pi/2N)$. If it does, the update rule stipulates that the state of the system is reset to $|0\rangle$. The evolution then starts all over again, and in the second measurement the probability of 'north'—and of another reset to $|0\rangle$—is again $\cos^2(\pi/2N)$. The probability that *all* N measurements yield 'north', and that, hence, even after time T the system is still stuck at the north pole is given by $[\cos^2(\pi/2N)]^N$. As the number of measurements increases, this probability approaches unity:

$$\left[\cos^2\left(\frac{\pi}{2N}\right)\right]^N \to 1 \quad (N \to \infty).$$

Thus, continuous monitoring (understood as the limit of infinitely fast repetitions of the same measurement) effectively prevents evolution; it 'freezes' the system in its initial

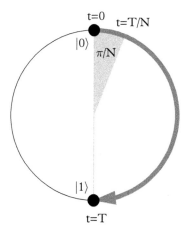

Fig. 3.5 *Quantum Zeno effect. The circle represents a two-dimensional section of the Bloch sphere. The thick grey curve from the north to the south pole marks the uniform time evolution of the state, uninhibited by measurement, between $t = 0$ and $t = T$. The total time is divided into N shorter intervals of duration T/N. During the first interval the state progresses by an angle $\Delta\theta = \pi/N$. When the system is monitored, measurements along the north–south axis are performed at the end of each interval. This triggers every time a state update. As the frequency of measurement increases, $N \to \infty$, the system becomes effectively 'frozen' in its initial state, $|0\rangle$.*

state. If Zeno's arrow were a quantum system, taking snapshots all the time would indeed make motion impossible!

Until now, all applications of the Hilbert space framework used only the angle θ, which parametrizes lines of longitude, or meridians, on the Bloch sphere. Pictorially speaking, our explanations have so far invoked only a great circle on the Bloch sphere, rather than its entire surface—more akin to the circle model (Fig. 2.5, left) than the sphere model. One may wonder, therefore, whether there is any experimental evidence that points unequivocally to the existence of the second angle, φ. Indeed, such evidence exists. In optical experiments one may employ particular components called 'quarter-wave plates' to build a modified beam splitter (Fig. 3.6). This new splitter is characterized by *two* angles, θ and α. Like the stand-alone polarizing beam splitter, it can be used in combination with detectors or a block to effect a binary measurement. Experiments confirm that this measurement is again reproducible, in the sense depicted in Fig. 1.6: when two measurements with identical parameter settings (now: θ, α) are performed in sequence, the second will confirm with certainty the result of the first. The set of measurements which can be realized by varying the two parameters θ and α includes the measurements effected with ordinary splitters as special cases. Indeed, the experiment sketched in Fig. 3.7 shows that when the first quarter-wave plate is rotated by the same angle as the polarizing beam splitter, $\alpha = \theta$, the combined setup performs the same measurement as a simple splitter without quarter-wave plates. The additional parameter is not redundant, however. There are values of α at which the measurement

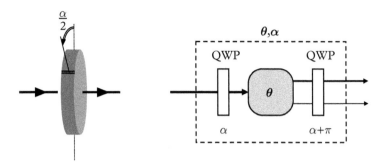

Fig. 3.6 *Exploring the full Bloch sphere. A quarter-wave plate (left) is an optical component made of a birefringent material, that is, a material whose refractive index depends on the polarization of the incoming light. It lets every photon pass but, in general, alters its polarization state; it converts linearly polarized into circularly (or, more generally, elliptically) polarized photons, and vice versa. The quarter-wave plate may be rotated about the beam axis, giving rise to a continuum of possible configurations parametrized by the rotation angle, $\alpha/2$. (The factor one half emphasizes the analogy with the definition of θ in Fig. 1.5.) The way in which the quarter-wave plate alters the state of a photon depends on the value of α. When a polarizing beam splitter is sandwiched between two quarter-wave plates (right), with respective parameter settings α and $\alpha + \pi$, the combined setup constitutes a modified beam splitter that is now controlled by two parameters, θ and α. By tuning both parameters, one can realize measurements along any diameter of the Bloch sphere; that is, arbitrary measurements on the two-level system.*

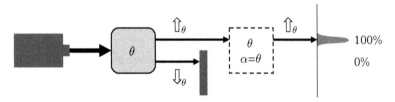

Fig. 3.7 *Special case $\alpha = \theta$. An ordinary polarizing beam splitter and subsequent blocking produce a beam of photons, $\Uparrow_\theta$. This beam is sent through a modified splitter set at the same θ and $\alpha = \theta$. The second measurement confirms with certainty the outcome of the first. Therefore, the two measurements are identical.*

effected by the combined setup genuinely differs from any measurement that can be realized with an ordinary beam splitter; one example is shown in Fig. 3.8. Finally, there is no way to construct even more general binary measurements on the polarization degrees of freedom of a photon that would be controlled by three or more parameters. Together, this experimental evidence confirms that an arbitrary measurement on the two-level system is specified by exactly two continuous parameters; and hence that the manifold of pure states has dimension two. We have already identified one of these parameters, θ, with the polar angle of the Bloch sphere. The other, α, is not directly related to the azimuthal angle, φ. However, there exists a functional relationship between these three angles. You will explore some aspects of this relationship in Exercise (3.13).

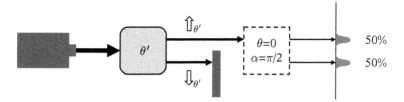

Fig. 3.8 *Independence of the parameters θ and α. An ordinary polarizing beam splitter and subsequent blocking produce a beam of photons, ⇑*$_{\theta'}$*. This beam is sent through a modified splitter with settings* $\theta = 0, \alpha = \pi/2$*. The latter will always split the beam in two, with equal intensity, for any value of* θ'*. This means that the second splitter must perform a measurement which genuinely differs from any measurement that can be realized with an ordinary beam splitter.*

One experimental finding which the theory of the single two-level system does *not* account for is the perfect anticorrelation depicted in Fig. 1.14. Its explanation will require a deeper understanding of composite quantum systems, to which we shall turn in Section 3.5.

3.4 Observables and Transformations

I introduce two further basic concepts of quantum theory: observables and transformations. With regard to observables, I discuss their expectation values and pertinent uncertainties, and I derive a fundamental result of quantum theory: the uncertainty relation. As for transformations, I consider the distinction between the Schrödinger and Heisenberg pictures, as well as between unitary and antiunitary transformations.

In physics, one is typically concerned not with testing propositions but with measuring observables. In general terms, an *observable*, A, is a set of pairs, $\{(x_i, a_i)\}$, where the $\{x_i\}$ are mutually exclusive propositions pertaining to the outcome of a measurement, and the $\{a_i\}$ are real numerical values (with the appropriate physical unit) assigned to these outcomes. For instance, when a die is rolled, x_i might constitute the proposition that the die shows i spots, and $a_i = i$ might be the number of spots itself. Together, they define the observable 'number of spots', $\{(x_i, i)\}$. Likewise, for a quantum system featuring discrete energy levels with respective energies $\{\epsilon_i\}$, the observable 'energy' is the set of pairs $\{(x_i, \epsilon_i)\}$, where x_i stands for the proposition 'measurement of the energy level yields level i'. In a given state, ρ, the *expectation value* of the observable is the probability-weighted average of its possible outcome values,

$$\langle A \rangle_\rho := \sum_i \text{prob}(x_i | \rho) \, a_i. \tag{3.70}$$

In Hilbert space, the state is represented by a statistical operator, $\hat{\rho}$, and the propositions pertaining to the measurement outcome are represented by mutually orthogonal projectors, $\{\hat{P}_i\}$. Then the expectation value is given by

$$\langle A \rangle_\rho = \sum_i \mathrm{tr}(\hat{\rho}\hat{P}_i)\, a_i = \mathrm{tr}(\hat{\rho}\hat{A}), \tag{3.71}$$

where

$$\hat{A} := \sum_i a_i \hat{P}_i. \tag{3.72}$$

This is the operator which represents the observable in Hilbert space. Just as we distinguished between a proposition, x, and its projector representation, $\hat{P}_x$, we shall make a distinction in our notation between an abstract observable, A (without a hat), and its representation as an operator on a Hilbert space, $\hat{A}$ (with a hat). The latter, in turn, is to be distinguished from its matrix representation in a particular basis, $\mathbf{A}$ (boldface). Where the state is pure, $\hat{\rho} = |\psi\rangle\langle\psi|$, the expectation value of the observable takes the form

$$\langle A \rangle_\psi = \langle\psi|\hat{A}|\psi\rangle, \tag{3.73}$$

analogous to the Born rule, Eq. (3.51).

The operator which represents an observable is Hermitian, and in Eq. (3.72) is already given in the form of its spectral decomposition. From the latter we glean that the possible measurement values, $\{a_i\}$, coincide with the eigenvalues of the Hermitian operator, and the projectors, $\{\hat{P}_i\}$, are those which project onto the associated eigenspaces. A pure state which lies in one of the eigenspaces,

$$\hat{A}|\psi\rangle = a_i|\psi\rangle,$$

is termed an *eigenstate* of the observable A. Two or more observables, $A = \{(x_i, a_i)\}$, $B = \{(y_k, b_k)\}$, $C = \{(z_l, c_l)\}, \ldots$, are *jointly measurable* if the respective propositions about the outcomes, $\{x_i\}, \{y_k\}, \{z_l\}, \ldots$, can be tested jointly. In Hilbert space, this means that the associated projectors commute. This is the case if and only if the Hermitian operators representing the observables pairwise commute:

$$[\hat{A}, \hat{B}] = [\hat{A}, \hat{C}] = [\hat{B}, \hat{C}] = \ldots = 0.$$

Observables which pairwise commute possess a joint eigenbasis. If this joint eigenbasis is unique (up to phase factors, $|k\rangle \to z_k|k\rangle$, with $|z_k| = 1$), the observables are said to form a *complete set of commuting observables*. Every member of the joint eigenbasis is then completely specified (again, up to phase factors) by a set of eigenvalues $a_i, b_k, c_l, \ldots$ pertaining to these observables, and may be labelled by $|a_i, b_k, c_l, \ldots\rangle$.

In an eigenstate of A the measurement of this observable will definitely yield the associated eigenvalue; there is no uncertainty about the measurement result. In an arbitrary state, however, there may be a range of possible outcomes, each with some non-zero probability. The width of this outcome distribution is characterized by the *variance*,

$$\operatorname{var}(A) := \langle (A - \langle A \rangle)^2 \rangle = \langle A^2 \rangle - \langle A \rangle^2, \tag{3.74}$$

or by its square root, the *standard deviation*,

$$\sigma(A) := \sqrt{\operatorname{var}(A)}. \tag{3.75}$$

Here the observable A^2 corresponds to the pairs $\{(x_i, a_i^2)\}$, and is represented in Hilbert space by the operator $\hat{A}^2$. Given two arbitrary observables, A and B, their standard deviations obey the *uncertainty relation*,

$$\sigma(A)\sigma(B) \geq \frac{1}{2} |\langle C_{A,B} \rangle|, \tag{3.76}$$

where the observable on the right-hand side, $C_{A,B}$, is represented in Hilbert space by the commutator (up to a factor $1/i$) of the previous two,

$$\hat{C}_{A,B} := \frac{1}{i}[\hat{A}, \hat{B}]. \tag{3.77}$$

The uncertainty relation may be derived as follows. We define the operator

$$\hat{T}_\lambda := (\hat{A} - \langle A \rangle) + i\lambda(\hat{B} - \langle B \rangle), \quad \lambda \in \mathbb{R},$$

where the parameter λ can take arbitrary real values. This operator is generally not Hermitian; however, the product $\hat{T}_\lambda \hat{T}_\lambda^\dagger$ is. The latter is, moreover, positive semi-definite. So its expectation value in any state must be non-negative:

$$\operatorname{tr}(\hat{\rho}\hat{T}_\lambda \hat{T}_\lambda^\dagger) \geq 0 \quad \forall \hat{\rho}, \lambda.$$

After inserting the definition of $\hat{T}_\lambda$, this yields the inequality

$$\operatorname{var}(B)\lambda^2 + \langle C_{A,B} \rangle \lambda + \operatorname{var}(A) \geq 0 \quad \forall \lambda.$$

It implies, first of all, that the variances of A and B cannot both vanish simultaneously unless $\langle C_{A,B} \rangle = 0$:

$$\operatorname{var}(A) = \operatorname{var}(B) = 0 \quad \Rightarrow \quad \langle C_{A,B} \rangle = 0.$$

Leaving this special case aside, we focus now on the case where at least one of the two variances does not vanish. Without loss of generality, we may assume this to be the variance of B, $\mathrm{var}(B) \neq 0$. Then we can divide the inequality by $\mathrm{var}(B)$ to obtain

$$\frac{\mathrm{var}(A)}{\mathrm{var}(B)} - \left[\frac{\langle C_{A,B}\rangle}{2\,\mathrm{var}(B)}\right]^2 \geq -\left[\lambda + \frac{\langle C_{A,B}\rangle}{2\,\mathrm{var}(B)}\right]^2 \quad \forall \lambda.$$

This must hold in particular when we choose λ such that the right-hand side vanishes. Therefore, the left-hand side must be non-negative, which is the case if and only if

$$\mathrm{var}(A)\,\mathrm{var}(B) \geq \frac{1}{4}\langle C_{A,B}\rangle^2.$$

Taking the square root then yields the uncertainty relation, Eq. (3.76).

   The uncertainty relation has profound implications for our ability to ascertain the values of two different observables. As long as the two observables are jointly measurable, and hence $C_{A,B} = 0$, the uncertainty relation imposes no particular constraints on the variances of A and B; both observables can be known with arbitrary precision. But if the observables are not jointly measurable, it is $C_{A,B} \neq 0$ and, in general, also $\langle C_{A,B}\rangle \neq 0$ (except perhaps in some special states). Then the uncertainty relation stipulates that neither variance can be zero; the value of neither observable can be known with perfect precision. Moreover, there is an inevitable trade-off: the higher the precision with which the value of one observable is known, the lower it must be for the other. We illustrate this trade-off in the simple case of a two-level system. The three Pauli operators, $\hat{X}$, $\hat{Y}$, and $\hat{Z}$, are all Hermitian and thus represent observables of the two-level system. Their mutual relationships, Eq. (3.64), imply that, for instance, $C_{X,Y} = 2Z$. Inserting the latter into the uncertainty relation, Eq. (3.76), yields the inequality

$$\sigma(X)\sigma(Y) \geq |\langle Z\rangle|,$$

and likewise for all other pairs of observables. At the same time, due to $\hat{X}^2 = \hat{Y}^2 = \hat{I}$, both standard deviations are bounded from above by one:

$$\sigma(X) \leq 1, \quad \sigma(Y) \leq 1.$$

When a two-level system is in an eigenstate of Z, with $\langle Z\rangle = \pm 1$, these inequalities can be satisfied simultaneously only if $\sigma(X) = \sigma(Y) = 1$. In other words, whenever the value of Z is known with perfect precision, the values of the other two observables, X and Y, are maximally uncertain.

   Next to 'state' and 'observable', another important concept in quantum theory is that of a *transformation*. A transformation may apply either to a state or to an observable, yielding a transformed state or observable, respectively. It is a bijective map either from the set of states onto itself or from the set of observables onto itself. If applied to an observable, a transformation does not change the possible measurement values of that observable, but it does alter the associated propositions. For instance, in a game of chance there might be an observable called 'gain (old)' which assigns different amounts of money to different outcomes of a die roll such as $\{(x_1, 2\$), (x_2, -1\$), \ldots\}$. A transformation may map this observable to another observable, 'gain (new)', which still features the same dollar amounts but assigns them differently to the various outcomes of the die roll. As another example, in a two-level system there might be a transformation mapping the Pauli observable X to the Pauli observable Y (Fig. 3.9). Both observables feature the same possible measurement values, $\{+1, -1\}$, yet the associated propositions—visualized by pairs of antipodes on the Bloch sphere—differ. More generally, a transformation, π, maps an observable, $A = \{(x_i, a_i)\}$, to another observable,

$$\pi : A \to \pi(A) = \{(\pi(x_i), a_i)\}. \tag{3.78}$$

This transformation must respect the fact that in both the old and the new observable the propositions associated with the outcomes are mutually exclusive. In other words, the transformation must preserve the mutual exclusion of propositions.

For each transformation of observables there exists a corresponding transformation of states. The transformed state, $\pi(\rho)$, is defined implicitly as that state which assigns to any transformed observable the same expectation value as does the original state to the original observable:

$$\langle \pi(A) \rangle_{\pi(\rho)} := \langle A \rangle_\rho \ \forall A. \tag{3.79}$$

This is equivalent to the requirement that the transformed state assign to all transformed propositions the same probabilities as does the original state to the original propositions. In other words, whenever states and observables (and hence propositions) are

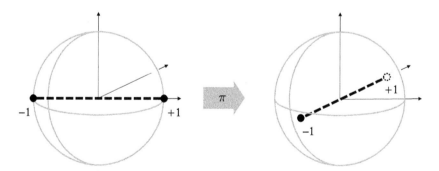

Fig. 3.9 *Observable and transformation. In a two-level system all measurements are binary. One may visualize a measurement as a diameter of the Bloch sphere (thick dashed line), its possible outcomes as the associated antipodes (dots), and an observable as such a pair of antipodes with numerical values attached (here: ±1). This example shows the Pauli observables X (left) and Y (right). Since they have the same set of measurement values, $\{+1, -1\}$, the former can be transformed into the latter by means of a transformation, π.*

transformed in sync, expectation values and probabilities remain the same. By contrast, when just the states or just the observables (or propositions) are transformed in isolation, expectation values and probabilities change. In this case a transformation of the state has the same effect on an expectation value as the *opposite* transformation of the pertinent observable; or, in terms of propositions and probabilities,

$$\mathrm{prob}(x|\pi(\rho)) = \mathrm{prob}(\pi^{-1}(x)|\rho). \tag{3.80}$$

Thus, any change of expectation values and probabilities that results from a transformation may be described in two alternative ways: either as stemming from a transformation of the state, called an *active* transformation (the so-called *Schrödinger picture*); or as stemming from the opposite transformation of the observable, called a *passive* transformation (*Heisenberg picture*).

In complex Hilbert space, the requirement that a transformation preserve the mutual exclusion of propositions translates into the requirement that it preserve orthogonality. If two propositions, x and y, are mutually exclusive and hence represented by subspaces of Hilbert space which are mutually orthogonal, then so must be their respective images. In terms of the associated projectors, this means that

$$\hat{P}_x \hat{P}_y = 0 \quad \Rightarrow \quad \hat{P}_{\pi(x)} \hat{P}_{\pi(y)} = 0. \tag{3.81}$$

Non-trivial transformations of projectors, $\hat{P}_x \rightarrow \hat{P}_{\pi(x)}$, which satisfy this requirement do exist; the most obvious example is a *unitary* transformation. More precisely, an abstract transformation of propositions might be represented by a unitary transformation of the associated projectors,

$$\hat{P}_{\pi(x)} = \hat{U}_\pi \hat{P}_x \hat{U}_\pi^\dagger, \quad \hat{U}_\pi \hat{U}_\pi^\dagger = \hat{U}_\pi^\dagger \hat{U}_\pi = \hat{I}. \tag{3.82}$$

Such a unitary transformation does indeed preserve the orthogonality of subspaces, Eq. (3.81). Another, less obvious example is a transformation of the kind

$$\hat{P}_{\pi(x)} = \hat{U}_\pi \hat{P}_x^* \hat{U}_\pi^\dagger, \tag{3.83}$$

where $\hat{U}_\pi$ is once again a unitary operator and $*$ denotes the complex conjugate (not the adjoint) of the projector. In contrast to the adjoint, the complex conjugate makes reference to a specific orthonormal basis in Hilbert space, $\{|i\rangle\}$. In terms of this basis, it is

$$\hat{P}^* = \sum_{ik} \langle i|\hat{P}|k\rangle^* |i\rangle\langle k|. \tag{3.84}$$

This complex conjugate is still a projector; it is still both Hermitian and idempotent. A transformation involving complex conjugation, of the form of Eq. (3.83), is called an *antiunitary* transformation. It preserves orthogonality, too. In fact, these two cases already exhaust all possibilities: according to one version of *Wigner's theorem*, every transformation of subspaces of a complex Hilbert space which preserves orthogonality, in the sense of Eq. (3.81), is either unitary or antiunitary. This version of the theorem uses only the preservation of logical structure, that is, orthogonality, and does not invoke probabilities. In this generality, it only holds for Hilbert spaces of dimension greater than two. If one requires that, in addition, the conditional probability of any most accurate proposition, e, given the truth of another most accurate proposition, f,

$$\text{prob}(e|f) = \text{tr}(\hat{P}_f \hat{P}_e), \tag{3.85}$$

be invariant under the joint transformation of both projectors, then Wigner's theorem in its original version stipulates that the transformation must be unitary or antiunitary for every Hilbert space dimension, including dimension two.

For a two-level system, unitary and antiunitary transformations can be visualized on the Bloch sphere. In Exercise (3.12) you will show that a unitary operator on two-dimensional Hilbert space can always be parametrized in the form

$$\hat{U} = \exp(i\beta)\hat{R}_{\hat{n}}(\gamma), \tag{3.86}$$

where

$$\hat{R}_{\hat{n}}(\gamma) := \exp\left(-i\frac{\gamma}{2}\hat{n}\cdot\hat{\sigma}\right) = \cos\left(\frac{\gamma}{2}\right)\hat{I} - i\sin\left(\frac{\gamma}{2}\right)\hat{n}\cdot\hat{\sigma}, \quad \gamma \in [0, 2\pi), \tag{3.87}$$

and $\hat{n}$ is a real three-dimensional unit vector, $|\hat{n}| = 1$. Since the phase factor, $\exp(i\beta)$, has no impact on the transformation of any object of interest—an observable, statistical operator, subspace, or projector—it may be disregarded. The remaining operator, $\hat{R}_{\hat{n}}(\gamma)$, effects on the Bloch sphere a rotation about $\hat{n}$ by the angle γ (Fig. 3.10). (You will verify this, too, in Exercise (3.12).) An antiunitary transformation, on the other hand, combines a unitary transformation—that is, rotation on the Bloch sphere—with complex conjugation. Provided complex conjugation is defined with respect to the standard basis, it effects a change of the relative phase angle, $\varphi \to (2\pi - \varphi)$, in Eq. (3.59). Thus, pictorially, complex conjugation amounts to a reflection through the x–z plane (Fig. 3.11). An arbitrary antiunitary transformation of a two-level system may then be visualized as a combination of a rotation with such a reflection. Rotations and reflections constitute the sole operations on a unit sphere which preserve all lengths and angles. Therefore, it appears plausible that beyond the unitary and antiunitary transformations there should be no further transformations.

Every unitary transformation can be divided into N incremental steps, each effected by its Nth root,

$$\hat{U} = \underbrace{\hat{U}^{1/N} \ldots \hat{U}^{1/N}}_{N}. \tag{3.88}$$

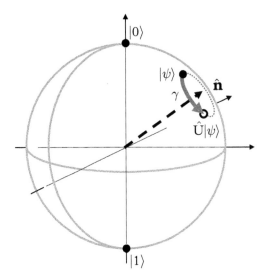

Fig. 3.10 *Unitary transformation on the Bloch sphere. Up to an unobservable phase factor, every unitary transformation of a two-level system corresponds to a rotation on the Bloch sphere about some axis, $\hat{n}$, by some angle, γ.*

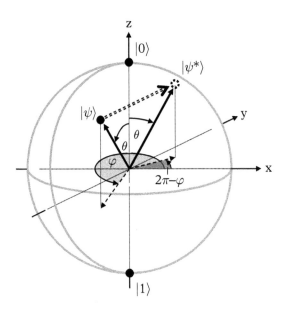

Fig. 3.11 *Complex conjugation on the Bloch sphere.*

This is similar in spirit to the division considered in the context of the quantum Zeno effect (Fig. 3.5). In the limit $N \to \infty$ the intermediate stages of this transformation, $\hat{U}^{M/N}$ $(0 \le M \le N)$, form a continuum, $\hat{U}^{\lambda}$ $(0 \le \lambda \le 1)$. This continuum connects the identity map ($\lambda = 0$) to the full transformation ($\lambda = 1$). As λ varies from zero to one and the state (Schrödinger picture) or observables (Heisenberg picture) are transformed via $\hat{U}^{\lambda}$ or $(\hat{U}^{\lambda})^{\dagger}$, respectively, expectation values and probabilities change in a continuous fashion. In this sense, every unitary transformation is connected smoothly to the identity. By contrast, an antiunitary transformation involves complex conjugation, which is *not* a continuous map. Upon complex conjugation, expectation values and probabilities may jump in a discontinuous manner. Therefore, it is impossible to build up an antiunitary transformation from a sequence of infinitesimal steps; an antiunitary transformation is necessarily discrete. In physics, most transformations of interest are of the former kind, to wit, they are transformations which can be connected smoothly to the identity: rotations, translations in space, and so on. These must be represented in Hilbert space by unitary transformations. Only a few discrete transformations—most notably, time reversal—are represented by antiunitary transformations.

3.5 Composite Systems

Quantum information processing works with multiple two-level systems, or 'qubits'. Together, these form a composite system. I discuss how to describe such a composite system mathematically. I examine how the state of a composite system is constructed given the—pure—states of its parts and, conversely, how to obtain the state of a part given the state of the whole. I pay particular attention to a genuinely quantum form of statistical correlation termed 'entanglement'.

We consider a quantum system that is composed of n constituents; such a system is also called an *n-partite system*. We will assume that its constituents are distinguishable, either by their different nature (for instance, a proton and an electron) or by other external features such as clearly distinct locations. Associated with each constituent, k, is some d_k-dimensional Hilbert space, $\mathcal{H}_k$. This Hilbert space has an orthonormal basis, $\{|i_k\rangle, i_k = 1, ..., d_k\}$. When all constituents, individually, have been prepared as accurately as possible, they are all, individually, in pure states. This amounts to a most accurate preparation of the composite system which, hence, must also be in a pure state. The latter is represented by a unit vector in some larger Hilbert space of the composite system, $\mathcal{H}$. When the first constituent is in the basis state $|i_1\rangle$, the second constituent is in the basis state $|i_2\rangle$, and so on, the resultant pure state of the composite system is denoted by $|i_1 i_2 ... i_n\rangle$. There are

$$d = \prod_{k=1}^{n} d_k \tag{3.89}$$

different such combinations of the constituent basis states. These combinations are linearly independent and, together, span the Hilbert space of the composite system. Hence the Hilbert space of the composite system has dimension d. This is consistent with our general product rule, Eq. (2.35).

Mathematically, the composite Hilbert space is the *tensor product* of the constituent Hilbert spaces,

$$\mathcal{H} = \bigotimes_{k=1}^{n} \mathcal{H}_k \equiv \mathcal{H}_1 \otimes \mathcal{H}_2 \otimes \cdots \otimes \mathcal{H}_n. \tag{3.90}$$

Its basis states, $|i_1 i_2 \ldots i_n\rangle$, are tensor products of the constituent basis states,

$$|i_1 i_2 \ldots i_n\rangle \equiv |i_1\rangle \otimes |i_2\rangle \otimes \cdots \otimes |i_n\rangle. \tag{3.91}$$

More generally, when the individual constituents are in arbitrary pure states, $\{|\psi_k\rangle \in \mathcal{H}_k, k = 1, \ldots, n\}$, the resultant pure state of the composite system is their tensor product,

$$|\psi\rangle = |\psi_1\rangle \otimes |\psi_2\rangle \otimes \cdots \otimes |\psi_n\rangle. \tag{3.92}$$

The tensor product is bilinear in the sense that, for example,

$$(a_1 |\psi_1\rangle + b_1 |\phi_1\rangle) \otimes (a_2 |\psi_2\rangle + b_2 |\phi_2\rangle)$$
$$= a_1 a_2 |\psi_1\rangle \otimes |\psi_2\rangle + a_1 b_2 |\psi_1\rangle \otimes |\phi_2\rangle + b_1 a_2 |\phi_1\rangle \otimes |\psi_2\rangle + b_1 b_2 |\phi_1\rangle \otimes |\phi_2\rangle. \tag{3.93}$$

The above equations are formulated for kets; analogous results hold for the bras. Switching between bras and kets commutes with taking the tensor product,

$$\{|\psi_k\rangle\} \leftrightarrow \{\langle\psi_k|\} \quad \Rightarrow \quad \bigotimes_{k=1}^{n} |\psi_k\rangle \leftrightarrow \bigotimes_{k=1}^{n} \langle\psi_k|.$$

The composite Hilbert space inherits from the constituent Hilbert spaces a scalar product,

$$((\langle\phi_1| \otimes \langle\phi_2| \otimes \cdots \otimes \langle\phi_n|)(|\psi_1\rangle \otimes |\psi_2\rangle \otimes \cdots \otimes |\psi_n\rangle)) = \prod_{k=1}^{n} \langle\phi_k|\psi_k\rangle. \tag{3.94}$$

With respect to this scalar product, the basis states of the composite Hilbert space are, indeed, orthonormal:

$$\langle i_1 i_2 \ldots i_n | j_1 j_2 \ldots j_n\rangle = \delta_{i_1 j_1} \delta_{i_2 j_2} \cdots \delta_{i_n j_n}.$$

Operators pertaining to the individual constituents, $\{\hat{A}_k\}$, can be combined into one operator pertaining to the composite system by taking their tensor product. The tensor product of operators which act on the individual constituent Hilbert spaces yields an operator that acts on the composite Hilbert space. It is defined by its action on arbitrary tensor product states,

$$\left(\hat{A}_1 \otimes \hat{A}_2 \otimes \cdots \otimes \hat{A}_n\right) \left(|\psi_1\rangle \otimes |\psi_2\rangle \otimes \cdots \otimes |\psi_n\rangle\right)$$
$$= (\hat{A}_1|\psi_1\rangle) \otimes (\hat{A}_2|\psi_2\rangle) \otimes \cdots \otimes (\hat{A}_n|\psi_n\rangle). \tag{3.95}$$

Finally, in Exercise (3.15) you will show that the trace of a tensor product of operators equals the product of the individual traces,

$$\operatorname{tr}\left(\bigotimes_{k=1}^{n} \hat{A}_k\right) = \prod_{k=1}^{n} \operatorname{tr}\left(\hat{A}_k\right). \tag{3.96}$$

For the logical AND operation which joins propositions about different systems, Eq. (2.33), we used the same symbol $\otimes$ as for the tensor product. This is not a coincidence. Indeed, if $\{x_1, \ldots, x_n\}$ are propositions pertaining to distinct constituents, represented in their respective Hilbert spaces by projectors $\{\hat{P}_1, \ldots, \hat{P}_n\}$, the joint proposition $x_1 \otimes \ldots \otimes x_n$ is represented in the composite Hilbert space by the tensor product of these projectors:

$$\hat{P}_{x_1 \otimes \ldots \otimes x_n} = \hat{P}_1 \otimes \ldots \otimes \hat{P}_n. \tag{3.97}$$

The law of distributivity for propositions, Eq. (2.34), then translates into

$$\hat{P}_x \otimes (\hat{P}_y + \hat{P}_z) = \hat{P}_x \otimes \hat{P}_y + \hat{P}_x \otimes \hat{P}_z, \quad \hat{P}_y \hat{P}_z = 0.$$

This is just a special case of the distributivity of the tensor product,

$$\hat{A} \otimes (\hat{B} + \hat{C}) = \hat{A} \otimes \hat{B} + \hat{A} \otimes \hat{C}. \tag{3.98}$$

There exists a similar correspondence for observables. We defined an observable, A, in abstract terms as a set of pairs, $\{(x_i, a_i)\}$, where the $\{x_i\}$ are mutually exclusive propositions pertaining to the outcome of a measurement and the $\{a_i\}$ are numerical values assigned to these outcomes. Where this observable and a second observable, $B = \{(y_k, b_k)\}$, pertain to two distinct constituents, we may define their product, $A \otimes B$, in abstract terms as the set of pairs $\{(x_i \otimes y_k, a_i b_k)\}$; this product is an observable pertaining to the composite. If in their respective constituent Hilbert spaces the observables A and B are represented by Hermitian operators $\hat{A}$ and $\hat{B}$, then the product observable, $A \otimes B$, is represented in the composite Hilbert space by the tensor product of these operators, $\hat{A} \otimes \hat{B}$.

Maximal knowledge about the constituents, individually, amounts to maximal knowledge about the whole. For example, once we ascertain the truth of the individual most accurate propositions 'the coin shows "Heads"' and 'the die shows "2"', we possess maximal knowledge about the composite system 'coin plus die'. Likewise, once we ascertain that 'the photon passes the horizontal polarization filter' and 'the atom is deflected upwards in the Stern–Gerlach apparatus', we have maximal knowledge about 'photon plus atom' (or, more precisely, about the composite of their polarization and magnetic moment degrees of freedom, respectively). In a classical theory without knowledge constraints, the converse also holds: maximal knowledge about the whole presupposes maximal knowledge about all its individual constituents. We can claim to possess maximal knowledge about 'coin plus die' only if we know which side the coin shows and how many spots the die shows; that is to say, only if we can make most accurate propositions about each constituent individually. In quantum theory, however, this is no longer true.

As an example, we consider a system composed of two two-level systems, or 'qubits' (Fig. 3.12). This is the simplest example of a *bipartite system*, which is a system composed of two constituents, A and B. Associated with each qubit is a two-dimensional Hilbert space, $\mathcal{H}_A$ or $\mathcal{H}_B$, respectively. The composite system is described in the tensor product space, $\mathcal{H} = \mathcal{H}_A \otimes \mathcal{H}_B$, of dimension $\dim \mathcal{H} = 2 \cdot 2 = 4$. This composite Hilbert space has four standard basis states, $|00\rangle$, $|01\rangle$, $|10\rangle$, and $|11\rangle$, where the first entry pertains to the state of qubit A and the second entry, to the state of qubit B. The allowed pure states of the composite system are arbitrary superpositions of these basis states. In particular, the composite system might be in one of the four so-called *Bell states*,

$$|\beta_{jk}\rangle := \frac{|0,k\rangle + (-1)^j |1,1-k\rangle}{\sqrt{2}}, \quad j,k \in \{0,1\}. \tag{3.99}$$

As you will verify in Exercise (3.17), these four states themselves constitute an (alternative) orthonormal basis in the composite Hilbert space, called the *Bell basis*. Having prepared the composite system in one of these Bell states, we now test a proposition, x, which pertains to the first qubit only. Let this proposition be represented in the Hilbert space of the first qubit, $\mathcal{H}_A$, by the projector $\hat{P}_x$. Since it is neutral as regards the second qubit, it is represented in the composite Hilbert space, $\mathcal{H}$, by the tensor product of $\hat{P}_x$

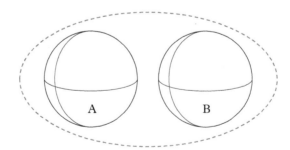

Fig. 3.12 *Bipartite system composed of two qubits.*

with the unit operator in $\mathcal{H}_B$, $\hat{I}_B$. The probability that x is revealed as true is given by the Born rule, Eq. (3.51):

$$
\begin{aligned}
\mathrm{prob}(x|\beta_{jk}) &= \langle \beta_{jk}|\hat{P}_x \otimes \hat{I}_B|\beta_{jk}\rangle \\
&= \frac{1}{2}\Big(\langle 0,k|\hat{P}_x \otimes \hat{I}_B|0,k\rangle + \langle 1,1-k|\hat{P}_x \otimes \hat{I}_B|1,1-k\rangle \\
&\quad + (-1)^j\,\langle 0,k|\hat{P}_x \otimes \hat{I}_B|1,1-k\rangle + (-1)^j\,\langle 1,1-k|\hat{P}_x \otimes \hat{I}_B|0,k\rangle \Big) \\
&= \frac{1}{2}\Big(\langle 0|\hat{P}_x|0\rangle + \langle 1|\hat{P}_x|1\rangle \Big) \\
&= \mathrm{tr}\left(\left[\frac{1}{2}\hat{P}_0 + \frac{1}{2}\hat{P}_1\right]\hat{P}_x \right).
\end{aligned}
$$

While the initial formula for the probability involved a scalar product in the composite Hilbert space, $\mathcal{H}$, the final expression contains only objects that pertain to the constituent Hilbert space, $\mathcal{H}_A$. It has the form of Gleason's theorem, Eq. (3.45), with the statistical operator given by

$$
\hat{\rho}_A = \frac{1}{2}\hat{P}_0 + \frac{1}{2}\hat{P}_1; \tag{3.100}
$$

in this particular example, the latter is proportional to the unit operator, $\hat{\rho}_A = \hat{I}_A/2$. This statistical operator is called the *reduced statistical operator* of the constituent A, and is labelled with the subscript A. It is the Hilbert space representation of the reduced state which we introduced, in more general terms, in Section 2.5. (Where the distinction between the abstract state and its Hilbert space representation does not matter, we will henceforth use the terms 'reduced state' and 'reduced statistical operator' interchangeably.) An analogous calculation yields the identical result for the second constituent, $\hat{\rho}_B = \hat{I}_B/2$. Evidently, both reduced states are not pure. In fact, they are states of total ignorance. So, indeed, we have a situation where we possess maximal knowledge about the composite system—described by a pure Bell state—but not about the individual constituents! This is an important and peculiar feature of quantum theory known as *entanglement*. In general, one says that a pure composite state is 'entangled' if there is at least one constituent whose reduced state is *not* pure. As you will show in Exercise (3.16), this is the case if and only if the pure composite state cannot be written as a product state, that is, in the form of Eq. (3.92). There also exist entangled states which are mixed, and you will discover how these are defined in Exercise (5.20); but in the main text we shall consider pure entangled states only.

Entanglement furnishes the last missing piece of explanation for the experimental evidence presented in Section 1.1. The perfect anticorrelation depicted in Fig. 1.14 means that, for arbitrary values of θ, the proposition $(\Uparrow_\theta \otimes \Downarrow_\theta) \oplus (\Downarrow_\theta \otimes \Uparrow_\theta)$ is true with certainty; or, equivalently, that the complementary proposition, $(\Uparrow_\theta \otimes \Uparrow_\theta) \oplus (\Downarrow_\theta \otimes \Downarrow_\theta)$, has probability zero. The latter requires that the composite state of the photon pair, $|\psi\rangle$, is an eigenstate of the pertinent projector, $\hat{P}_{\Uparrow_\theta} \otimes \hat{P}_{\Uparrow_\theta} + \hat{P}_{\Downarrow_\theta} \otimes \hat{P}_{\Downarrow_\theta}$, with eigenvalue zero:

$$
(\hat{P}_{\Uparrow_\theta} \otimes \hat{P}_{\Uparrow_\theta} + \hat{P}_{\Downarrow_\theta} \otimes \hat{P}_{\Downarrow_\theta})\,|\psi\rangle = 0.
$$

This is possible only if each summand vanishes individually,

$$(\hat{P}_{\Uparrow\theta} \otimes \hat{P}_{\Uparrow\theta})|\psi\rangle = (\hat{P}_{\Downarrow\theta} \otimes \hat{P}_{\Downarrow\theta})|\psi\rangle = 0. \tag{3.101}$$

A non-entangled state, $|\psi\rangle = |\psi_A\rangle \otimes |\psi_B\rangle$, can achieve that if $|\psi_A\rangle$ and $|\psi_B\rangle$, individually, are eigenstates of $\hat{P}_{\Uparrow\theta}$ and $\hat{P}_{\Downarrow\theta}$: either $|\psi_A\rangle = |\Uparrow\theta\rangle$ and $|\psi_B\rangle = |\Downarrow\theta\rangle$, or vice versa. In other words, for any specific, fixed value of θ there does exist a product state which exhibits the observed anticorrelation. However, every parameter setting would require its own tailor-made product state; there is no single product state that would yield the perfect anticorrelation for *all* values of θ. But how can the nonlinear crystal in Fig. 1.14 know which product state it ought to produce? Should it not always produce the same state? Unless we are willing to attribute prophetic powers to the crystal, we are forced to conclude that the two-photon state produced by the crystal is an entangled state.

In fact, the nonlinear crystal produces photon pairs in one particular Bell state:

$$|\beta_{11}\rangle = \frac{|01\rangle - |10\rangle}{\sqrt{2}}. \tag{3.102}$$

This Bell state is special in that it is invariant under synchronized unitary transformations in both single-qubit Hilbert spaces:

$$(\hat{U} \otimes \hat{U})|\beta_{11}\rangle = |\beta_{11}\rangle \quad \forall \hat{U}. \tag{3.103}$$

That is to say, if on the right-hand side of Eq. (3.102) we replace the standard basis states, $\{|0\rangle, |1\rangle\}$, by rotated orthonormal basis states,

$$|i\rangle \rightarrow |i'\rangle = \hat{U}|i\rangle, \quad i \in \{0, 1\},$$

the result is still the same Bell state:

$$|\beta_{11}\rangle = \frac{|0'1'\rangle - |1'0'\rangle}{\sqrt{2}};$$

the Bell state maintains its form in any orthonormal basis. (By contrast, the other three Bell states are invariant only under certain selected, not all, basis changes. You will investigate these invariance properties in Exercise (3.17).) In particular, we can choose as basis states the states associated with the two possible outcomes of the beam splitter,

$$|\beta_{11}\rangle = \frac{|\Uparrow\theta\Downarrow\theta\rangle - |\Downarrow\theta\Uparrow\theta\rangle}{\sqrt{2}}, \tag{3.104}$$

where θ may take any value. Written in this form, it is evident that the Bell state does indeed satisfy Eq. (3.101) for *every* value of θ, as is needed to explain the observed perfect anticorrelation. In quantum computation, sensing, and communication such perfect anticorrelation (or perfect correlation) is often created deliberately and put to use to improve the efficiency, accuracy, or security of the respective protocols. This is the reason why entangled states, in particular the Bell states defined above, will play a prominent role in many of the protocols that we shall discuss in Chapters 4 and 5.

Finally, we expand some more on the concept of the reduced statistical operator, which we introduced earlier in the special setting of a qubit pair in a Bell state. The calculation which led to the reduced statistical operator of one of the qubits, Eq. (3.100), is readily generalized to arbitrary bipartite systems in arbitrary composite states. A bipartite system is composed of two constituents which, again, we label by A and B. The composite system might be in a pure state, as in our earlier example, or in a mixed state. In order to account for both possibilities, we shall now describe the state of the composite system with a statistical operator, $\hat{\rho}_{AB}$. Again, we consider the probability of some proposition, x, that pertains to constituent A only. As before, this proposition is represented by a projector, $\hat{P}_x$, on the Hilbert space of constituent A, $\mathcal{H}_A$, or by the tensor product $\hat{P}_x \otimes \hat{I}_B$ on the composite Hilbert space, $\mathcal{H} = \mathcal{H}_A \otimes \mathcal{H}_B$. According to Gleason's theorem, Eq. (3.45), its probability is given by

$$\mathrm{prob}(x|\rho_{AB}) = \mathrm{tr}_{AB}\left[\hat{\rho}_{AB}(\hat{P}_x \otimes \hat{I}_B)\right]$$

$$= \sum_{jklm} \langle jk|\hat{\rho}_{AB}|lm\rangle \langle lm|\hat{P}_x \otimes \hat{I}_B|jk\rangle$$

$$= \sum_{jklm} \langle jk|\hat{\rho}_{AB}|lm\rangle \langle l|\hat{P}_x|j\rangle \delta_{mk}$$

$$= \mathrm{tr}_A(\hat{\rho}_A \hat{P}_x),$$

with the reduced statistical operator

$$\hat{\rho}_A = \mathrm{tr}_B(\hat{\rho}_{AB}) := \sum_{jkl} \langle jk|\hat{\rho}_{AB}|lk\rangle \, |j\rangle\langle l|. \qquad (3.105)$$

While in the original expression for the probability the trace is taken in the composite Hilbert space, tr_{AB}, the trace featuring at the end is taken in the Hilbert space of the constituent of interest, A, only. The transition from the composite to the reduced state has effectively eliminated ('traced out') the degrees of freedom of the unconsidered constituent, B. Indeed, the formula for the reduced statistical operator, Eq. (3.105), features a *partial trace*, tr_B, of the composite state over the disregarded degrees of freedom. The above result for the probability confirms that tracing out degrees of freedom is consistent with Gleason's theorem, in the following sense. Whenever the probabilities for a composite system have the form stipulated by Gleason's theorem, Eq. (3.45), so do the probabilities for all its constituents, only with the composite state replaced by the respective reduced state. The latter explains why we have always assumed that Gleason's theorem applies to two-level systems as well, even though, strictly speaking, two-dimensional Hilbert spaces are exempt from the original statement of the theorem. After all, a two-level system never exists in perfect isolation; ultimately, it is always a constituent of some higher-dimensional composite system.

In order to illustrate how the reduced statistical operator is calculated when the composite state is mixed, we consider the example of two qubits in the state

$$\hat{\rho}_{AB} = \frac{3}{4}|00\rangle\langle00| + \frac{1}{4}|\beta_{00}\rangle\langle\beta_{00}|,$$

where

$$|\beta_{00}\rangle = \frac{|00\rangle + |11\rangle}{\sqrt{2}}$$

is one of the Bell states, Eq. (3.99). The reduced statistical operator of constituent A is given by

$$\hat{\rho}_A = \frac{3}{4}\sum_{jkl}\langle jk|00\rangle\,\langle00|lk\rangle\,|j\rangle\langle l| + \frac{1}{4}\sum_{jkl}\langle jk|\beta_{00}\rangle\,\langle\beta_{00}|lk\rangle\,|j\rangle\langle l|$$

$$= \frac{3}{4}\sum_{jkl}\langle jk|00\rangle\,\langle00|lk\rangle\,|j\rangle\langle l| + \frac{1}{8}\sum_{jkl}\langle jk|00\rangle\,\langle00|lk\rangle\,|j\rangle\langle l| + \frac{1}{8}\sum_{jkl}\langle jk|11\rangle\,\langle11|lk\rangle\,|j\rangle\langle l|$$

$$= \frac{7}{8}|0\rangle\langle0| + \frac{1}{8}|1\rangle\langle1|.$$

In the second line of this calculation there appeared no mixed terms containing, say, the product $\langle jk|00\rangle\,\langle11|lk\rangle$, because whatever value k takes, one of the two factors in these mixed terms always vanishes. Due to the symmetry of the composite state, the reduced state of constituent B has the same form as the reduced state of constituent A,

$$\hat{\rho}_B = \frac{7}{8}|0\rangle\langle0| + \frac{1}{8}|1\rangle\langle1|.$$

However, the bras and kets now pertain to the Hilbert space of constituent B rather than to the Hilbert space of constituent A.

If the constituents are statistically independent, then, by definition, the composite state is a product state, $\rho_{AB} = \rho_A \otimes \rho_B$ (Eq. (2.41)). Such a product state is represented in the composite Hilbert space by the tensor product of the respective reduced statistical operators,

$$\rho_{AB} = \rho_A \otimes \rho_B \quad \Rightarrow \quad \hat{\rho}_{AB} = \hat{\rho}_A \otimes \hat{\rho}_B. \tag{3.106}$$

Indeed, thanks to the tensor product representation of joint propositions, Eq. (3.97), and the factorization of the trace, Eq. (3.96), it is

$$\text{prob}(x \otimes y|\rho_{AB}) = \text{tr}\left[(\hat{\rho}_A \otimes \hat{\rho}_B)(\hat{P}_x \otimes \hat{P}_y)\right]$$
$$= \text{tr}\left[(\hat{\rho}_A \hat{P}_x) \otimes (\hat{\rho}_B \hat{P}_y)\right]$$
$$= \text{tr}\left(\hat{\rho}_A \hat{P}_x\right) \cdot \text{tr}\left(\hat{\rho}_B \hat{P}_y\right)$$
$$= \text{prob}(x|\rho_A) \cdot \text{prob}(y|\rho_B),$$

mirroring Eq. (2.41). In the special case where both reduced states are pure, $\hat{\rho}_i = |\psi_i\rangle\langle\psi_i|$, their tensor product is pure, too:

$$|\psi_A\rangle\langle\psi_A| \otimes |\psi_B\rangle\langle\psi_B| = (|\psi_A\rangle \otimes |\psi_B\rangle)(\langle\psi_A| \otimes \langle\psi_B|).$$

This is the counterpart, in terms of statistical operators, of the tensor product structure of the unit vector, Eq. (3.92). When the two constituents are statistically independent and a measurement is performed on just one of them, yielding the truth of, say, the proposition x^A pertaining to constituent A, the composite state must be updated according to Lüders' rule, Eq. (3.56):

$$[\hat{\rho}_A \otimes \hat{\rho}_B]_{|x^A} = \frac{(\hat{P}_{x^A} \otimes \hat{I}_B)(\hat{\rho}_A \otimes \hat{\rho}_B)(\hat{P}_{x^A} \otimes \hat{I}_B)}{\text{tr}[(\hat{\rho}_A \otimes \hat{\rho}_B)(\hat{P}_{x^A} \otimes \hat{I}_B)]} = \frac{\hat{P}_{x^A}\hat{\rho}_A\hat{P}_{x^A}}{\text{tr}(\hat{\rho}_A\hat{P}_{x^A})} \otimes \frac{\hat{I}_B\hat{\rho}_B\hat{I}_B}{\text{tr}(\hat{\rho}_B\hat{I}_B)} = \hat{\rho}_{A|x^A} \otimes \hat{\rho}_B.$$

As we would expect, the measurement on A triggers an update of the reduced state of A only; the reduced state of the other constituent, B, remains unaffected. This is in agreement with our earlier result, Eq. (2.42).

If the constituents are not statistically independent, then the composite state cannot be written as a tensor product, $\hat{\rho}_{AB} \neq \hat{\rho}_A \otimes \hat{\rho}_B$. In this case there exists at least one joint measurement that will reveal statistical correlations between the constituents. These correlations might be of the same kind as the correlations found in classical probability theory, or they might be genuine quantum correlations due to entanglement.

3.6 The Issue of Reality

In a classical world, every object is in some—possibly unknown—definite configuration. Measurements reveal this configuration, wholly or in part, but do not influence it. I present two pieces of evidence, the Kochen–Specker theorem and the Bell experiment, which cast serious doubt on whether such an underlying configuration exists in the quantum realm.

In our operational approach we have taken pains to work only with propositions which can be tested—that is, propositions which pertain to the outcomes of measurements that can actually be performed. We never speculated about outcomes of measurements that cannot be performed. Moreover, we have steered clear of assuming anything about the object of a measurement. Indeed, provided they are used in their strictly operational sense, notions like proposition, logical implication, mutual exclusion, or probability all pertain to the *outcomes*—rather than the *object*—of a measurement. We have seen that as long as we stick to this purely operational approach, these concepts apply equally well to classical and quantum theory. It is only when additional assumptions are made as to the presumed object of measurement that the profound differences between classical and quantum theory emerge.

In classical theory, one assumes that the object of measurement—a coin, a die, an arrangement of cars and goats—constitutes a *preexisting reality*. The object is presumed to be in some possibly unknown, yet definite configuration which uniquely determines the outcomes of all measurements; and the measurements then merely reveal, rather than influence, this preexisting configuration. This conforms perfectly with our everyday experience: a die is as it is, and it shows a definite number, whether we look at it or not. Since it seems so obvious, the assumption of a preexisting reality is usually taken for granted and rarely spelt out explicitly. It also extends to situations where our knowledge is constrained and hence the underlying configuration can be revealed only partially. For instance, in the 3×3 problem discussed in Section 2.1 the rules of the game preclude that a candidate ever learns the locations of all three cars. Nevertheless, we continue to assume that at every stage of the game the three cars must be in one of the six configurations shown in Eq. (2.1). The configuration, known or unknown, uniquely determines whether the candidate will find behind a given door a car or a goat; and opening a door will reveal some aspect of, but not affect in any way, this preexisting configuration.

The assumption of a preexisting reality has wide-ranging implications. If indeed the outcomes of all measurements are predetermined by an underlying configuration, then (at least in theory, if not in practice) it must be possible to assign to all testable propositions definite truth values, 0 or 1, with no truth value left undetermined. As soon as an omniscient observer knows the underlying configuration, he is able to predict with certainty the outcomes of all measurements. In other words, he then knows the definite truth values, 0 or 1, of all propositions pertaining to these outcomes. For example, if in the 3×3 problem the game show host knows the full configuration of cars and goats, he can predict with certainty the outcome whenever the candidate opens a door. Thus, in a given configuration, every proposition has a preordained truth value. This truth value is 'objective', in the sense that it does not depend on whether the proposition is actually being tested or how it is being tested. Clearly, such an assignment of objective truth values must be consistent with the logical relationships between the testable propositions.

In quantum theory, this idea of a preexisting reality runs into serious difficulties. This is exemplified by two important mathematical results: the Kochen–Specker theorem and the violation of Bell's inequality. Of the *Kochen–Specker theorem* we shall give only a very simplified version, which, nevertheless, captures the essence of the argument. We shall pick a special set of observables in a Hilbert space of dimension four (which might be, say, the Hilbert space associated with two qubits). They are the following nine observables, arranged in a 3×3 scheme called the 'Peres–Mermin square':

$$
\begin{pmatrix} 0 & \hat{I} \\ \hat{I} & 0 \end{pmatrix} \quad \begin{pmatrix} \hat{X} & 0 \\ 0 & \hat{X} \end{pmatrix} \quad \begin{pmatrix} 0 & \hat{X} \\ \hat{X} & 0 \end{pmatrix} \quad \rightarrow 1
$$

$$
\begin{pmatrix} \hat{Y} & 0 \\ 0 & \hat{Y} \end{pmatrix} \quad \begin{pmatrix} 0 & -i\hat{I} \\ i\hat{I} & 0 \end{pmatrix} \quad \begin{pmatrix} 0 & -i\hat{Y} \\ i\hat{Y} & 0 \end{pmatrix} \quad \rightarrow 1 \qquad (3.107)
$$

$$
\begin{pmatrix} 0 & \hat{Y} \\ \hat{Y} & 0 \end{pmatrix} \quad \begin{pmatrix} 0 & -i\hat{X} \\ i\hat{X} & 0 \end{pmatrix} \quad \begin{pmatrix} \hat{Z} & 0 \\ 0 & -\hat{Z} \end{pmatrix} \quad \rightarrow 1
$$

$$
\quad\quad\;\; \downarrow \quad\quad\quad\quad\quad \downarrow \quad\quad\quad\quad\;\; \downarrow
$$

$$
\quad\quad\;\; 1 \quad\quad\quad\quad\quad 1 \quad\quad\quad\quad -1
$$

Here the entries of the various 2×2 matrices—the Pauli operators $\hat{X}$, $\hat{Y}$, and $\hat{Z}$, as well as the unit operator, $\hat{I}$—are themselves operators in a two-dimensional Hilbert space, so that the nine operators in the square are indeed defined in a four-dimensional Hilbert space. First of all, these nine operators are all Hermitian and thus, indeed, observables. All nine have eigenvalues ± 1; so a measurement of any of these can yield only the outcomes $+1$ or -1. Secondly, the arrangement in the square is such that within any given row and within any given column, the observables commute. (By contrast, observables which are neither in the same row nor in the same column need not commute.) Finally, as the arrows indicate, the product of the three observables in the same row or column yields the unit operator, with the important exception of the third column, where this product yields an extra minus sign. Now let us suppose that:

- there is a preexisting reality, and hence it is logically possible to assign to all nine observables definite values $+1$ or -1

- the value assigned to a particular observable does not depend on which other commuting observables one might choose to measure with it. In particular, the value does not depend on whether the observable is measured jointly with the other observables in the same row or jointly with the other observables in the same column. This assumption is known as *non-contextuality*.

In order to be logically consistent, any such assignment of definite values must reflect the functional relationships between the observables; in particular, the fact that the product of the observables in any given row yields the unit operator. Correspondingly, multiplying the assigned values in any given row must yield $+1$; and so must multiplying all nine values, row by row. The same argument can be made for the columns. Here, however, the product in the third column must yield -1 rather than $+1$; and so, when multiplied column by column, the product of the nine values must be -1. But this is a logical contradiction: it cannot be that the product of all nine values yields different results if multiplied row by row, versus column by column. Thus we are forced to conclude that our initial assumption—a non-contextual assignment of preordained values to all observables—must have been wrong! Kochen and Specker have shown in their general theorem that similar logical contradictions arise in Hilbert spaces of arbitrary dimension greater than two.

Another way to see how the classical idea of a preexisting reality breaks down in quantum theory is by considering a so-called *Bell experiment* (Fig. 3.13). Two experimenters, Alice and Bob, are sufficiently far apart that their respective actions cannot influence each other. (To be on the safe side, their laboratories might be so far apart that any communication or other causal influence between them would require superluminal signals, in violation of the theory of relativity.) A pair of two-level systems, or 'qubits', is prepared in some composite state and subsequently split between Alice and Bob. Alice measures on her qubit one of two observables, Q or R, represented, like in Fig. 3.9, as diameters of the Bloch sphere. Bob, on the other hand, measures one of S or T, likewise represented as a pair of diameters. Each pair of diameters is assumed to be orthogonal, and both pairs are assumed to lie in the same two-dimensional section of the Bloch sphere. However, Bob's pair is rotated with respect to Alice's by $3\pi/4$ (counterclockwise). Each of the four observables in question may take the values $+1$ or -1. By combining their measurement results afterwards, Alice and Bob together can measure exactly one of the four product observables $Q \otimes S, Q \otimes T, R \otimes S$, or $R \otimes T$. Out of these four product observables one may construct another composite observable,

$$A := Q \otimes S + R \otimes S + R \otimes T - Q \otimes T = (Q + R) \otimes S + (R - Q) \otimes T. \qquad (3.108)$$

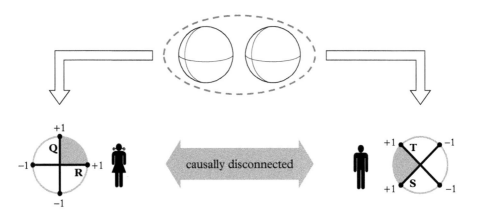

Fig. 3.13 *Bell experiment. A pair of two-level systems, or 'qubits', is prepared in a composite state. Its members are subsequently sent to two widely separated locations where, independently of each other, they are subjected to measurement. In each location the respective experimenter, Alice (left) or Bob (right), may measure one of two observables. Alice may measure either the observable Q or the observable R, whereas Bob may measure either S or T. Each observable has the possible measurement values +1 or −1. The two observables available to a given experimenter are represented as orthogonal diameters of the Bloch sphere, with the possible measurement values attached to their ends. The pair pertaining to Bob lies in the same two-dimensional section of the Bloch sphere as the pair pertaining to Alice, yet is rotated against the latter counterclockwise by 3π/4; this is indicated by the relative orientation of the grey quarter circles.*

This observable cannot be measured directly in the experiment just described. However, by repeating the experiment multiple times, it is possible to determine its expectation value in the given composite state. In each run the identical composite state is prepared initially, and then, randomly, one of the four product observables is measured. In this fashion, after sufficiently many runs, Alice and Bob are able to determine the expectation values of all four product observables separately. A simple linear combination of these then yields the expectation value of A,

$$\langle A \rangle = \langle Q \otimes S \rangle + \langle R \otimes S \rangle + \langle R \otimes T \rangle - \langle Q \otimes T \rangle.$$

Now let us suppose that:

- there is a preexisting reality; to wit, in each run all four observables, Q, R, S, and T, have preordained values which are merely revealed, but not influenced, by the measurements
- given that Alice and Bob are causally disconnected, the values of Q and R do not depend on whether they are measured jointly with S or T. This second assumption, which is similar in spirit to non-contextuality, is termed *locality*.

Under these assumptions, the observable A can have only two possible measurement values, $+2$ or -2. Indeed, on the right-hand side of Eq. (3.108) the values of Q and R are either equal, in which case it is $(Q + R) = \pm 2$ and $(R - Q) = 0$, or they are different, in which case it is $(Q + R) = 0$ and $(R - Q) = \pm 2$. In either case, subsequent multiplication with $S = \pm 1$ or $T = \pm 1$, respectively, yields $A = \pm 2$. This means that for arbitrary states the expectation value of A must lie between -2 and $+2$; or, equivalently,

$$|\langle A \rangle| \leq 2. \tag{3.109}$$

The latter inequality is a particular variant of *Bell's inequality*, known as the Clauser–Horne–Shimony–Holt (in short: CHSH) inequality. Yet experimental evidence shows that there exist states for which the CHSH inequality is violated. Once again, our premise—in this case, a local assignment of preordained values to all observables—must have been wrong!

One particular state in which the CHSH inequality is violated is the Bell state $|\beta_{11}\rangle$ produced by the nonlinear crystal in Fig. 1.14 and given mathematically in Eqs (3.102) and (3.104), respectively. In this state, the joint probability that a beam splitter with parameter setting θ for Alice's qubit and a second beam splitter with setting $\theta + \Delta\theta$ for Bob's qubit will both yield 'up' equals

$$\langle \beta_{11} | \hat{P}_{\Uparrow_\theta} \otimes \hat{P}_{\Uparrow_{\theta+\Delta\theta}} | \beta_{11} \rangle = \frac{1}{2} \langle \Downarrow_\theta | \hat{P}_{\Uparrow_{\theta+\Delta\theta}} | \Downarrow_\theta \rangle = \frac{1}{2} [1 - \text{prob}(\Uparrow_{\theta+\Delta\theta} \mid \Uparrow_\theta)].$$

The single-qubit probability in the last expression is given by Eq. (3.68), which leads to

$$\langle \beta_{11}|\hat{P}_{\Uparrow\theta} \otimes \hat{P}_{\Uparrow\theta+\Delta\theta}|\beta_{11}\rangle = \frac{1}{2}\sin^2\left(\frac{\Delta\theta}{2}\right). \tag{3.110}$$

Like the single-qubit probability, the joint probability depends only on the relative orientation of the splitters, $\Delta\theta$. It does not depend on θ, to wit, on the orientation of the splitters with respect to some external reference frame. This is a peculiar feature of the chosen Bell state and reflects its invariance under arbitrary rotations of the single-qubit basis, Eq. (3.103). The result for the joint probability implies the perfect anticorrelation depicted in Fig. 1.14 as a special case: whenever the splitters have identical parameter settings, $\Delta\theta = 0$, the probability that both yield 'up' vanishes. Moreover, the result allows us to calculate the sought-after expectation values of the four product observables $Q \otimes S$, $Q \otimes T$, $R \otimes S$, and $R \otimes T$. Without loss of generality, we assume that the diameter associated with the observable Q coincides with the north–south axis of the Bloch sphere, with the measurement value $+1$ situated at the north pole ($\theta = 0$) and -1 situated at the south pole ($\theta = \pi$). Then, for instance, the Hermitian operator representing $Q \otimes S$ is given by

$$\hat{Q} \otimes \hat{S} = +1 \cdot \hat{P}_0 \otimes \hat{P}_{3\pi/4} - 1 \cdot \hat{P}_0 \otimes \hat{P}_{\pi/4} - 1 \cdot \hat{P}_\pi \otimes \hat{P}_{3\pi/4} + 1 \cdot \hat{P}_\pi \otimes \hat{P}_{\pi/4},$$

where $\hat{P}_\theta \equiv \hat{P}_{\Uparrow\theta}$. Its expectation value in the Bell state follows directly from the formula for the joint probability, Eq. (3.110):

$$\langle \beta_{11}|\hat{Q} \otimes \hat{S}|\beta_{11}\rangle = \frac{1}{2}\left[\sin^2\left(\frac{3\pi}{8}\right) - \sin^2\left(\frac{\pi}{8}\right) - \sin^2\left(\frac{\pi}{8}\right) + \sin^2\left(\frac{3\pi}{8}\right)\right] = \frac{1}{\sqrt{2}}.$$

In a similar fashion we obtain the other three expectation values,

$$\langle \beta_{11}|\hat{R} \otimes \hat{T}|\beta_{11}\rangle = \langle \beta_{11}|\hat{R} \otimes \hat{S}|\beta_{11}\rangle = \frac{1}{\sqrt{2}}, \quad \langle \beta_{11}|\hat{Q} \otimes \hat{T}|\beta_{11}\rangle = -\frac{1}{\sqrt{2}}.$$

In combination they yield

$$\langle \beta_{11}|\hat{A}|\beta_{11}\rangle = 2\sqrt{2}, \tag{3.111}$$

which does indeed violate the CHSH inequality, Eq. (3.109). As you will verify in Exercise (3.25), $2\sqrt{2}$ is in fact the maximum value that the expectation value of A can attain in any quantum state; this maximum value is known as the 'Tsirelson bound'.

The most straightforward, and also the most radical, conclusion from these findings is to throw out the idea of a preexisting reality. This is the conclusion advocated by Niels Bohr and Werner Heisenberg in their classic 'Copenhagen interpretation' of quantum theory. It is in line with Bohr's broader philosophical view that physics ought to be regarded not so much as the study of something a priori given but rather as the development of methods for ordering and surveying human experience. In this view, there is no longer a preexisting reality that is merely revealed, but not influenced, by the act of measurement. Rather, the image of reality that emerges through acts of measurement reflects as much the history of intervention as the external world (Fig. 3.14). The essence of this way of thinking is captured beautifully in a metaphor by Wheeler (1983):

> About the game of twenty questions. You recall how it goes—one of the after-dinner party sent out of the living room, the others agreeing on a word, the one fated to be questioner returning and starting his questions. 'Is it a living object?' 'No.' 'Is it here on Earth?' 'Yes.' So the questions go from respondent to respondent around the room until at length the word emerges: victory if in twenty tries or less; otherwise, defeat.
>
> Then comes the moment when we are fourth to be sent from the room. We are locked out unbelievably long. On finally being readmitted, we find a smile on everyone's face, sign of a joke or a plot. We innocently start our questions. At first the answers come quickly. Then each question begins to take longer in the answering—strange, when the answer itself is only a simple 'yes' or 'no'. At length, feeling hot on the trail, we ask, 'Is the word "cloud"?' 'Yes', comes the reply, and everyone bursts out laughing. When we were out of the room, they explain, they had agreed not to agree in advance on any word at all. Each one around the circle could respond 'yes' or 'no' as he pleased to whatever question we put to him. But however he replied he had to have a word in mind compatible with his own reply—and with all the replies that went before. No wonder some of those decisions between 'yes' and 'no' proved so hard!

In Wheeler's game the word does not already exist 'out there'. Rather, information about the word is only brought into being through the questions raised. While the answers given are internally consistent, and eventually converge to some word, the outcome is also influenced by the questioner: a different sequence of questions will generally lead to a different word.

Fig. 3.14 *Conceptual illustration of the contrasting world views in classical theory (left) versus the Copenhagen interpretation of quantum theory (right). Classically, there is a preexisting reality which is independent of the observer. Measurements merely reveal, rather than influence, this preexisting reality. By contrast, in quantum theory the image of reality that emerges through acts of measurement reflects as much the history of intervention as the external world.*

An alternative conclusion is to maintain the notion of a preexisting reality—'realism'—and give up locality and non-contextuality instead. This alternative would be equally disturbing, however, since it would call into question a fundamental tenet of the scientific method: that it ought to be possible to learn about the world by isolating small parts of it and studying them in a laboratory, confined in space and time. In other words, we are caught between a rock and a hard place; we must either forgo realism or renounce a crucial part of our scientific method.

3.7 Classical Limit

> Macroscopic systems behave classically. I examine how this classical behaviour
> emerges as a system grows larger, and how the structure of the emergent classical
> theory is related to that of the underlying quantum theory.

Our everyday experience in the macroscopic world is governed by the rules of classical, not quantum, theory. It should be possible, therefore, to recover the classical from the quantum theory in the limit of a large number of constituents. In order to make the latter more precise, we consider a system composed of an arbitrary number, L, of constituents. We focus on the special case where these constituents are all of the same kind and have all been prepared in exactly the same way. However, we shall continue to assume that they are distinguishable. Such a situation may be realized, for example, by a sequence of photons emitted, one by one, by the same attenuated laser (as in Fig. 1.3) and subsequently filtered by the same polarizers, or by an array of elementary magnets inside a homogeneous magnetic material. As long as the laser is sufficiently weak, the photons, albeit prepared identically, may be distinguished by their temporal order; and if the elementary magnets reside, say, on definite sites of a crystal lattice, they can be distinguished by their locations. As we discussed in Section 2.6, a composite system of this kind constitutes an *exchangeable assembly*. Consequently, the L-partite state of such a composite system, $\rho^{(L)}$, has the de Finetti form, Eq. (2.50). It is a mixture of product states, $\rho^{\otimes L}$, weighted with some probability density function, $\mathrm{pdf}(\rho)$, on the manifold of normalized single-constituent states. In the following, we shall:

1. Assume that we know exactly how the constituents were prepared—for example, which source they came from and which subsequent filters were applied. As a result, the probability density function is a δ-function and the L-partite state a product state, $\rho^{(L)} = \rho^{\otimes L}$. However, we do not require the preparation to be most accurate, so ρ need not be pure.
2. Let the number of constituents become very large, $L \gg 1$.
3. Consider only macroscopic observables, to wit, totals pertaining to the assembly as a whole rather than observables pertaining to individual constituents. For instance, in a measurement we count only the total number of photons hitting a certain area on a screen, or measure only the total magnetic moment of a magnetic material, rather than investigating individual photons or elementary magnets, respectively.

We will show that under these circumstances the relevant totals attain quasi sharp values. This is the *classical limit*.

Let $\hat{A}^{(i)}$ represent some single-constituent observable, A, referring to the ith constituent. The corresponding total is then represented by the sum over all constituents,

$$\hat{\mathcal{A}} := \sum_{i=1}^{L} \hat{A}^{(i)}, \tag{3.112}$$

which, strictly speaking, is a shorthand notation for

$$\hat{\mathcal{A}} = (\hat{A}^{(1)} \otimes \hat{I} \otimes \ldots) + (\hat{I} \otimes \hat{A}^{(2)} \otimes \hat{I} \otimes \ldots) + \ldots + (\ldots \otimes \hat{I} \otimes \hat{A}^{(L)}).$$

In a product state, $\rho^{\otimes L}$, the expectation value of this total simply equals L times the expectation value of the single-constituent observable,

$$\langle \mathcal{A} \rangle_{\rho^{\otimes L}} = L \langle A \rangle_{\rho}. \tag{3.113}$$

The square of the total is represented by

$$\hat{\mathcal{A}}^2 = \sum_{i \neq k} \hat{A}^{(i)} \otimes \hat{A}^{(k)} + \sum_{i} (\hat{A}^{(i)})^2.$$

Since in a product state the constituents are statistically independent, its expectation value is given by

$$\langle \mathcal{A}^2 \rangle_{\rho^{\otimes L}} = \sum_{i \neq k} \langle A \rangle_{\rho} \langle A \rangle_{\rho} + \sum_{i} \langle A^2 \rangle_{\rho} = (L^2 - L) \langle A \rangle_{\rho}^2 + L \langle A^2 \rangle_{\rho}.$$

This yields the variance of the total,

$$\mathrm{var}(\mathcal{A}) = \langle \mathcal{A}^2 \rangle_{\rho^{\otimes L}} - \langle \mathcal{A} \rangle_{\rho^{\otimes L}}^2 = L \left(\langle A^2 \rangle_{\rho} - \langle A \rangle_{\rho}^2 \right) = L \,\mathrm{var}(A), \tag{3.114}$$

and the associated standard deviation,

$$\sigma(\mathcal{A}) = \sqrt{L}\,\sigma(A). \tag{3.115}$$

While the expectation value of the total, Eq. (3.113), scales linearly with the number of constituents, its standard deviation scales only with the square root of L. Thus, as the system gets larger, the *relative* size of the uncertainty about the total diminishes:

$$\frac{\sigma(\mathcal{A})}{\langle \mathcal{A} \rangle} \sim \frac{\sqrt{L}}{L} \to 0 \quad (L \to \infty). \tag{3.116}$$

(Here we disregard special states in which $\langle A \rangle$ is precisely equal to zero.) In relative terms, therefore, the uncertainty about the total becomes negligible. Another way of expressing this relative sharpness is in terms of the ratio

$$\frac{\langle A^2 \rangle}{\langle A \rangle^2} = 1 + \frac{1}{L}\frac{\text{var}(A)}{\langle A \rangle^2},$$

which approaches one as $L \to \infty$. In other words, it is asymptotically

$$\langle A^2 \rangle \sim \langle A \rangle^2.$$

More generally, in this asymptotic limit, 'taking a function' and 'taking the expectation value' commute,

$$\langle f(A) \rangle \sim f(\langle A \rangle), \tag{3.117}$$

for every sufficiently smooth function, f. It is in this sense that the total attains a quasi sharp value.

Next, we consider an entire set of macroscopic totals, $\{\mathcal{B}_\mu\}$. We assume that they are closed under application of the commutator bracket, in the sense that every commutator,

$$\hat{C}_{\mu\nu} := \frac{1}{i}[\hat{\mathcal{B}}_\mu, \hat{\mathcal{B}}_\nu], \tag{3.118}$$

is again a linear combination of these totals,

$$\hat{C}_{\mu\nu} = \sum_\sigma c_{\mu\nu}{}^\sigma \hat{\mathcal{B}}_\sigma + \eta_{\mu\nu}\hat{I}, \tag{3.119}$$

with certain *structure constants*, $c_{\mu\nu}{}^\sigma$ and $\eta_{\mu\nu}$. One example is the set $\{\mathcal{X}, \mathcal{Y}, \mathcal{Z}\}$ pertaining to an assembly of two-level systems. These macroscopic totals are defined as

$$\mathcal{X} := \sum_i X^{(i)}, \quad \mathcal{Y} := \sum_i Y^{(i)}, \quad \mathcal{Z} := \sum_i Z^{(i)}, \tag{3.120}$$

where $X^{(i)}, Y^{(i)}, Z^{(i)}$ are the Pauli observables pertaining to the ith two-level system. Thanks to the interrelationships between the single-constituent Pauli operators, Eq. (3.64), the totals are indeed closed under application of the commutator bracket:

$$C_{\mathcal{X},\mathcal{Y}} = 2\mathcal{Z}, \quad C_{\mathcal{Y},\mathcal{Z}} = 2\mathcal{X}, \quad C_{\mathcal{Z},\mathcal{X}} = 2\mathcal{Y}. \tag{3.121}$$

As long as there is this kind of closure, the expectation values of the commutator brackets,

$$\Omega_{\mu\nu} := \langle \mathcal{C}_{\mu\nu} \rangle = \sum_{\sigma} c_{\mu\nu}{}^{\sigma} \langle \mathcal{B}_{\sigma} \rangle + \eta_{\mu\nu}, \tag{3.122}$$

are linear combinations of the expectation values of the macroscopic totals. We illustrate this again for the collection of two-level systems. Denoting the expectation values of the pertinent macroscopic totals, Eq. (3.120), by

$$x := \langle \mathcal{X} \rangle, \quad y := \langle \mathcal{Y} \rangle, \quad z := \langle \mathcal{Z} \rangle,$$

the commutators of these totals, Eq. (3.121), have the expectation values (in compact matrix form)

$$\Omega(x, y, z) = \begin{pmatrix} 0 & 2z & -2y \\ -2z & 0 & 2x \\ 2y & -2x & 0 \end{pmatrix}. \tag{3.123}$$

Indeed, these are linear functions of the expectation values x, y, and z.

A classical, macroscopic system is typically characterized by the quasi sharp values of just a small set of 'relevant' macroscopic observables, which may be totals or averages. For instance, a macroscopic magnet might be characterized by the three components of its total magnetic moment, which is the sum of the magnetic moments of the many constituent elementary magnets; or a little iron ball (approximating the concept of a classical 'point particle') might be characterized by its momentum and position, which are really the sum or average of the momenta and positions of the constituent iron atoms, respectively. Since macroscopic averages are related to macroscopic totals by a simple division by the number of constituents, we may limit ourselves, without loss of generality, to the case where the relevant set comprises only totals, $\{\mathcal{B}_{\mu}\}$. We denote their quasi sharp values by lower-case letters,

$$b_{\mu} := \langle \mathcal{B}_{\mu} \rangle. \tag{3.124}$$

The state of a classical system may then be visualized as a point in a manifold which has these quasi sharp values as its coordinates. Physical properties of the classical system—for example, its energy—are functions of its state and hence functions on this manifold. Let $f(\{b_{\mu}\})$ and $g(\{b_{\nu}\})$ be two such functions. We can define a bracket which maps these two functions to another function on the manifold,

$$\{f, g\} := \sum_{\mu\nu} \Omega_{\mu\nu} \frac{\partial f}{\partial b_{\mu}} \frac{\partial g}{\partial b_{\nu}}, \tag{3.125}$$

where $\Omega_{\mu\nu}$ are the 'fundamental brackets' defined in Eq. (3.122) and $\partial \ldots / \partial \ldots$ are partial derivatives of the functions with respect to the coordinates. As we discussed earlier, the fundamental brackets are themselves functions on the manifold, $\Omega_{\mu\nu}(\{b_\sigma\})$, as long as the set of macroscopic totals is closed. In our example of a collection of two-level systems, the pertinent fundamental brackets, Eq. (3.123), allow us to calculate the brackets of arbitrary functions $f(x,y,z)$ and $g(x,y,z)$; for instance, the bracket

$$\{x^2, y\} = 4xz. \tag{3.126}$$

The bracket exhibits four important properties:

1. It is antisymmetric,

$$\{f,g\} = -\{g,f\}. \tag{3.127}$$

2. It is linear in both arguments; for instance, it is

$$\{f, c_1 g + c_2 h\} = c_1\{f,g\} + c_2\{f,h\}, \tag{3.128}$$

    where c_1 and c_2 are arbitrary constants.
3. It satisfies the Leibniz rule,

$$\{f, gh\} = \{f,g\}h + g\{f,h\}. \tag{3.129}$$

4. It satisfies the Jacobi identity,

$$\{\{f,g\},h\} + \{\{h,f\},g\} + \{\{g,h\},f\} = 0. \tag{3.130}$$

This is the only property whose proof requires a bit more thought. By definition of the bracket, Eq. (3.125), the Jacobi identity is tantamount to a constraint on the fundamental brackets,

$$\sum_\sigma \left(\frac{\partial \Omega_{\mu\nu}}{\partial b_\sigma} \Omega_{\sigma\tau} + \frac{\partial \Omega_{\tau\mu}}{\partial b_\sigma} \Omega_{\sigma\nu} + \frac{\partial \Omega_{\nu\tau}}{\partial b_\sigma} \Omega_{\sigma\mu} \right) = 0. \tag{3.131}$$

Thanks to Eq. (3.122) and its partial derivative,

$$\frac{\partial \Omega_{\mu\nu}}{\partial b_\sigma} = c_{\mu\nu}{}^\sigma,$$

the first of the three summands yields the contribution

$$\sum_\sigma \frac{\partial \Omega_{\mu\nu}}{\partial b_\sigma} \Omega_{\sigma\tau} = \sum_\sigma c_{\mu\nu}{}^\sigma \left(\sum_\omega c_{\sigma\tau}{}^\omega b_\omega + \eta_{\sigma\tau} \right).$$

Up to a sign, this is the expectation value of a double commutator, $[[\hat{\mathcal{B}}_\mu, \hat{\mathcal{B}}_\nu], \hat{\mathcal{B}}_\tau]$. Likewise, the other two summands contribute expectation values of double commutators, with the indices permuted cyclically. Thus the constraint on the fundamental brackets, Eq. (3.131), amounts to the requirement that the cyclic sum of double commutators vanishes. By the Jacobi identity for the ordinary commutator bracket (which you will prove in Exercise (3.2)), this is indeed the case.

These four properties define a *Poisson bracket*, akin to the Poisson bracket used in classical mechanics. Any manifold endowed with a Poisson bracket is called a *Poisson manifold*. As we just verified, the manifold of classical states is such a Poisson manifold.

The Poisson brackets on the manifold of classical states are closely related to the commutator brackets of quantum observables. First of all, by the very definition of the Poisson bracket, Eq. (3.125), in conjunction with Eqs (3.118) and (3.122), the commutator brackets of the selected macroscopic totals have expectation values which equal the Poisson brackets of the associated coordinates,

$$\langle \mathcal{C}_{\mu\nu} \rangle = \{b_\mu, b_\nu\}; \tag{3.132}$$

this explains why we called these expectation values the 'fundamental brackets'. Secondly, thanks to the linearity of both the commutator and the Poisson brackets, this equality extends to arbitrary linear functions of the macroscopic totals,

$$f(\{\mathcal{B}_\mu\}) = \sum_\mu m^\mu \mathcal{B}_\mu + c, \quad g(\{\mathcal{B}_\nu\}) = \sum_\nu n^\nu \mathcal{B}_\nu + d,$$

and their respective counterparts on the manifold of classical states, $f(\{b_\mu\})$ and $g(\{b_\nu\})$:

$$\langle \mathcal{C}_{f(\{\mathcal{B}_\mu\}), g(\{\mathcal{B}_\nu\})} \rangle = \{f, g\}. \tag{3.133}$$

Finally, in the classical limit this correspondence between commutator and Poisson brackets extends even to *arbitrary*, not just linear, functions f and g. Rather than giving a full proof, we will illustrate this with a specific example. We consider again an assembly of two-level systems, with the macroscopic totals defined as in Eq. (3.120). For the two functions we choose $f(\mathcal{X}, \mathcal{Y}, \mathcal{Z}) = \mathcal{X}^2$ and $g(\mathcal{X}, \mathcal{Y}, \mathcal{Z}) = \mathcal{Y}$, of which the first is not linear. In order to calculate their commutator, we use the commutator brackets of the totals, Eq. (3.121), and subsequently break the result down into single-constituent contributions. We thus obtain a double sum,

$$\mathcal{C}_{\mathcal{X}^2, \mathcal{Y}} = 4 \sum_{i \neq k} X^{(i)} Z^{(k)}.$$

In a product state, $\rho^{\otimes L}$, the expectation value of each summand factorizes:

$$\langle X^{(i)} Z^{(k)} \rangle = \langle X^{(i)} \rangle \langle Z^{(k)} \rangle = \frac{\langle \mathcal{X} \rangle}{L} \frac{\langle \mathcal{Z} \rangle}{L} = \frac{xz}{L^2} \quad \forall i \neq k.$$

Thus we obtain the expectation value of the commutator,

$$\langle \mathcal{C}_{\mathcal{X}^2, \mathcal{Y}} \rangle = 4(L^2 - L)\frac{xz}{L^2} = 4xz \left(1 - \frac{1}{L} \right).$$

Asymptotically, as $L \to \infty$, this expectation value does indeed approach the Poisson bracket $\{x^2, y\}$, given in Eq. (3.126).

The one-to-one correspondence between Poisson brackets and commutator brackets in Hilbert space,

$$\frac{1}{i}[\cdot, \cdot] \leftrightarrow \{\cdot, \cdot\}, \tag{3.134}$$

in the sense of Eq. (3.133), is often employed as a heuristic 'quantization rule'. It is used in situations where the classical limit is known and one seeks a description of the underlying quantum system that is compatible with the known classical limit. Reading the correspondence from right to left, classical functions are replaced by Hilbert space operators in such a manner that their commutation relations exhibit the same structure as the known Poisson brackets. The paradigmatic example is a mechanical system which, classically, is described with functions on phase space. In the simplest case of a particle in one spatial dimension, this phase space has dimension two. Its coordinates are the particle's position, q, and momentum, p. The phase space constitutes a Poisson manifold, with the fundamental brackets given by

$$\{q, q\} = \{p, p\} = 0, \quad \{q, p\} = 1.$$

Guided by the correspondence discussed earlier, the 'quantization' of such a system consists in representing position and momentum by Hermitian operators, $\hat{q}$ and $\hat{p}$, in a Hilbert space, where these operators are defined such that they satisfy the commutation relations

$$\frac{1}{i\hbar}[\hat{q}, \hat{q}] = \frac{1}{i\hbar}[\hat{p}, \hat{p}] = 0, \quad \frac{1}{i\hbar}[\hat{q}, \hat{p}] = \hat{I}. \tag{3.135}$$

Here the commutation relations feature an additional constant,

$$\hbar \approx 1.05 \times 10^{-34}\,\text{Js},\tag{3.136}$$

called the *reduced Planck constant* (and pronounced 'h-bar'), which relates the physical units of momentum and energy to those of length and time, respectively. In contrast to the finite-dimensional Hilbert spaces considered so far, that particular Hilbert space has to be infinite-dimensional. (You will discover the reason in Exercise (3.29).) We shall not delve deeper into this example, because doing so would lead us to the quantum theory of matter, which is outside the scope of this book.

In the classical limit, propositions and their probabilities obey the rules of classical logic and probability theory, in the following sense. The classical counterpart of a most accurate proposition is a proposition that specifies quasi sharp values for all relevant macroscopic observables of the system at hand; for instance, 'this little iron ball has position q and momentum p'. It corresponds to a point in the Poisson manifold, or—in classical mechanics—phase space. (A more careful analysis would show that the correspondence is actually with an infinitesimal neighbourhood of the point rather than the point itself, but we will not go into these details here.) In other words, the set of 'classical' most accurate propositions coincides with the Poisson manifold. These propositions are mutually exclusive: when one of them is true, corresponding to a particular assignment of quasi sharp values to the macroscopic observables, every alternative assignment of values, and thus every other classical most accurate proposition, must be false. This, in turn, implies that all these propositions can be tested jointly, and hence that on this classical level, there are no knowledge constraints. Consequently, the classical set picture applies: the Poisson manifold as a whole plays the role of a classical sample space, and every classical proposition corresponds to a subset, or region, thereof. As for the state, we argued earlier that in the classical limit we are dealing with systems composed of many identically prepared constituents; that is, exchangeable assemblies. We focused on the case where we knew exactly how the constituents were prepared. This did not mean that the state was pure, but that in its de Finetti representation, Eq. (2.50), the probability density function was a δ-function. We have seen that such a state is associated with a point in the Poisson manifold, just like a classical most accurate proposition; it is, therefore, the classical version of a state of maximal knowledge. If we cease to assume complete knowledge of the preparation procedure but work with a generic de Finetti state instead, with an arbitrary probability density function, then the macroscopic properties of the system are no longer described by a point but by a probability distribution on the Poisson manifold. Being defined on a classical sample space, this distribution yields probabilities which obey all rules of classical probability theory. A state of maximal knowledge is the special case where this probability density function becomes a δ-function. The various correspondences between concepts in quantum theory and their counterparts in the classical limit are summarized in Table 3.4.

Table 3.4 *Concepts in quantum theory and their counterparts in the classical limit.*

Operational concept	Quantum theory	Classical limit
overarching proposition	Hilbert space	Poisson manifold
testable proposition	subspace thereof	region thereof
most accurate proposition	ray	point
observable	Hermitian operator	function
	commutator	Poisson bracket
state	statistical operator	probability density function
state of maximal knowledge	pure state	δ-function

Chapter Summary

- A quantum system is described in complex Hilbert space. Propositions about measurement outcomes correspond to subspaces of the Hilbert space or to the projectors thereon. They exhibit a non-classical logical structure.

- The state of a quantum system is represented by a statistical operator. It is Hermitian, positive semi-definite, and (if normalized) has unit trace.

- The state is pure if and only if the statistical operator is a projector; otherwise it is mixed. A pure state may also be represented by a unit vector, which is defined up to an undetermined phase factor.

- The probability of a proposition is calculated with the help of Gleason's theorem. Upon measurement, the statistical operator is updated according to Lüders' rule. For pure states represented by unit vectors, the former becomes the Born rule, and the latter amounts to a simple projection.

- The simplest quantum system is a two-level system, or 'qubit'. Its pure states can be visualized as points on the surface of the Bloch sphere, its mixed states, as points in the interior.

- The mirror image of lossless steering is the quantum Zeno effect. Continuous monitoring—infinitely fast repetitions of the same measurement—effectively prevents evolution; it 'freezes' a system in its initial state.

- An observable is represented by a Hermitian operator. Its eigenvalues correspond to possible measurement values, its eigenspaces, to the associated propositions. Its expectation value equals the probability-weighted average of the possible measurement values.

- Two observables are jointly measurable if and only if they commute. Where they do not, their measurement uncertainties obey an uncertainty relation: the higher the precision with which one observable is known, the lower it must be for the other.

- A transformation applies to states (Schrödinger picture) or observables (Heisenberg picture). It is either unitary or antiunitary. A unitary transformation can be connected smoothly to the identity, whereas an antiunitary transformation is discrete. In physics, most transformations of interest are unitary.

- On the Bloch sphere, an observable is visualized as a diameter with numerical values attached to its endpoints, a unitary transformation as a rotation, and an antiunitary transformation as a rotation combined with a reflection through a plane.

- When several quantum systems form a larger composite system, the latter is described in the tensor product Hilbert space.

- The reduced state of an individual constituent is the partial trace of the composite state over the disregarded degrees of freedom.

- If all constituents, individually, are in pure states, then the composite state is pure, too. The converse is not true: it may be that a composite state is pure but the reduced states are not. This phenomenon is termed entanglement.

- The Bell states are entangled states of two qubits. One of them yields universal anticorrelation; that is, opposite measurement results for the two qubits, whatever the orientation of the measuring device.

- The Kochen–Specker theorem and the Bell experiment show that in the quantum realm one must either abandon the idea of a preexisting reality, or abandon non-contextuality and locality.

- The classical limit regards selected macroscopic totals or averages of a large assembly. These attain quasi sharp values and may be considered the coordinates of a manifold of macroscopic states. A point in that manifold corresponds to a classical pure state, a probability density function thereon to a classical mixed state.

- The manifold is endowed with a Poisson bracket. There is a correspondence between Poisson and commutator brackets, often employed as a heuristic quantization rule.

Further Reading

Most textbooks on quantum theory contain a chapter on the mathematics of Hilbert space, often more detailed than the concise summary that I have given here. There are many excellent books that you can consult; I personally like those by Cohen-Tannoudji *et al.* (1991), Bohm (1993), and Ballentine (1998) for their clarity and precision, as well as the previously mentioned text by Schumacher and Westmoreland (2010) for its modern approach which takes recent developments in quantum information processing into account. These are also good references for quantum theory in general. The correspondence between subspaces of Hilbert space and propositions about measurement outcomes is at the heart of operational statistics, quantum logic, and related approaches to quantum theory, for which I have already provided reading suggestions in Chapters 1 and 2. For fundamental results such as the theorems of Gleason, Wigner, and Kochen and Specker, as well as Bell's inequality, an excellent place to start is the book by Peres (1995). Additional details about Wigner's theorem and its various generalizations may be found in the review by Chevalier (2007). The results of Kochen, Specker, and Bell have prompted intense debate and countless research publications. It is instructive to consult the original work by John Bell, which is collected in Bell (2004). Moreover, I recommend the lucid review by Mermin (1993) as well as the witty analysis of Bell's inequality by Peres (1978). The simple proof of the Kochen–Specker theorem in four-dimensional Hilbert space with the help of the Peres–Mermin square is due to Peres (1990), Mermin (1990), and Mermin (1995). The quantum Zeno effect is not often presented in a standard textbook, and you can find more information about it in the original article by Misra and Sudarshan (1977). Poisson structures are known mainly in classical mechanics. You will find more information about these in books by Abraham *et al.* (1988), Arnold (1997), or Öttinger (2005).

EXERCISES

3.1. Scalar product and norm

(a) Prove the *Cauchy–Bunyakovsky–Schwarz inequality*,

$$|(|u\rangle, |v\rangle)| \leq \| |u\rangle \| \, \| |v\rangle \|.$$

(b) Prove the triangle inequality, Eq. (3.15).

3.2. Jacobi identity

Prove the Jacobi identity for the commutator bracket,

$$[[\hat{A},\hat{B}],\hat{C}] + [[\hat{C},\hat{A}],\hat{B}] + [[\hat{B},\hat{C}],\hat{A}] = 0.$$

3.3. Trace

(a) Verify that the definition of the trace, Eq. (3.25), is independent of the chosen orthonormal basis, to wit, invariant under arbitrary unitary transformations of that basis.

(b) Show that

$$\mathrm{tr}(\hat{A}\hat{B}) = \mathrm{tr}(\hat{B}\hat{A})$$

and therefore the trace of a commutator vanishes:

$$\mathrm{tr}([\hat{A}, \hat{B}]) = 0.$$

This result holds only in the finite-dimensional Hilbert spaces that we have been considering. It cannot be extended to infinite-dimensional Hilbert spaces.

3.4. Projector

Let $\{|u_i\rangle\}$ be mutually orthogonal unit vectors that span a subspace of Hilbert space. Show that the projector onto that subspace can be written as

$$\hat{P} = \sum_i |u_i\rangle\langle u_i|.$$

Verify that it has all the properties mentioned in the text: it is Hermitian and idempotent, and its trace equals the dimension of the subspace spanned by $\{|u_i\rangle\}$.

3.5. Distributivity in Hilbert space

(a) Show that, in general, the binary operators $\wedge$ and $\vee$, defined in Hilbert space as the intersection and sum of subspaces, respectively, do not satisfy the distributivity property, Eq. (2.30). Give a specific counterexample. This is in line with the qualitative conclusion that we drew from the experimental evidence in Exercise (2.7).

(b) Show that, on the other hand, the distributivity property *is* satisfied if the propositions involved can be tested jointly, to wit, if the associated subspaces share a joint orthogonal basis.

3.6. Logical structure in terms of projection operators

Verify the one-to-one correspondence between the representations of the logical structure in terms of subspaces (Table 3.1) and projection operators (Table 3.2). In particular, show that the various mathematical expressions in the right column of Table 3.2 do indeed mirror the corresponding properties of, or operations on, the subspaces.

3.7. Statistical operator

(a) Show that every statistical operator obeys the inequality

$$\hat{\rho}^2 \leq \hat{\rho}.$$

Show that a state is mixed if and only if the statistical operator possesses at least one eigenvalue which is neither zero nor one.

(b) Show that in a Hilbert space of dimension d, a state represents total ignorance, $\hat{\rho} = \hat{I}/d$, if and only if

$$\text{tr}(\hat{\rho}^2) = \frac{1}{d}.$$

3.8. Real Hilbert space

One might formulate an alternative quantum theory in real, rather than complex, Hilbert space. The statistical operator in real Hilbert space has the same properties as in complex Hilbert space, only that in a given basis it is represented by a real symmetric, rather than Hermitian, matrix.

(a) Show that two-dimensional real Hilbert space—describing a 'real qubit', or 'rebit'—corresponds to the circle model depicted in Fig. 2.5 (left).
(b) Derive the number of state parameters, $S(d)$. Show that it satisfies Eq. (2.71) but not the power law, Eq. (2.39).

3.9. Lüders' versus Bayes' rule

Let a quantum system be initially in a state described by the statistical operator $\hat{\rho}$, and let $\hat{A}$, $\hat{B}$ be two commuting observables, $[\hat{A}, \hat{B}] = 0$, with respective spectra $\{a_i\}$ and $\{b_j\}$. Whenever one of the observables is measured, the state must be updated according to Lüders' rule, Eq. (3.56). When $\hat{A}$ is measured, yielding outcome a_i, the post-measurement state is denoted by $\hat{\rho}_{|a_i}$; when $\hat{B}$ is measured, yielding outcome b_j, the post-measurement state is denoted by $\hat{\rho}_{|b_j}$. Show that

$$\text{prob}(a_i|\hat{\rho}_{|b_j}) = \frac{\text{prob}(b_j|\hat{\rho}_{|a_i})\,\text{prob}(a_i|\hat{\rho})}{\text{prob}(b_j|\hat{\rho})}.$$

This equation has the same structure as the classical Bayes' rule, Eq. (1.39). In this sense, Lüders' rule may be regarded as a generalization of the classical Bayes' rule.

3.10. Cartesian coordinates of the Bloch sphere

According to Eq. (3.65), the statistical operator of a two-level system may be written as a linear combination of the unit and the three Pauli operators.

(a) Prove that the coefficients in front of the latter are indeed given by Eq. (3.66).
(b) For a pure state, $|\vec{r}| = 1$, express these coefficients in terms of the angles θ and φ (Eq. (3.59)). Interpret their relationship.
(c) Show that for two pure states, $|\psi\rangle$ and $|\chi\rangle$, described by unit vectors $\vec{r}$ and $\vec{s}$, respectively, it is

$$|\langle\chi|\psi\rangle|^2 = \frac{1}{2}(1 + \vec{r}\cdot\vec{s}).$$

This furnishes an alternative expression, in terms of Cartesian coordinates, for the conditional probability in Eq. (3.68).

3.11. Pauli operators

Show that for any two traceless operators, $\mathrm{tr}(\hat{A}) = \mathrm{tr}(\hat{B}) = 0$, in a two-dimensional complex Hilbert space it is

$$\frac{1}{2}\sum_{i=1}^{3}\mathrm{tr}(\hat{\sigma}_i\hat{A})\,\mathrm{tr}(\hat{\sigma}_i\hat{B}) = \mathrm{tr}(\hat{A}\hat{B}),$$

where $\{\hat{\sigma}_i\}$ are the three Pauli operators.

3.12. Unitary transformation on the Bloch sphere

Verify that a unitary transformation of a two-level system can always be parametrized as in Eqs (3.86) and (3.87), and that its effect on the Bloch sphere is the rotation depicted in Fig. 3.10.

3.13. Quarter-wave plate

In the Bloch sphere picture, a polarizing beam splitter at parameter setting θ, in combination with a pair of detectors, performs a binary measurement along the axis $\hat{m} = (\sin\theta, 0, \cos\theta)$. A quarter-wave plate (Fig. 3.6) at parameter setting α effects a unitary transformation and thus a rotation on the Bloch sphere. The axis of rotation, $\hat{n}$, depends on the value of α: it lies in the x–z plane, with components $\hat{n}(\alpha) = (\sin\alpha, 0, \cos\alpha)$. The rotation angle equals $\pi/2$ (which is a quarter of 2π, whence the name).

(a) Show that in the standard basis the Bloch sphere rotation associated with a quarter-wave plate has the matrix representation

$$\mathbf{R}_{\hat{n}(\alpha)}\left(\frac{\pi}{2}\right) = \frac{1}{\sqrt{2}}\begin{pmatrix} 1 - i\cos\alpha & -i\sin\alpha \\ -i\sin\alpha & 1 + i\cos\alpha \end{pmatrix}.$$

(b) Consider the experimental setup depicted in Fig. 1.6. Place a quarter-wave plate in between the two beam splitters. How does that change the observed probabilities? Calculate the relative intensities of upper and lower beam as a function of the parameters θ and α.

(c) Explain the experimental evidence shown in Figs 3.7 and 3.8.

(d) Prove that the combination of a polarizing beam splitter with quarter-wave plates as shown in Fig. 3.6 exhausts all possible measurements on a two-level system.

(e) Which values must you choose for θ and α in order to effect a measurement along the axis $(\cos\varphi, \sin\varphi, 0)$?

3.14. Elitzur–Vaidman bomb tester

A beam splitter alone does not perform a measurement; it is the combination with a screen or a pair of detectors, as in Figs 1.1 and 1.4, which does. The splitter *per se* effects a unitary transformation, resulting in a superposition of 'up' (upper beam) and 'down' (lower beam) states. In the main text we considered special splitters which do so in such a way that internal degrees of freedom of the system (quantized magnetic moment in the case of a Stern–Gerlach apparatus, polarization in the case of a polarizing beam splitter) become correlated—entangled—with the beam direction. Yet there also exist 'ordinary' splitters which do no more than create this superposition of 'up' and 'down' states without any coupling to internal degrees of freedom. For photons, such ordinary splitters can be realized in the laboratory with half-silvered plates. In general, these have not just one but two incoming beams, which are then transformed, unitarily, into the two outgoing beams. One particular setup, of a type known as a *Mach–Zehnder interferometer*, is shown in Fig. 3.15. It features two splitters (dotted diagonal lines), two mirrors (solid diagonal lines), and two detectors, D_0 and D_1. Beams are directed either to the right or upwards. Associated with these two directions are two orthogonal kinetic states, $|0\rangle$ and $|1\rangle$, which serve as the standard basis. In this basis, the unitary transformation effected by each beam splitter is assumed to have the matrix representation

$$\mathbf{U}_{\mathrm{BS}} = \frac{1}{\sqrt{2}} \begin{pmatrix} 1 & i \\ i & 1 \end{pmatrix}.$$

The two mirrors in concert effect a unitary transformation, too, which is modelled to have the matrix representation

$$\mathbf{U}_{\mathrm{M}} = \begin{pmatrix} 0 & i \\ i & 0 \end{pmatrix}.$$

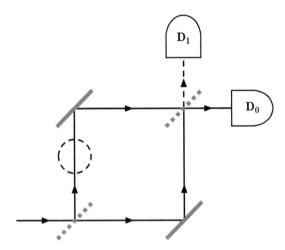

Fig. 3.15 *Mach–Zehnder type interferometer.*

(a) Each of the unitaries effects a rotation on the Bloch sphere. Determine the respective axes and angles.

(b) Show that the net effect of the beam splitters and mirrors depicted in Fig. 3.15 is simply the identity map: the state of the outgoing photon equals that of the incoming photon (up to an irrelevant phase factor). Consequently, the photon coming in from the left, in the state $|0\rangle$, will certainly exit in the state $|0\rangle$ and make the detector D_0 click.

(c) Imagine that an object (dashed circle) blocks the left path. How does that affect the probabilities with which the detectors click?

(d) Avshalom Elitzur and Lev Vaidman (Elitzur and Vaidman (1993)) proposed the following gedankenexperiment:

> Consider a stock of bombs with a sensor of a new type: if a single photon hits the sensor, the bomb explodes. Suppose further that some of the bombs in the stock are out of order: a small part of their sensor is missing so that photons pass through the sensor's hole without being affected in any way, and the bomb does not explode. Is it possible to find out which bombs are still in order?
>
> Of course, we can direct some light at each bomb. If it does not explode, it is not good. If it does, it *was* good. But we are interested in finding a good bomb without destroying it. The trouble is that the bomb is designed in such a way that *any* interaction with light, even a very soft photon bouncing on the bomb's sensor, causes an explosion. The task therefore seems to be impossible, and in classical physics it surely is.

Can you use the setup in Fig. 3.15 to achieve the 'impossible' task—also termed an 'interaction-free measurement'—with at least some probability?

(e) Imagine that you are free to test each individual bomb as often as you like with the setup in Fig. 3.15. By making optimal use of this capacity, which fraction (on average) of the 'good' bombs can you certify as such without destroying them? (Answer: one third)

(f) You can do even better than that by modifying the beam splitters. Rather than having both splitters in Fig. 3.15 perform the same unitary transformation, you may have one splitter effect a Bloch sphere rotation by an angle θ (about the same axis as before) and the other, by $(\pi - \theta)$. Verify that, without an obstacle, the net effect of the modified setup is still the identity map. Calculate the fraction of 'good' bombs that you can certify, undetonated, with the modified setup, as a function of θ. Show that with a suitable choice of θ you can bring that fraction arbitrarily close to one half.

(g) Actually, you can do better still and bring the fraction arbitrarily close to one. The trick is to use the setup multiple times, not with different photons but with the *same* photon. You can realize this (at least on paper) by feeding the beams exiting from the second splitter into the setup once again, bypassing the first splitter; for instance, by using additional mirrors, as sketched in Fig. 3.16. Assume that the first splitter effects a Bloch sphere rotation by $\theta = \pi/N$

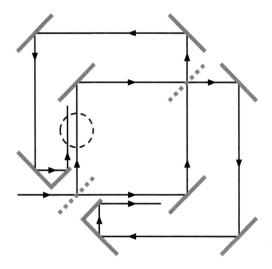

Fig. 3.16 *Testing a bomb multiple times with the same photon.*

and the second, a rotation by $(\pi - \theta) = \pi(N - 1)/N$. Feed the photon back into the setup a total of N times (in addition to its initial passage), and then measure with the two detectors. Show that if the bomb is out of order, detector D_1 will click with certainty. For a 'good' bomb, calculate the joint probability that it will not explode and detector D_0 will click. Show that as $N \to \infty$, this probability approaches one. Discuss the connection with the quantum Zeno effect.

3.15. Tensor product

(a) Calculate the matrix representation (in the standard basis) of the tensor product $\hat{X} \otimes \hat{Y}$, where $\hat{X}$ and $\hat{Y}$ are Pauli operators.

(b) Prove that the trace of a tensor product factorizes,

$$\text{tr}(\hat{A} \otimes \hat{B}) = \text{tr}(\hat{A}) \cdot \text{tr}(\hat{B}).$$

3.16. Entangled versus product states

Show that a pure composite state is entangled if and only if it cannot be written in the form of Eq. (3.92), that is, as a product state.

3.17. Bell states

(a) Show that the four Bell states, Eq. (3.99), constitute an orthonormal basis in the composite Hilbert space of two qubits.

(b) Verify the invariance of the Bell state $|\beta_{11}\rangle$ under arbitrary rotations of the single-qubit basis, Eq. (3.103). Investigate whether the other Bell states are also invariant under arbitrary, or perhaps only selected, basis changes.

3.18. Maximally entangled states

The composite state of two qubits is called 'maximally entangled' if it represents maximal knowledge about the whole but total ignorance about its parts. That is to say, the composite state is pure, but the reduced states of both constituents are states of total ignorance. Examples of such maximally entangled states are the four Bell states, Eq. (3.99).

(a) Show that all two-qubit states of the form

$$|\gamma_{AB}\rangle = \frac{|\psi\rangle \otimes |\chi\rangle + e^{i\varphi} |\bar{\psi}\rangle \otimes |\bar{\chi}\rangle}{\sqrt{2}}$$

are maximally entangled. Here $|\bar{\psi}\rangle$ and $|\bar{\chi}\rangle$ denote the states orthogonal to $|\psi\rangle$ and $|\chi\rangle$, respectively.

(b) Show that an arbitrary pure two-qubit state,

$$|\psi_{AB}\rangle = \frac{1}{\sqrt{2}} \sum_{jk} \Gamma_{jk} |jk\rangle,$$

is maximally entangled if and only if the coefficient matrix, Γ, is unitary. Calculate the coefficient matrix for $|\gamma_{AB}\rangle$ and verify that it is unitary.

(c) Show that, up to an irrelevant global phase factor, every maximally entangled state can be cast in the form

$$|\psi_{AB}\rangle = \frac{|0\rangle \otimes |\chi'\rangle + e^{i\varphi'} |1\rangle \otimes |\bar{\chi}'\rangle}{\sqrt{2}},$$

where, once again, $|\chi'\rangle$ and $|\bar{\chi}'\rangle$ are orthogonal.

(d) Count the number of parameters needed to specify an arbitrary

   i. pure state
   ii. pure product state
   iii. maximally entangled state

of two qubits.

3.19. Two-photon experiment

Consider the experimental setup depicted in Fig. 1.14, with the photon pair produced by the crystal in the Bell state $|\beta_{11}\rangle$. Imagine that the measurements on the two photons are handled by two different agents, Alice and Bob. They are hidden from each other's view and are not allowed to communicate with each other.

(a) Imagine further that both Alice and Bob remove their respective upper detector. What is the probability that of the remaining detectors
    i. both will click
    ii. one will click
    iii. none will click?

(b) Imagine another scenario where Bob is free to change the setting of his splitter from θ to θ', remove the splitter altogether, remove one or both detectors, or block the photon at any point; Alice's setup (with both detectors) stays as it is. Do any of Bob's actions alter the probabilities with which Alice's detectors click?

3.20. Reduced state

(a) A system composed of three two-level systems is in the state

$$|\psi\rangle = \frac{1}{\sqrt{2}}\left(|000\rangle + |111\rangle\right).$$

(This is a so-called 'Greenberger–Horne–Zeilinger state'.) Determine the reduced state of a subsystem composed of two out of the three two-level systems.

(b) Consider a bipartite system, AB, and an arbitrary orthonormal basis in the Hilbert space of constituent B, $\{|k\rangle\}_{k=0}^{d_B-1}$. Show that the operators

$$\hat{U}_i := \hat{I}_A \otimes \sum_{k=0}^{d_B-1} |k\rangle\langle k+i\,(\mathrm{mod}\,d_B)|, \quad i = 0,\ldots,d_B-1,$$

defined on the composite Hilbert space, are unitary.

(c) Show that for any normalized composite state, $\hat{\rho}_{AB}$, it is

$$\sum_{i=0}^{d_B-1} \hat{U}_i \hat{\rho}_{AB} \hat{U}_i^\dagger = \hat{\rho}_A \otimes \hat{I}_B.$$

This is an alternative (albeit not very practical) way of extracting the reduced state of constituent A from the composite state.

3.21. Measuring and discarding constituents

(a) Consider a bipartite system in some arbitrary composite state, ρ_{AB}. Show that ascertaining the truth of a most accurate proposition, e^A, about constituent A destroys all prior correlations with constituent B. That is to say, the post-measurement state is a product state,

$$\frac{(\hat{P}_{e^A} \otimes \hat{I}_B)\hat{\rho}_{AB}(\hat{P}_{e^A} \otimes \hat{I}_B)}{\mathrm{tr}[(\hat{P}_{e^A} \otimes \hat{I}_B)\hat{\rho}_{AB}(\hat{P}_{e^A} \otimes \hat{I}_B)]} = \hat{P}_{e^A} \otimes \hat{\rho}_{B|e^A}.$$

Show that unless the prior state was already a product state, the posterior reduced state of B, $\rho_{B|e^A}$, generally differs from the prior reduced state, ρ_B.

(b) Explain why the preceding result implies that, more generally, whenever we have maximal knowledge about a constituent, it cannot be correlated with the other; the composite state must be a product state:

$$\rho_A \text{ pure} \quad \Rightarrow \quad \rho_{AB} = \rho_A \otimes \rho_B.$$

(c) Argue that discarding a constituent, say, B, corresponds mathematically to taking the partial trace over the associated degrees of freedom, tr_B, to wit, going from the composite state to the reduced state of the remaining constituent, ρ_A.

(d) Show that it does not matter whether a measurement on A, confirming the truth of some proposition, x^A, is performed before or after B is discarded; both the probability of that outcome and the ensuing state update are unaffected by the timing of the disposal. Argue that, in mathematical terms, this invariance is reflected in the equation

$$\mathrm{tr}_B\big[(\hat{P}_{x^A} \otimes \hat{I}_B)\hat{\rho}_{AB}(\hat{P}_{x^A} \otimes \hat{I}_B)\big] = \hat{P}_{x^A}[\mathrm{tr}_B(\hat{\rho}_{AB})]\hat{P}_{x^A},$$

and prove this equation for any composite state $\hat{\rho}_{AB}$. Show that this result implies Eq. (2.42) as a special case.

3.22. Measurement process

In the main text we took measurement to be a primitive of the theory and did not worry about the inner workings of the measurement process. In reality, however, measurement involves various systems and happens in various stages. To begin with, the observable of interest is often not measured directly but is coupled to another degree of freedom where different measurement values can be discerned more easily. For instance, the magnetic moment of a silver atom or polarization of a photon, respectively, is not measured directly but is first coupled, via a splitter, to kinetic degrees of freedom—the direction of motion—of the particle. It is the latter which is subsequently ascertained with a screen or detector (Figs 1.1 and 1.4). A simple model considers the degrees of freedom of interest (say, polarization) and their proxy (beam direction) as two two-level systems, A and B. We denote the two basis states of polarization by $\{|0_A\rangle, |1_A\rangle\}$ and the two basis states of beam direction, or 'kinetic' states (say, to the right and upwards), by $\{|0_B\rangle, |1_B\rangle\}$. We assume that if the polarization state is $|0_A\rangle$, a polarizing beam splitter lets the photon pass without changing direction, whereas if the polarization state is $|1_A\rangle$, its direction of motion is changed from rightwards to upwards and vice versa. In sum, the splitter effects the transformation

$$|00\rangle \mapsto |00\rangle, \ |01\rangle \mapsto |01\rangle, \ |10\rangle \mapsto |11\rangle, \ |11\rangle \mapsto |10\rangle,$$

which in the standard basis is described by the unitary matrix

$$
\mathbf{U}_{BS} = \begin{pmatrix} 1 & 0 & 0 & 0 \\ 0 & 1 & 0 & 0 \\ 0 & 0 & 0 & 1 \\ 0 & 0 & 1 & 0 \end{pmatrix}.
$$

After this transformation the measurement is performed on B rather than A.

(a) Suppose a photon enters the splitter depicted in Fig. 1.4 in the polarization state $|+\rangle$ (Eq. (3.60)). Assume that the two detectors are such that they assert the passage of a photon without destroying it, deflecting it, or altering its polarization state. (Such detectors exist.) Thus, they serve to ascertain the photon's direction of flight but do not stop that flight. Consider four separate instants during the measurement process:

    i. before the photon enters the splitter

    ii. after the photon has passed the splitter, but before it reaches a detector

    iii. after it has passed a detector, but before you know which detector it was

    iv. after you have learnt which detector it passed.

Determine the composite state $\hat{\rho}_{AB}$, which describes both polarization and kinetic degrees of freedom, and the reduced state $\hat{\rho}_A$, which describes polarization only, at each instant.

(b) One might add yet another layer to this description. For instance, the detectors might be connected to a device with a little lamp that shines either blue or red (corresponding to, say, two different level transitions of an atom). If the photon passes the detector to the right, this colour does not change; if the photon passes the upper detector, on the other hand, the lamp changes its colour. In a gross simplification we might treat the lamp as yet another two-level system, C, whose two basis 'colour' states, $\{|0_C\rangle, |1_C\rangle\}$, correspond to 'blue' and 'red'. The device then effects the transformation

$$
|x00\rangle \mapsto |x00\rangle,\ |x01\rangle \mapsto |x01\rangle,\ |x10\rangle \mapsto |x11\rangle,\ |x11\rangle \mapsto |x10\rangle,\ x \in \{0,1\},
$$

where the first entry refers to the photon polarization (which does not affect the lamp colour), the second to the beam direction, and the third to the lamp colour. After this transformation the measurement is performed on C rather than A or B. Repeat the previous analysis of the states at various instants during the measurement process, taking into account that there are now additional instants (for instance, after coupling the detectors to the device but before ascertaining the colour of the lamp) and states ($\hat{\rho}_{ABC}$, $\hat{\rho}_{AB}$, and $\hat{\rho}_A$) to consider.

(c) Compare the two scenarios with each other and with the situation assumed in the main text, where the measurement on A is simply taken as a primitive. Do they differ in the post-measurement state of A? In other words, what difference does it make where the dividing line (sometimes called the 'Heisenberg cut') is drawn between the quantum realm and the realm of classical measuring devices? Discuss.

(d) In the initial scenario with two systems, A and B, imagine that with some probability, p, both detectors fail. In that case, it is as if they were absent. Which state $\hat{\rho}_{AB}$ would you assign after the photon has passed a (possibly defective) detector but before learning which detector it was? What if just one of them failed?

3.23. Purification

Consider a system composed of two constituents, A and B, described in Hilbert spaces of respective dimensions d_A and d_B. The composite system is in a pure state, $\hat{\rho}_{AB} = |\psi_{AB}\rangle\langle\psi_{AB}|$. The reduced states of the constituents, which may be pure or mixed, are described by $\hat{\rho}_A$ and $\hat{\rho}_B$, respectively. Each of these reduced statistical operators possesses a spectral decomposition. In particular, for constituent A it is

$$\hat{\rho}_A = \sum_{i=1}^{d_A} \rho_i |i_A\rangle\langle i_A|.$$

(a) In the Hilbert space of constituent B, choose an arbitrary orthonormal basis, $\{|k_B'\rangle\}$, which need not be an eigenbasis of $\hat{\rho}_B$. Then define the vectors

$$|i_B''\rangle := \sum_{k=1}^{d_B} \langle i_A k_B'|\psi_{AB}\rangle \, |k_B'\rangle$$

for all $i = 1, \ldots, d_A$; in general, these do not constitute a basis, nor are they normalized. Show that it is

$$|\psi_{AB}\rangle = \sum_{i=1}^{d_A} |i_A\rangle \otimes |i_B''\rangle.$$

Starting from this representation of the composite state, calculate the reduced state $\hat{\rho}_A$ and compare with the spectral decomposition given above. Deduce that

$$\langle i_B''|j_B''\rangle = \delta_{ij}\rho_i.$$

For those i where $\rho_i \neq 0$, define the rescaled vectors

$$|i_B\rangle := \frac{1}{\sqrt{\rho_i}} |i_B''\rangle.$$

These are normalized and mutually orthogonal (but need not necessarily constitute a basis). The pure composite state can then be written in the form

$$|\psi_{AB}\rangle = \sum_{i=1}^{d_A} \sqrt{\rho_i}\, |i_A\rangle \otimes |i_B\rangle.$$

This is called the *Schmidt decomposition* of the composite state, and the coefficients $\{\sqrt{\rho_i}\}$ are termed the *Schmidt coefficients*.

(b) Calculate the reduced state of constituent B. Show that the $\{|i_B\rangle\}$ are eigenvectors of $\hat{\rho}_B$. Show that $\hat{\rho}_B$ has the same non-zero eigenvalues (with the same multiplicities) as $\hat{\rho}_A$.

(c) You are given an arbitrary state $\hat{\rho}_A$ of a system A. Prove that it is always possible to add a second system, B, and define a pure state of the composite, $|\psi_{AB}\rangle$, such that the latter yields $\hat{\rho}_A$ as the reduced state of A,

$$\hat{\rho}_A = \text{tr}_B(|\psi_{AB}\rangle\langle\psi_{AB}|).$$

This process of going from the arbitrary state of a system to a pure state of some larger system is known as *purification*.

3.24. Symmetry versus exchangeability

In order for a composite state to be exchangeable, it must satisfy two conditions. First, the state must be symmetric, Eq. (2.43). Secondly, it must be possible to add another constituent and assign to the enlarged system a state which is equally symmetric and satisfies Eq. (2.44).

(a) Show that, expressed in terms of statistical operators, Eq. (2.44) amounts to

$$\hat{\rho}^{(L)} = \text{tr}_{L+1}(\hat{\rho}^{(L+1)}),$$

where the partial trace is taken over the added $(L+1)$th copy. Thus, the second condition states that it must be possible to write the composite state as a reduced state, starting from an equally symmetric state with one added constituent.

(b) Show that the Bell states are symmetric but not exchangeable.

(c) Prove that, more generally, a pure entangled state cannot be exchangeable.

3.25. Tsirelson bound

(a) For a Hermitian operator, $\hat{A}$, with eigenvalues $\{a_i\}$, one defines its 'spectral radius' as the largest modulus of an eigenvalue,

$$\|\hat{A}\| := \max_i |a_i|.$$

The Hermitian operators themselves form a vector space. Show that in this vector space the spectral radius represents a norm, to wit, it satisfies Eqs (3.13) to (3.15). Show, moreover, that in a tensor product space it satisfies

$$\|\hat{A} \otimes \hat{B}\| = \|\hat{A}\| \, \|\hat{B}\|.$$

(b) Use the properties of the norm to show that for the observable defined by Eq. (3.108), it is

$$\|\hat{A}\| \leq 2\sqrt{2}.$$

Argue that this implies the Tsirelson bound,

$$\langle A \rangle \leq 2\sqrt{2}.$$

(c) Show that the observable in question has eigenvalues $\pm 2\sqrt{2}$ (non-degenerate) and zero (with multiplicity two).
Hint: Show that $\hat{A}^2/8$ is a projector, and consider the traces of $\hat{A}$ and $\hat{A}^2$.

(d) Verify that the Bell state $|\beta_{11}\rangle$ is the *only* state in which the Tsirelson bound is saturated (Eq. (3.111)). In other words, whenever a Bell-type experiment reveals that $\langle A \rangle = 2\sqrt{2}$, we can be sure that the qubit pair is in this particular Bell state.

3.26. Specker's parable revisited

Consider the parable in Exercise (2.1) and the associated Hasse diagram. Does this model exhibit non-contextual realism? Discuss.

3.27. Quantum cakes

To illustrate a phenomenon known as *Hardy's paradox*, Paul Kwiat and Lucien Hardy (Kwiat and Hardy (2000)) tell the following story about a fictitious 'quantum kitchen' (with my omissions and clarifications in square brackets):

> We have a kitchen with two opposing doors, out of which come conveyor belts, and on the belts come pairs of ovens, one to each side. There is an experimenter on each side, call them Lucy (left side) and Ricardo (right), who will make measurements on the ovens; later the two will come together to compare their results. In particular, there are two types of measurements that can be made on a given oven. The tester could wait until the oven reaches the end of the conveyor belt before opening it. Inside, he/she finds a cake, which can then be tested to see whether it tastes Good or Bad. This is one observable, the taste of the cake. Alternatively, the tester can open the oven midway on its journey, to see whether or not the batter has Risen or Not Risen early, the second observable. Assuming we have some sort of soufflé, it is easy to justify why these measurements might

be noncommuting—re-closing the oven in the middle will cause the cake to collapse, and the result will always be a poor cake (perhaps even worse than it would naturally have been). Hence, only *one* of these qualities can be measured on a given cake.

Each experimenter will randomly decide which measurement they will make, and record the results obtained. Comparing the records later on will reveal the strangeness which arises if the cakes are quantum mechanically correlated. There are three main classes to consider, depending on whether Lucy and Ricardo both opened their respective ovens in the middle, one waited until the end to do so, or both did. Below, we describe the results which would be obtained [...].

#1 In cases where Lucy and Ricardo both checked their ovens midway, they find that 9% of the time, both cakes rose early. The rest of the time, only one or neither did.

#2 In cases where one checked midway and the other waited: whenever Lucy's cake rose early, Ricardo's tasted good; and whenever Ricardo's cake rose early, Lucy's tasted good. [...]

#3 [In cases where both Lucy and Ricardo performed taste-tests on their respective cakes:] *both cakes NEVER taste good*! That is, at least one of the cakes always tastes bad. [...]

In view of the first two observations, the third is striking. For consider the 9% of cases where both cakes would have been seen to rise early, had Lucy and Ricardo made those measurements instead. Since (in this 9% of cases) Lucy's cake would have risen early, the second observation implies that Ricardo's cake will taste good. Likewise, since (again in this 9% of cases) Ricardo's cake would have risen early, the said observation implies that Lucy's cake will taste good. Hence, on the basis of this reasoning, we would have expected that both cakes will taste good in at least 9% of cases.

(a) Explain why this story exhibits a violation of local realism.

(b) Show that the following entangled state of an oven pair explains all observations:

$$|\psi\rangle = \frac{1}{2}|BB\rangle - \sqrt{\frac{3}{8}}(|BG\rangle + |GB\rangle),$$

where G and B are the Good- and Bad-tasting eigenstates, which are related to the R (Risen) and N (Not Risen) eigenstates by

$$|G\rangle = \sqrt{\frac{2}{5}}|R\rangle - \sqrt{\frac{3}{5}}|N\rangle,$$

$$|B\rangle = \sqrt{\frac{3}{5}}|R\rangle + \sqrt{\frac{2}{5}}|N\rangle.$$

3.28. **Basis-independence of the Poisson bracket**

The definition of the Poisson bracket, Eq. (3.125), makes reference to a particular set of basis observables, $\{\hat{\mathcal{B}}_\mu\}$. Show that the Poisson bracket is in fact independent of these basis observables. That is to say, the Poisson bracket is invariant under an arbitrary change of basis in the linear space of Hermitian operators,

$$\hat{\mathcal{B}}_\mu \rightarrow \hat{\mathcal{B}}'_\mu = \sum_\nu J_\mu{}^\nu \hat{\mathcal{B}}_\nu.$$

3.29. **Hilbert space of a particle in one spatial dimension**

Show that the Hilbert space in which the position and momentum operators, $\hat{q}$ and $\hat{p}$, satisfy the commutation relations given by Eq. (3.135) must have infinite dimension.

Hint: Remember Exercise (3.3).

4

Computation

4.1 Gates and Circuits

I review elementary concepts of classical computing—bits, logic gates, circuits—
and show how these are transferred to the quantum realm. I discuss the
most important quantum gates and explain the basic stages of a quantum
computation. I highlight important differences from the classical case such as the
impossibility of copying a qubit or the often probabilistic nature of a quantum
computation.

In very broad terms, a *computer* is a machine which receives an input, processes
it according to some set of rules, and produces an output. In the case of a classical
computer, input and output consist of classical *information*, represented by strings of
symbols. Commonly, these symbols are classical *bits*, which take one of two possible
values, 0 or 1. In general, it is not specified what the input information refers to; it might
refer to *data* (say, some set of numbers), a *program*—instructions as to their handling
('calculate the sum')—or a combination of both. If, given a particular input, the computer
will always pass through the same sequence of states, always leading to the same output,
it is termed *deterministic*. A deterministic computer maps every allowed input bit string
to a unique output bit string. Thus, in essence, it computes a function. By contrast,
a *probabilistic* computer processes the input in a way which may involve randomness.
The same input might trigger different sequences of intermediate states of the machine,
yield different final outputs, or both, according to some probability distribution. In other
words, in different runs with the same input a probabilistic computer might exhibit
varying runtimes, produce varying outputs, or both. In cases where several different
outputs are possible, only one of them will constitute the desired, 'correct' function of
the input. In order to identify this correct output with a high degree of confidence, it may
be necessary to repeat the computation several times.

While there exist classical 'randomized' algorithms which introduce randomness on
purpose, the underlying classical machine works in a deterministic fashion. (We shall
ignore the fact that, in practice, there is always some probability of error.) Thus, every
single bit in its output is a unique function of the input bit string. Such a function,
$f : \{0,1\}^n \to \{0,1\}$, which maps an input bit string of length n to a single output bit,
is called a *Boolean function*. Any Boolean function may be expressed as a combination of

Quantum Theory: An Information Processing Approach. Jochen Rau, Oxford University Press (2021). © Jochen Rau.
DOI: 10.1093/oso/9780192896308.003.0004

logical operations ($\wedge, \vee, \neg$) on the input bits, assuming that '1' represents 'true' and '0' represents 'false'. For example, the Boolean function of two bits defined by

x_1	x_2	$f(x_1, x_2)$
0	0	1
0	1	1
1	0	0
1	1	1

can be written as

$$f(x_1, x_2) = (\neg x_1 \wedge \neg x_2) \vee (\neg x_1 \wedge x_2) \vee (x_1 \wedge x_2).$$

The same also works for Boolean functions of more than two bits (even though the number of logical operations required quickly becomes very large). Thanks to this correspondence, all Boolean functions can be implemented by manipulating bits one at a time ($\neg$) or in pairs ($\wedge, \vee$). Consequently, a classical computer can always be constructed from elementary *logic gates*, each acting on just one or two bits at a time. This is the *Boolean circuit model* of classical computation. The three logic gates used here—$\wedge$, $\vee$, and $\neg$—suffice; one says that they constitute a *universal set*. In fact, there exist other universal sets which are even smaller, containing just one type of logic gate; for instance, the NAND gate. The binary operation NAND is defined as a logical AND followed by a negation,

$$\text{NAND}(x_1, x_2) := \neg(x_1 \wedge x_2).$$

Indeed, all logic gates from the previous universal set can be implemented with NAND gates only:

$$x_1 \wedge x_2 = \text{NAND}(\text{NAND}(x_1, x_2), \text{NAND}(x_1, x_2)),$$
$$x_1 \vee x_2 = \text{NAND}(\text{NAND}(x_1, x_1), \text{NAND}(x_2, x_2)),$$

and

$$\neg x = \text{NAND}(x, x).$$

All these constructions tacitly assume that a classical bit may be copied as often as desired and inputted multiple times into the same or several different gates. Such limitless copying is not obvious and, in fact, will turn out to be severely restricted in a quantum computer. Logic gates with two inputs and just one output, such as $\wedge$, $\vee$, or NAND, entail a loss of information: given their output, one can in general not reconstruct their input. In this sense, classical computation with such gates is *irreversible*.

Even though in computer science information is treated as an abstract entity, it is ultimately physical. In the real world, information always resides on a physical substrate; for instance, as blobs of ink on a sheet of paper, holes punched in a card, little magnets pointing up or down, or different voltage levels in a semiconductor. Likewise, the processing of information always involves processes which are physical, such as flipping magnets or moving electrons around. Therefore, the study of the inner workings of a computer can never be completely dissociated from the laws of physics governing the underlying hardware. When the hardware is miniaturized down to atomic scale and sufficiently cooled and insulated, its quantum nature will come to the fore, inevitably affecting the way a computer will function.

In a *quantum computer*, the quantum nature of its physical constituents is actively exploited. This gives rise to novel computational capabilities which are beyond the reach of a classical computer; we will encounter a number of examples later in this chapter and in Chapter 5. In a quantum computer, input and output consist not of classical information but of quantum states, or some combination of classical information and quantum states. As we discussed in Sections 2.2 and 3.2, a 'state' in general, and a quantum state in particular, is also a form of information—namely, about the prior preparation and measurement history of the system at hand. Like classical information, it requires a physical carrier; for instance, an atom, a nuclear spin, or a photon. In principle, the quantum states being processed by a quantum computer might be defined in a Hilbert space of arbitrary (finite) dimension, d—just as the symbols representing classical information might be taken from an alphabet of arbitrary (finite) size. In practice, however, analogous to the representation of classical information in terms of bits, the Hilbert space is commonly that of an assembly of *qubits*. On these qubits, a quantum computer performs two basic types of operations: *unitary transformations* and *measurements*.

In the same way that Boolean functions of an arbitrary number of bits may be implemented with one- and two-bit gates only, giving rise to the classical Boolean circuit model, unitary transformations on an arbitrary number of qubits can be effected with elementary *quantum logic gates* which each act on just one or two qubits at a time; this is the *quantum circuit model*. (We shall prove this universality of the quantum circuit model in Section 4.2.) In the following, we investigate the elementary quantum logic gates more closely. We start with the simplest case, a quantum logic gate acting on a single qubit. Classically, there exists only one interesting single-bit gate—the NOT gate ($\neg$)—but in a quantum computer there are many more possibilities: the gate might perform an arbitrary unitary transformation, $\hat{U}$, in the Hilbert space of the qubit,

$$|\psi\rangle -\boxed{U}- \hat{U}|\psi\rangle. \tag{4.1}$$

Like in a classical circuit, the graphical representation of the respective operation should be read from left to right. An input qubit in some state, $|\psi\rangle$, enters the gate, which then performs a unitary transformation. The output is a qubit in the transformed state, $\hat{U}\,|\psi\rangle$. In addition to unitary single-qubit gates, there also exists a *measurement gate*,

$$|\psi\rangle \ -\!\!\boxed{\measuredangle}\!\!= x. \tag{4.2}$$

This gate performs a measurement on the qubit in the standard basis, $\{|0\rangle,|1\rangle\}$, which we had introduced in Section 3.3. The output is now not a qubit but a classical bit representing the measurement outcome, $x \in \{0,1\}$. All quantum computing algorithms feature such measurement gates because eventually the result of a quantum computation will have to be expressed in terms of classical bits. Qubits and classical bits are distinguished graphically by different types of wires: a qubit is represented by a single line, whereas a classical bit is represented by a double line. While a unitary transformation constitutes a *reversible* operation (which can always be undone by applying its inverse, $\hat{U}^{\dagger}$), a measurement is *irreversible*. Unless the input state, $|\psi\rangle$, is constrained a priori to be one of the two basis states, the outcome of the measurement will not allow a reconstruction of the input state. Moreover, the outcome of the measurement is random. This inherent randomness renders the typical quantum computer probabilistic, rather than deterministic.

As for the unitary transformation of a single qubit, $\hat{U}$, there are myriad possibilities. Some special transformations are employed frequently and therefore have their own symbols. To begin with, there are single-qubit gates associated with the three Pauli operators, $\hat{X}$, $\hat{Y}$, and $\hat{Z}$, defined in Eq. (3.62), which are all unitary. When acting on a basis state, $|0\rangle$ or $|1\rangle$, the *Pauli-X* gate has the effect

$$|t\rangle \ -\!\boxed{X}\!-\ |t\oplus 1\rangle, \quad t\in\{0,1\}, \tag{4.3}$$

where $\oplus$ denotes addition modulo two. In other words, the basis state $|0\rangle$ is transformed into the basis state $|1\rangle$, and vice versa. In this sense, the Pauli-X gate may be regarded as the quantum version of a classical NOT gate. The *Pauli-Z* gate, on the other hand, effects a selective sign change,

$$|t\rangle \ -\!\boxed{Z}\!-\ (-1)^{t}\,|t\rangle, \quad t\in\{0,1\}; \tag{4.4}$$

such an operation has no classical analogue. In addition to the Pauli gates, there are three further often-used single-qubit gates: the *Hadamard gate*, the *phase gate*, and the $\pi/8$ *gate*, represented by the letters H, S, and T, respectively. In the standard basis, $\{|0\rangle,|1\rangle\}$, these effect the respective unitary transformations

$$\mathbf{H} := \frac{1}{\sqrt{2}}\begin{pmatrix} 1 & 1 \\ 1 & -1 \end{pmatrix}, \quad \mathbf{S} := \begin{pmatrix} 1 & 0 \\ 0 & i \end{pmatrix}, \quad \mathbf{T} := \begin{pmatrix} 1 & 0 \\ 0 & e^{i\pi/4} \end{pmatrix}. \tag{4.5}$$

In particular, the Hadamard gate transforms a basis state into one of two superposition states,

$$|0\rangle \;-\boxed{H}-\; |+\rangle\,, \qquad |1\rangle \;-\boxed{H}-\; |-\rangle\,, \tag{4.6}$$

where $|+\rangle$ and $|-\rangle$ are the rotated basis states defined in Eq. (3.60). This can be succinctly summarized as

$$\hat{H}\,|t\rangle = \frac{1}{\sqrt{2}}\sum_{z\in\{0,1\}} (-1)^{tz}\,|z\rangle\,, \quad t\in\{0,1\}. \tag{4.7}$$

The various single-qubit gates are not independent of each other; between them there exist numerous relationships. For instance,

$$\hat{H}^2 = \hat{I}, \quad \hat{H}\hat{X}\hat{H} = \hat{Z}, \quad \hat{H}\hat{Z}\hat{H} = \hat{X}, \quad \hat{S}^2 = \hat{Z}, \quad \hat{T}^2 = \hat{S}, \tag{4.8}$$

or, graphically,

$$
\begin{array}{rcl}
-\boxed{H}\boxed{H}- &=& - \\[2pt]
-\boxed{H}\boxed{X}\boxed{H}- &=& -\boxed{Z}- \\[2pt]
-\boxed{H}\boxed{Z}\boxed{H}- &=& -\boxed{X}- \\[2pt]
-\boxed{S}\boxed{S}- &=& -\boxed{Z}- \\[2pt]
-\boxed{T}\boxed{T}- &=& -\boxed{S}-
\end{array}
\tag{4.9}
$$

Next, we turn to two-qubit gates. Pictorially, these gates have two qubits going in and two qubits going out:

$$|\psi\rangle \left\{ \;\boxed{\;U\;}\; \right\} \hat{U}\,|\psi\rangle \tag{4.10}$$

Here $|\psi\rangle$ denotes an arbitrary composite state of the two qubits; this might be a product state or an entangled state like a Bell state, Eq. (3.99). In case it is a product state, for instance, $|01\rangle$, the first entry (here: '0') refers to the state of the upper qubit and the second entry (here: '1') to that of the lower qubit in the graph. The gate effects a unitary transformation, $\hat{U}$, in the four-dimensional composite Hilbert space of the two qubits. This Hilbert space has as its standard basis the four distinct tensor products of the constituent basis states, $|00\rangle$, $|01\rangle$, $|10\rangle$, and $|11\rangle$. The entries in these basis states may also be regarded as binary representations of the natural numbers $0,1,2,3$. Therefore, an alternative notation for these basis states reads

$$|0\rangle_2 \equiv |00\rangle\,, \quad |1\rangle_2 \equiv |01\rangle\,, \quad |2\rangle_2 \equiv |10\rangle\,, \quad |3\rangle_2 \equiv |11\rangle\,, \tag{4.11}$$

where the subscript, '2', indicates that the states pertain to two qubits. In this standard basis, a two-qubit gate is represented by a unitary 4×4 matrix. One example of particular importance is the gate associated with the unitary matrix

$$\mathbf{U}_{\text{CNOT}} = \begin{pmatrix} 1 & 0 & 0 & 0 \\ 0 & 1 & 0 & 0 \\ 0 & 0 & 0 & 1 \\ 0 & 0 & 1 & 0 \end{pmatrix} = \begin{pmatrix} \mathbf{I} & \mathbf{0} \\ \mathbf{0} & \mathbf{X} \end{pmatrix}, \tag{4.12}$$

where on the right-hand side the entries are themselves 2×2 matrices. It maps the basis states $|00\rangle$ and $|01\rangle$ to themselves, whereas it maps $|10\rangle$ to $|11\rangle$ and vice versa. In short, whenever the first qubit is in the basis state $|0\rangle$, the state of the second qubit remains unchanged; but when the first qubit is in the basis state $|1\rangle$, the state of the second qubit undergoes a NOT operation. For this reason the gate is called the CNOT ('controlled-NOT') gate. It is represented graphically by

$$\begin{array}{c} |c\rangle \quad\longrightarrow\quad |c\rangle \\ |t\rangle \quad\longrightarrow\quad |t \oplus c\rangle \end{array} \quad , \quad c, t \in \{0, 1\}. \tag{4.13}$$

The first, upper qubit is termed the 'control qubit'. Its state (provided it is a basis state) is preserved. By contrast, the state of the second, lower qubit—termed the 'target qubit'—may flip, depending on the state of the control qubit. This controlled change is indicated by the mapping

$$t \mapsto t \oplus c, \tag{4.14}$$

where $c, t \in \{0, 1\}$ and, as before, $\oplus$ denotes addition modulo two. When the control qubit is in an arbitrary superposition state rather than a basis state, its state need no longer be preserved. For instance, if the control qubit is in the state $|+\rangle$, defined in Eq. (3.60), and the target qubit is in the basis state $|0\rangle$, the CNOT gate will produce an entangled state:

$$\begin{array}{c} |+\rangle \quad\longrightarrow \\ |0\rangle \quad\longrightarrow \end{array} \Big\} \; \frac{1}{\sqrt{2}} (|00\rangle + |11\rangle) \tag{4.15}$$

After the action of the gate neither qubit is in a pure state on its own, let alone in the same state as before.

There are other, similar two-qubit gates which also feature a control and a target qubit, yet where the controlled operation on the target qubit is not a flip but, say, an application of the Pauli operator $\hat{Z}$:

$$\begin{array}{c} |c\rangle \quad\longrightarrow\quad |c\rangle \\ |t\rangle \quad\boxed{Z}\quad (-1)^{ct} |t\rangle \end{array} \quad , \quad c, t \in \{0, 1\}. \tag{4.16}$$

This controlled-Z gate has the matrix representation

$$\mathbf{U}_{\text{C}-Z} = \begin{pmatrix} 1 & 0 & 0 & 0 \\ 0 & 1 & 0 & 0 \\ 0 & 0 & 1 & 0 \\ 0 & 0 & 0 & -1 \end{pmatrix} = \begin{pmatrix} \mathbf{I} & \mathbf{0} \\ \mathbf{0} & \mathbf{Z} \end{pmatrix}. \tag{4.17}$$

Thanks to the first two relationships in Eq. (4.8), the controlled-Z gate can be constructed from CNOT and Hadamard gates. In matrix representation it is

$$\begin{pmatrix} \mathbf{I} & \mathbf{0} \\ \mathbf{0} & \mathbf{Z} \end{pmatrix} = \begin{pmatrix} \mathbf{H} & \mathbf{0} \\ \mathbf{0} & \mathbf{H} \end{pmatrix} \begin{pmatrix} \mathbf{I} & \mathbf{0} \\ \mathbf{0} & \mathbf{X} \end{pmatrix} \begin{pmatrix} \mathbf{H} & \mathbf{0} \\ \mathbf{0} & \mathbf{H} \end{pmatrix}, \tag{4.18}$$

which pictorially translates into

$$\tag{4.19}$$

Using three CNOT gates in a row, it is possible to *swap* two qubits; the resulting 'swap gate' has its own graphical symbol:

$$\tag{4.20}$$

Indeed, the three successive applications of CNOT have the net effect of swapping the qubit states:

$$|a,b\rangle \mapsto |a, b \oplus a\rangle \mapsto |(b \oplus a) \oplus a, b \oplus a\rangle \mapsto |(b \oplus a) \oplus a, (b \oplus a) \oplus (b \oplus a) \oplus a\rangle = |b,a\rangle,$$

since $x \oplus x = 0$ for any $x \in \{0,1\}$.

There are limits, however, to what can be done with a two-qubit gate. In particular, it is impossible to build a 'quantum copying machine' which would simply duplicate one of the incoming qubit states. Such a hypothetical machine would have to transform states as follows:

$$\tag{4.21}$$

where $\hat{U}$ is some unitary two-qubit operator. (The auxiliary input $|s\rangle$ is necessary in order to have the same number of qubits going in and out.) Alas, there exists no $\hat{U}$ which achieves such a transformation for arbitrary input states, $|\psi\rangle$. To see this, suppose that such a $\hat{U}$ did exist. This operator would then have to satisfy

$$\hat{U}(|\psi\rangle \otimes |s\rangle) = |\psi\rangle \otimes |\psi\rangle, \quad \hat{U}(|\varphi\rangle \otimes |s\rangle) = |\varphi\rangle \otimes |\varphi\rangle \tag{4.22}$$

for arbitrary $|\psi\rangle$ and $|\varphi\rangle$. Taking the respective scalar products of the left-hand sides and of the right-hand sides of these equations yields the condition

$$\langle\psi|\varphi\rangle = (\langle\psi|\varphi\rangle)^2. \tag{4.23}$$

This can only be satisfied if the scalar product is either zero or one; that is, if the two states are either orthogonal or identical. In other words, it *is* possible to duplicate states, yet only as long as these are limited to the orthogonal basis states. For all other states, duplication is not possible. This finding is known as the *no-cloning theorem*. It contrasts sharply with classical computing, where the copying of a classical bit posed no problem at all. The no-cloning theorem is deeply linked to the basic structure of quantum theory. In quantum theory it is impossible to measure the state of an individual quantum system. Whenever one measures some observable of a single system, the post-measurement state becomes an eigenstate of this observable, regardless of the state prior to measurement. This entails an unavoidable and irretrievable loss of information about the prior state which, therefore, can never be reconstructed. If, prior to a measurement, the state could be cloned, one would be able to retain a copy of the system in its original state, thereby circumventing this fundamental limitation on the measurement of individual quantum states.

One may also define gates which act on more than two qubits. However, as we already noted, the effects of such gates can always be emulated by one- and two-qubit gates. While the general proof of this assertion will have to wait until Section 4.2, we consider here a specific example, the so-called *Toffoli gate*, which acts on a total of three qubits. In the standard basis, $\{|000\rangle, |001\rangle, \ldots, |111\rangle\}$, it has the 8×8 matrix representation

$$\mathbf{U}_{\text{Toffoli}} = \begin{pmatrix} \mathbf{I} & & \\ & \mathbf{I} & \\ & & \begin{matrix} \mathbf{I} & \\ & \mathbf{X} \end{matrix} \end{pmatrix}. \tag{4.24}$$

As before, the entries of this matrix are themselves 2×2 matrices; the empty spaces are all zeros. This gate has the graphical representation

$$\begin{array}{l} |c_1\rangle \quad\longrightarrow\quad |c_1\rangle \\ |c_2\rangle \quad\longrightarrow\quad |c_2\rangle \\ |t\rangle \quad\longrightarrow\quad |t \oplus c_1 c_2\rangle \end{array} \quad , \quad c_1, c_2, t \in \{0,1\}. \tag{4.25}$$

It features two control qubits and one target qubit. Basis states of the control qubits are preserved, whereas a basis state of the target qubit will flip if and only if both control qubits are in the state $|1\rangle$. This is the simplest example of a 'multiply-controlled' NOT gate, where the NOT operation on the target qubit is contingent upon the states of not just one but several control qubits—in contrast to the ordinary CNOT gate, Eq. (4.13), which features a single control qubit only. The Toffoli gate can, indeed, be constructed from single-qubit and CNOT gates. The required combination is not straightforward; it turns out that the following sequence of six CNOT and nine single-qubit gates does the job:

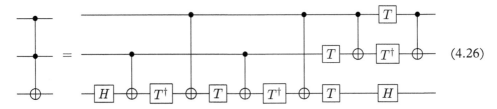

$$(4.26)$$

This corresponds to the fact that the matrix representing the Toffoli gate, Eq. (4.24), equals the product of the following 15 matrices:

$$
\begin{pmatrix} \mathbf{I} & & \\ & \mathbf{I} & \\ & & \mathbf{I} \\ & & & \mathbf{X} \end{pmatrix} = \begin{pmatrix} \mathbf{I} & & \\ & \mathbf{I} & \\ & & \mathbf{I} \\ & & & \mathbf{I} \end{pmatrix} \begin{pmatrix} \mathbf{H} & & \\ & \mathbf{H} & \\ & & \mathbf{H} \\ & & & \mathbf{H} \end{pmatrix} \begin{pmatrix} \mathbf{I} & & \\ & e^{-i\pi/4} & \\ & & \mathbf{I} \\ & & & e^{-i\pi/4} \end{pmatrix} \begin{pmatrix} \mathbf{I} & & \\ & \mathbf{I} & \\ & & e^{i\pi/4} \\ & & & e^{i\pi/4} \end{pmatrix}
$$

$$
\begin{pmatrix} \mathbf{I} & & \\ & \mathbf{I} & \\ & & \mathbf{I} \end{pmatrix} \begin{pmatrix} \mathbf{T} & & \\ & \mathbf{T} & \\ & & \mathbf{T} \end{pmatrix} \begin{pmatrix} \mathbf{I} & & \\ & e^{i\pi/4} & \\ & & \mathbf{I} \\ & & & e^{i\pi/4} \end{pmatrix} \begin{pmatrix} \mathbf{I} & & \\ & \mathbf{I} & \\ & & \mathbf{X} \\ & & & \mathbf{X} \end{pmatrix}
$$

$$
\begin{pmatrix} \mathbf{T}^\dagger & & \\ & \mathbf{T}^\dagger & \\ & & \mathbf{T}^\dagger \\ & & & \mathbf{T}^\dagger \end{pmatrix} \begin{pmatrix} \mathbf{I} & & \\ & \mathbf{X} & \\ & & \mathbf{I} \\ & & & \mathbf{X} \end{pmatrix} \begin{pmatrix} \mathbf{T} & & \\ & \mathbf{T} & \\ & & \mathbf{T} \\ & & & \mathbf{T} \end{pmatrix} \begin{pmatrix} \mathbf{I} & & \\ & \mathbf{I} & \\ & & \mathbf{X} \\ & & & \mathbf{X} \end{pmatrix}
$$

$$
\begin{pmatrix} \mathbf{T}^\dagger & & \\ & \mathbf{T}^\dagger & \\ & & \mathbf{T}^\dagger \\ & & & \mathbf{T}^\dagger \end{pmatrix} \begin{pmatrix} \mathbf{I} & & \\ & \mathbf{X} & \\ & & \mathbf{I} \\ & & & \mathbf{X} \end{pmatrix} \begin{pmatrix} \mathbf{H} & & \\ & \mathbf{H} & \\ & & \mathbf{H} \\ & & & \mathbf{H} \end{pmatrix}. \qquad (4.27)
$$

Once again, all components of these matrices are themselves 2×2 matrices, and empty spaces are zeros. The exponentials are short for multiples of the 2×2 unit matrix, $\exp(\pm i\pi/4)\mathbf{I}$.

The fact that the controlled-Z gate (Eq. (4.19)), the swap gate (Eq. (4.20)), and now the Toffoli gate can all be built from single-qubit and CNOT gates is not a coincidence. In Section 4.2 we will show not only that arbitrary unitary transformations on an arbitrary number of qubits can be effected with combinations of one- and two-qubit gates, but also, more stringently, that it is sufficient to use combinations of one-qubit gates and just a single type of two-qubit gate—indeed, the CNOT gate. In other words, one-qubit gates and the CNOT gate together form a universal set; they play the same role in quantum computing that the NAND gate or similar universal sets played in classical computing. This reduces the diversity of the gate zoo significantly and opens the prospect that practically useful quantum algorithms can be implemented with a manageable number of different gate types.

There exists an alternative paradigm for quantum computing that dispenses with unitary gates and instead works with measurement gates only; this is called 'measurement-based computation'. It is equally universal: by means of a suitable sequence of measurements, it is possible to emulate arbitrary unitary transformations on an arbitrary number of qubits. Its practical implementation is difficult, however, because it requires the preparation of rather complex initial states or measurement gates, or both. Still, since it is so markedly different from classical computing, it is quite interesting conceptually. For this reason, we shall devote to measurement-based computation a brief interlude in Section 4.3.

Quantum circuits may be designed to help solve specific computational problems; for instance, to factorize integers, or to find the ground state of a complex molecule. The typical computation involves three stages:

1. The *input* of the problem is encoded in the initial state of the qubits. This encoding may happen either directly or indirectly. For example, if the input is the integer '21', which has binary representation 10101, one might directly prepare the qubits in the initial basis state $|0\ldots010101\rangle$. (The preceding zeros in the basis state account for the possibility that there are more than five qubits.) Alternatively, the qubits could be initialized to the state $|0\ldots00\rangle$ and the input used to control subsequent unitary transformations of these qubits. In the given example, each qubit might be subjected to a Pauli-X gate, contingent upon the pertinent bit in the classical input bit string having the value '1'. After these transformations, the state would again be $|0\ldots010101\rangle$. The gates controlled by the input might also effect more general unitary transformations, producing a state that need no longer be a basis state. Regardless of the specific implementation, the number of qubits being thus initialized is a measure of the *length* of the input.

2. Next, the quantum computation proper is realized as a unitary transformation in the composite Hilbert space of all the input qubits. This unitary transformation is effected by a concatenation of elementary quantum logic gates. At this stage the circuit might make use of additional 'ancilla' (or 'auxiliary', or 'work') qubits which at the beginning and end of the computation are in the basis state $|0\rangle$, yet may change their states in the interim.

3. Finally, the *output* of the computation consists in the final state of some or all of the outgoing qubits, classical information in terms of bits, or some combination thereof. In order to extract classical information, one performs measurements in the standard basis on some or all of the output qubits. In general, the output qubits are not exactly in a basis state, so the outcomes of these measurements will not be certain. Rather, there will be various possible outcomes, subject to some probability distribution, of which one represents the correct solution to the computational problem.

These three stages of the computation may have to be repeated several times, and they might be complemented by some form of pre- or post-processing on a classical computer. All steps together constitute the *algorithm* for solving the problem at hand.

The gates in the circuit, the ancilla qubits, and the time employed for repeated runs represent precious computational resources. The resources needed to solve a given computational problem grow with the length of the input: for instance, a circuit which can factorize integers with up to ten digits must contain more gates or run more often (or both) than a circuit able to factorize integers with just up to five digits. How exactly the requisite resources scale with the input length depends on the problem at hand; the pertinent scaling is a measure of the problem's *complexity*. Categorizing computational problems into 'complexity classes' and studying the relationships between these classes is a major area of research in theoretical computer science, on which we shall not elaborate here. (If you want to learn more about this, you will find a suggestion for further reading at the end of this chapter.) Suffice it to say that a computational problem is deemed *tractable* if it can be solved with resources that scale at most polynomially with the input length, and intractable, or simply *hard*, if instead the resource requirements grow exponentially. In order to demonstrate that a particular problem is tractable on a quantum computer, it is sufficient to find one specific quantum algorithm that solves it with polynomial resources. Such an algorithm is called *efficient*.

The great promise of quantum computing rests on the discovery that for some computational problems which are intractable on a classical computer, efficient algorithms do exist on a quantum computer. In Section 4.4 we will encounter a simple toy example which, albeit not very useful in practical terms, will illustrate the basic idea behind this 'quantum advantage'. In terms of applications with more practical relevance, a prominent case in point is the aforementioned factorization of large integers. While this is a hard problem on a classical computer, requiring resources that grow exponentially with the number of digits of the integer, there is indeed an efficient quantum algorithm. Alas, a detailed discussion of this particular algorithm, called 'Shor's algorithm', is beyond the scope of this book, but you will find some reading suggestions at the end of this chapter. Another example, also mentioned earlier, is the determination of the ground state of a molecule. Here we shall sketch a pertinent algorithm in Section 4.5, and we will find that the same algorithm may also be used to tackle hard classical optimization problems. Finally, one of the most natural applications of quantum computing is to simulate the time evolution of *other* quantum systems. Whenever such systems contain many interacting particles, this is again hard on a classical computer. In Section 4.6 we will see that on a quantum computer it can be done efficiently, provided the interactions between the particles are local.

Quantum circuits can be employed not just for computation and simulation but also in other technological fields, notably metrology and communication. In metrology, the precision of measurements can be enhanced by the clever use of quantum entanglement. This enhancement may be described succinctly in the pictorial language of quantum circuits; we shall do so, for a simple example, in Section 4.7. Lastly, it is in the area of communication that quantum circuits perform perhaps their greatest feats. There, they accomplish certain tasks which would be completely *impossible* by classical means; for instance, 'teleporting' a quantum state from one place to another without physically moving the carrier of that state. This particular and a few other prominent examples, along with general considerations about the transmission of classical and quantum information, merit a chapter of their own: Chapter 5.

4.2   Universality

I claimed in Section 4.1 that single-qubit gates and the CNOT gate form a universal set, to wit, that these gates suffice to effect arbitrary unitary transformations on any number of qubits. Here I furnish a proof of this assertion. Should you trust the final result and wish to fast-forward to practical applications of quantum information processing, you may skip this section.

We consider a quantum system described in a Hilbert space of arbitrary finite dimension, d. The Hilbert space is endowed with an orthonormal basis, with the basis states labelled by $|0\rangle, \ldots, |d-1\rangle$. The natural numbers labelling these basis states can also be written in binary form,

$$|0\rangle_n \equiv |0\ldots00\rangle$$
$$|1\rangle_n \equiv |0\ldots01\rangle$$
$$|2\rangle_n \equiv |0\ldots10\rangle$$
$$\vdots$$

(4.28)

This binary representation must have n digits, where $2^{n-1} < d \leq 2^n$. Formally, then, the quantum system may be thought of as an n-qubit system. Where the original Hilbert space dimension, d, is not exactly equal to the Hilbert space dimension of an n-qubit system, 2^n, one may nevertheless think of the system as a collection of n qubits whose allowed states have been restricted to some subspace of their Hilbert space. Without loss of generality, therefore, we shall assume that the system is composed of n qubits.

The most general reversible operation on such a system is a unitary transformation, $\hat{U}$, represented in the given basis by a unitary $d \times d$ matrix. We want to prove that any such unitary transformation can be written as a product of unitary transformations representing single-qubit and CNOT operations only. We break this down into five smaller results, which we shall prove separately:

1. Every unitary transformation in d-dimensional Hilbert space can be written as a product of unitaries involving two levels only.

2. Every such two-level transformation can be written as a product of level swaps and a multiply-controlled unitary transformation.

3. A level swap can be effected by a sequence of multiply-controlled NOT operations.

4. A multiply-controlled NOT operation can be decomposed into single-qubit and CNOT operations. A more general multiply-controlled unitary transformation can be decomposed into single-qubit operations, CNOTs, and a controlled unitary transformation with just a single control qubit.

5. A controlled unitary transformation with a single control qubit may be effected by a sequence of single-qubit and CNOT operations.

We begin with the first step. For the purposes of our calculation, it will be convenient to write the $d \times d$ unitary matrix in the form

$$
\mathbf{U}_{d \times d} = \begin{pmatrix} x_0 \\ \vdots & \boxed{} \\ x_{d-1} \end{pmatrix},
$$

where the box is a placeholder for the remaining $d \times (d-1)$ entries. Since the matrix is unitary, the first column vector, $(x_0, \ldots, x_{d-1})^{\mathrm{T}}$, must be a unit vector. In the following, we will describe an algorithm which transforms this general unitary matrix into the unit matrix, via successive multiplication with unitary matrices that each pertain to two levels only:

$$
\mathbf{U}_{2 \times 2}^{(1)} \ldots \mathbf{U}_{2 \times 2}^{(r)} \mathbf{U}_{d \times d} = \mathbf{I}_{d \times d}. \tag{4.29}
$$

A simple inversion of this equation then shows that $\mathbf{U}_{d \times d}$ can be written as a product of unitary transformations which each involve two levels only. In the algorithm we distinguish two cases:

- The first column vector comprises two or more non-zero entries. In this case we pick the uppermost entry which is not zero, x_i, and some other non-zero entry, x_k, with $k > i$. Then we multiply the matrix from the left with a unitary transformation involving only the levels (basis states) $|i\rangle$ and $|k\rangle$. The matrix elements of this two-level transformation can always be chosen such that in the transformed unitary matrix the kth component of the first column vector becomes zero:

$$
\begin{pmatrix}
\ddots & & & & & & \\
& 1 & & & & & \\
& & \hat{x}_i^* & & & \hat{x}_k^* & \\
& & & 1 & & & \\
& & & & \ddots & & \\
& & & & & 1 & \\
& & -\hat{x}_k & & & \hat{x}_i & \\
& & & & & & 1 \\
& & & & & & & \ddots
\end{pmatrix}
\begin{pmatrix}
\vdots \\
0 \\
x_i \\
x_{i+1} \\
\vdots \\
x_{k-1} \\
x_k \\
x_{k+1} \\
\vdots
\end{pmatrix}
=
\begin{pmatrix}
\vdots \\
0 \\
x_i' \\
x_{i+1} \\
\vdots \\
x_{k-1} \\
0 \\
x_{k+1} \\
\vdots
\end{pmatrix},
$$

where

$$
\hat{x}_i := \frac{x_i}{\sqrt{|x_i|^2 + |x_k|^2}}
$$

(and likewise $\hat{x}_k$), and

$$
x_i' = \sqrt{|x_i|^2 + |x_k|^2}
$$

is positive and real. (In the two-level unitary transformation all off-diagonal elements vanish except those explicitly shown.) In this fashion we have reduced the number of non-zero entries in the first column vector by one. We can iterate the procedure until the ith component is the only non-zero entry left. This remaining non-zero component is, moreover, positive and real; and since the column vector must be a unit vector, it is, in fact, equal to one. If this '1' is the first component of the column vector (that is, in our labelling convention, $i = 0$), we are done. Otherwise, we move on to the second case.

- The first column vector has only a single non-zero entry, x_i, with $|x_i| = 1$. In this case we multiply the matrix from the left with a unitary transformation which again involves only two levels, this time the levels $|0\rangle$ and $|i\rangle$. The result is yet another unitary matrix where the first column vector has the form $(1, 0, \ldots, 0)^{\mathrm{T}}$:

$$
\begin{pmatrix}
0 & & & & x_i^* & \\
& 1 & & & & \\
& & \ddots & & & \\
& & & 1 & 0 & \\
-x_i & & & 0 & & \\
& & & & 1 & \\
& & & & & \ddots
\end{pmatrix}
\begin{pmatrix}
0 \\
0 \\
\vdots \\
0 \\
x_i \\
0 \\
\vdots
\end{pmatrix}
=
\begin{pmatrix}
1 \\
0 \\
\vdots \\
0 \\
0 \\
0 \\
\vdots
\end{pmatrix}. \qquad (4.30)
$$

In either case, our procedure yields a matrix of the form given on the right-hand side of the last equation. Being unitary, not only its first column vector but also its first row vector must be a unit vector. This implies that, in fact, in the box of $d \times (d-1)$ unspecified entries the first row consists only of zeros. Therefore, the right-hand side of Eq. (4.30) is a unitary matrix whose dimension has, in effect, been reduced to $(d-1) \times (d-1)$. We can repeat the procedure $(d-1)$ times, each time reducing the effective dimension of the unitary matrix by one, $\mathbf{U}_{d \times d} \to \mathbf{U}_{(d-1) \times (d-1)} \to \ldots$, until we arrive at the identity matrix on the right-hand side of Eq. (4.29).

In the second step, we have to show that every two-level unitary transformation can be written as a product of level swaps and a multiply-controlled unitary transformation. A general two-level transformation has the matrix representation

$$
\mathbf{U}_{2 \times 2} =
\begin{pmatrix}
\ddots & & & & \\
& a & & b & \\
& & \ddots & & \\
& c & & d & \\
& & & & \ddots
\end{pmatrix},
\tag{4.31}
$$

where all diagonal elements other than a and d are equal to one, and all off-diagonal elements other than b and c are zero. If the non-trivial entries are situated in the two rows and columns labelled s and t $(t > s)$, respectively, this transformation relates the two levels $|s\rangle$ and $|t\rangle$. It may then be written in the form

$$
\hat{U}_{2 \times 2} = a|s\rangle\langle s| + b|s\rangle\langle t| + c|t\rangle\langle s| + d|t\rangle\langle t| + \sum_{i \neq s,t} |i\rangle\langle i|.
\tag{4.32}
$$

The two labels, s and t, each have a binary representation; for example, $s = 010$ and $t = 100$. Being the smaller of the two numbers, s features in its binary representation at least one '0'; let's say, its kth digit is '0'. Then there exists another number, t' $(t' > s)$, whose binary representation is identical to that of s except for the kth digit, which is '1' rather than '0'. In our example there are two possibilities, corresponding to $k = 1$ or $k = 3$. If we opt for the latter, we have $t' = 011$. With the definition

$$
\hat{U}_{t \leftrightarrow t'} := |t\rangle\langle t'| + |t'\rangle\langle t| + \sum_{j \neq t,t'} |j\rangle\langle j|
\tag{4.33}
$$

we can now write the two-level transformation as

$$
\hat{U}_{2 \times 2} = \hat{U}_{t \leftrightarrow t'} \hat{U}'_{2 \times 2} \hat{U}_{t \leftrightarrow t'}.
\tag{4.34}
$$

Here the unitary transformation $\hat{U}_{t \leftrightarrow t'}$ is a *level swap*, mapping $|t\rangle$ to $|t'\rangle$ and vice versa. The unitary operator in the middle, $\hat{U}'_{2 \times 2}$, has the same form as the original two-level transformation, Eq. (4.32), except that the levels involved are now $|s\rangle$ and $|t'\rangle$ (rather than $|t\rangle$). Since $|s\rangle$ and $|t'\rangle$ differ in the state of the kth qubit only, this is now a transformation which acts on the kth qubit only. The transformation is conditioned on a particular state of the other qubits. In our example, the states involved in the unitary transformation are $|s\rangle = |010\rangle$ and $|t'\rangle = |011\rangle$. That is to say, the transformation alters the state of the third qubit (the 'target' qubit), *conditioned* on the first two qubits (the 'control' qubits) being in the states 0 and 1, respectively. In a quantum circuit such an operation is represented graphically by the gate

$$
\begin{array}{c}
|c_1\rangle \longquad|c_1\rangle \\
|c_2\rangle \longquad|c_2\rangle \\
|t\rangle \boxed{U'} \hat{U}'^{(1-c_1)c_2}|t\rangle
\end{array}
\qquad , \qquad c_1, c_2, t \in \{0, 1\},
\qquad (4.35)
$$

where U' is the single-qubit transformation with matrix representation

$$
\mathbf{U}' = \begin{pmatrix} a & b \\ c & d \end{pmatrix}.
\qquad (4.36)
$$

The entries in this matrix are the same as in the original two-level transformation, Eq. (4.31). In the diagram the open circle indicates that a control qubit has to be '0', and the full dot indicates that a control qubit has to be '1', in order for the target qubit state to undergo the unitary transformation. Like the Toffoli gate, Eq. (4.25), this gate is multiply controlled. Since the operation on the target is an arbitrary unitary transformation, U', it is called a *multiply-controlled unitary transformation*.

The third step consists in showing that every level swap, $\hat{U}_{t \leftrightarrow t'}$, can be effected by a sequence of multiply-controlled NOT operations. The labels of the two levels involved, t and t', each have a binary representation: in our example, $t = 100$ and $t' = 011$. We can find a sequence of binary numbers which interpolates between t and t', in the sense that each member of the sequence differs from its predecessor in one bit only. This sequence is not unique. For the given example, one possibility might be the sequence

$$
t = 100 \rightarrow 101 \rightarrow 111 \rightarrow 011 = t'.
\qquad (4.37)
$$

Such a sequence is called a 'Gray code'. The swap $t \leftrightarrow t'$ can then be written as a product of nearest-neighbour swaps, where 'nearest neighbour' refers to the position in the Gray code. In our example, the swap $t \leftrightarrow t'$ is effected by the following sequence of nearest-neighbour swaps (from top to bottom):

$$
\begin{array}{cccc}
100 & 101 & 111 & 011 \\
\leftrightarrow & & & \\
 & \leftrightarrow & & \\
 & & \leftrightarrow & \\
 & \leftrightarrow & & \\
\leftrightarrow & & &
\end{array}
\tag{4.38}
$$

Each nearest-neighbour swap amounts to a NOT operation on one of the qubits, conditioned on particular states of the other qubits. For instance, the first nearest-neighbour swap in this sequence, $100 \leftrightarrow 101$, is a NOT operation on the third qubit (the 'target' qubit), conditioned on the first two qubits (the 'control' qubits) being in the states 1 and 0, respectively. In a quantum circuit this particular operation is represented graphically by the multiply-controlled gate

$$
\begin{aligned}
|c_1\rangle &\longrightarrow |c_1\rangle \\
|c_2\rangle &\longrightarrow |c_2\rangle \\
|t\rangle &\longrightarrow |t \oplus (c_1(1-c_2))\rangle
\end{aligned}
\qquad , \qquad c_1, c_2, t \in \{0,1\}.
\tag{4.39}
$$

Once again, the open circle indicates that a control qubit has to be '0', and the full dot indicates that a control qubit has to be '1', in order for the target qubit state to flip. This is an example of a *multiply-controlled* NOT *operation*. It constitutes a special case of a multiply-controlled unitary transformation, with the unitary transformation, U', being a Pauli-X.

In the fourth step, we begin by showing that multiply-controlled NOT operations can be decomposed into single-qubit and CNOT operations. Without loss of generality, we may limit ourselves to multiply-controlled NOT operations where all the control qubits have to be '1' in order for the target to flip; all other multiply-controlled NOT operations can be related to this type by adding Pauli-X gates. For instance, in our example it is

$$
\tag{4.40}
$$

If the multiply-controlled NOT operation features exactly two control qubits, like in our example, it coincides with the Toffoli gate, Eq. (4.25). Of this gate, we already know that it can be constructed from a combination of single-qubit and CNOT gates (Eq. (4.26)). If the number of control qubits is larger than two, the multiply-controlled NOT operation can be constructed from CNOT and Toffoli gates. The pertinent construction involves additional ancilla qubits which at the beginning and end of the process are in the basis state $|0\rangle$ yet may change their states at intermediate stages. Graphically, the circuit

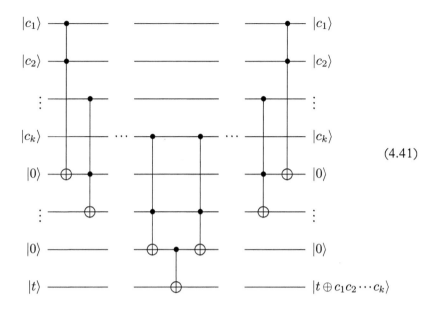

$$(4.41)$$

with k control qubits, $(k-1)$ ancilla qubits, and one target qubit, achieves the desired outcome. The circuit comprises a total of one CNOT and $2(k-1)$ Toffoli gates. While the first $(k-1)$ Toffoli gates serve to impose the additional controls on the target qubit flip, the last $(k-1)$ Toffoli gates serve to reset all the ancilla qubits to their original default state, $|0\rangle$. Each Toffoli gate, in turn, can be constructed from a fixed number of single-qubit and CNOT gates (Eq. (4.26)). The circuit shown in Eq. (4.41) may also be used to implement a multiply-controlled rotation, with the sole difference being that the CNOT gate at its centre must be replaced by a controlled unitary transformation,

$$\begin{array}{c} |c\rangle \longrightarrow\!\!\!\bullet\!\!\!\longrightarrow |c\rangle \\ |t\rangle \longrightarrow\!\boxed{U'}\!\longrightarrow \hat{U}'^c|t\rangle \end{array} \qquad , \qquad c,t \in \{0,1\}. \qquad (4.42)$$

Then it only remains to be shown that for any rotation U' such a controlled transformation, with a single control qubit, can be effected by a sequence of single-qubit and CNOT operations. This is what we shall do in the fifth and final step.

An arbitrary unitary transformation on a single-qubit can always be decomposed into a product of Bloch-sphere rotations about the y and z axes,

$$\hat{U}' = \exp(i\alpha)\hat{R}_z(\beta)\hat{R}_y(\gamma)\hat{R}_z(\delta), \qquad (4.43)$$

for some angles $\alpha, \beta, \gamma, \delta$, and with the Bloch-sphere rotations defined by Eq. (3.87). Indeed, in matrix representation,

$$
\begin{aligned}
\mathbf{R}_z(\beta)\mathbf{R}_y(\gamma)\mathbf{R}_z(\delta) &= \begin{pmatrix} e^{-i\beta/2} & 0 \\ 0 & e^{i\beta/2} \end{pmatrix} \begin{pmatrix} \cos(\gamma/2) & -\sin(\gamma/2) \\ \sin(\gamma/2) & \cos(\gamma/2) \end{pmatrix} \begin{pmatrix} e^{-i\delta/2} & 0 \\ 0 & e^{i\delta/2} \end{pmatrix} \\
&= \begin{pmatrix} e^{-i(\beta+\delta)/2}\cos(\gamma/2) & -e^{-i(\beta-\delta)/2}\sin(\gamma/2) \\ e^{i(\beta-\delta)/2}\sin(\gamma/2) & e^{i(\beta+\delta)/2}\cos(\gamma/2) \end{pmatrix}
\end{aligned}
\tag{4.44}
$$

has the most general form of a unitary matrix with determinant equal to one. This decomposition can be cast in an alternative form that will prove particularly useful for our purposes. We define the unitary operators

$$
\hat{A} := \hat{R}_z(\beta)\hat{R}_y\left(\frac{\gamma}{2}\right), \quad \hat{B} := \hat{R}_y\left(-\frac{\gamma}{2}\right)\hat{R}_z\left(-\frac{\delta+\beta}{2}\right), \quad \hat{C} := \hat{R}_z\left(\frac{\delta-\beta}{2}\right), \tag{4.45}
$$

which satisfy

$$
\hat{A}\hat{B}\hat{C} = \hat{I}. \tag{4.46}
$$

Thanks to the properties of the Pauli operators, Eq. (3.64), and the definition of the Bloch-sphere rotations, Eq. (3.87), it is

$$
\hat{X}\hat{R}_y\left(-\frac{\gamma}{2}\right)\hat{X} = \hat{R}_y\left(\frac{\gamma}{2}\right), \quad \hat{X}\hat{R}_z\left(-\frac{\delta+\beta}{2}\right)\hat{X} = \hat{R}_z\left(\frac{\delta+\beta}{2}\right),
$$

and therefore

$$
\hat{X}\hat{B}\hat{X} = \hat{R}_y\left(\frac{\gamma}{2}\right)\hat{R}_z\left(\frac{\delta+\beta}{2}\right).
$$

This means that the single-qubit transformation may also be written in the form

$$
\hat{U}' = \exp(i\alpha)\hat{A}\hat{X}\hat{B}\hat{X}\hat{C}. \tag{4.47}
$$

The latter is called the '*A-B-C* decomposition' of a single-qubit transformation. With its help, the matrix representation of a controlled unitary transformation, Eq. (4.42), can be expressed as a product of five matrices:

$$
\begin{pmatrix} I & 0 \\ 0 & U' \end{pmatrix} = \begin{pmatrix} A & 0 \\ 0 & e^{i\alpha}A \end{pmatrix}\begin{pmatrix} I & 0 \\ 0 & X \end{pmatrix}\begin{pmatrix} B & 0 \\ 0 & B \end{pmatrix}\begin{pmatrix} I & 0 \\ 0 & X \end{pmatrix}\begin{pmatrix} C & 0 \\ 0 & C \end{pmatrix}. \tag{4.48}
$$

Pictorially, this product translates into

$$\text{(circuit diagram)} \qquad (4.49)$$

where, according to Eq. (4.5), $T^{4\alpha/\pi}$ has the matrix representation

$$\mathbf{T}^{4\alpha/\pi} = \begin{pmatrix} 1 & 0 \\ 0 & e^{i\alpha} \end{pmatrix}. \qquad (4.50)$$

Thus, indeed, the controlled unitary transformation is effected by a sequence of single-qubit and CNOT gates.

In sum, we have verified that single-qubit gates and the CNOT gate do indeed constitute a universal set of quantum gates. Actually, this set may be reduced even further. It turns out that the effect of any single-qubit gate can be emulated, to arbitrary (finite) accuracy, by a sequence of just two types of single-qubit gates: the Hadamard gate, H, and the $\pi/8$ gate, T. Thus, there exists a universal set comprising just three different gates: CNOT, H, and T. The number of Hadamard and $\pi/8$ gates required to approximate an arbitrary single-qubit transformation to some given accuracy, ϵ, scales as $(\log(1/\epsilon))^c$, where $c \approx 2$. This result is known as the *Solovay–Kitaev theorem*.

4.3 Measurement-Based Computation

In principle, it is possible to do without unitary gates and implement a quantum algorithm with measurement gates only. This is termed 'measurement-based' computation. In this brief interlude I give a simple example that illustrates the basic idea.

Instead of a sequence of unitary gates, one may also employ a sequence of measurements to effect a computation. This is equally universal and called *measurement-based quantum computation*. It has no classical analogue, because only in quantum theory is it possible for measurements to have an effect on pure states. For this reason, measurement-based quantum computation exhibits perhaps the starkest contrast to classical computation.

There are various ways of implementing measurement-based computation. Some of them involve the preparation of particular highly entangled initial states, the so-called 'cluster states'; others make use of ancilla qubits and of algorithms which are probabilistic. We will not delve into the details of these various possibilities, instead providing a very simple example that illustrates the basic principle. We wish to build a quantum circuit that employs measurements only, and that succeeds in emulating an arbitrary unitary transformation, $\hat{U}$, on a single qubit. Specifically, it must transform an arbitrary input state, $|\psi\rangle$, into the output state $\hat{U}|\psi\rangle$. We shall show that this can be achieved with the following simple circuit:

$$\text{(4.51)}$$

(repeat until outcome is '1')

There are two incoming qubits: an ancilla qubit in its default state, $|0\rangle$, and the input qubit proper in an arbitrary state, $|\psi\rangle$, which we want to transform by means of measurements only. The protocol then proceeds as follows:

1. Perform a joint measurement on both qubits. The observable to be measured is the composite observable

$$\hat{B} := |1\rangle\langle 0| \otimes \hat{U} + |0\rangle\langle 1| \otimes \hat{U}^{\dagger}, \tag{4.52}$$

where the first factor in a tensor product pertains to the ancilla, whereas the second factor pertains to the input qubit. The operator $\hat{U}$ is the unitary transformation which we wish to emulate. This composite observable is indeed a Hermitian operator, $\hat{B}^{\dagger} = \hat{B}$. Moreover, it has the following two properties:

(a) It satisfies $\hat{B}^2 = \hat{I}$. It is, therefore, a 'binary' observable, in the sense that it can only have the measurement values $+1$ or -1.

(b) Its effect on the composite state of the incoming qubits,

$$\hat{B}(|0\rangle \otimes |\psi\rangle) = |1\rangle \otimes \hat{U}|\psi\rangle,$$

means that its expectation value is zero:

$$(\langle 0| \otimes \langle\psi|)\hat{B}(|0\rangle \otimes |\psi\rangle) = 0.$$

This, in turn, implies that the two allowed measurement values, ± 1, occur with equal probability.

After the measurement of $\hat{B}$, the composite state of ancilla and input qubit is projected onto the eigenspace associated with the measured eigenvalue. Since $\hat{B}$ is binary, the respective projectors are given by

$$\hat{P}_{\pm 1} = \frac{1}{2}(\hat{I} \pm \hat{B}). \tag{4.53}$$

Thus, depending on the measurement outcome, the initial state is updated to

$$(\hat{I} \pm \hat{B})(|0\rangle \otimes |\psi\rangle) = |0\rangle \otimes |\psi\rangle \pm |1\rangle \otimes \hat{U} |\psi\rangle \qquad (4.54)$$

(up to a normalization factor), where the plus and minus signs occur with equal probability. While the state before measurement was a product state, $|0\rangle \otimes |\psi\rangle$, the post-measurement state (with either sign) is entangled.

2. Perform a standard measurement on the ancilla qubit. This will yield either '0' or '1'. Regardless of the sign in Eq. (4.54), and hence regardless of the outcome of the previous measurement of $\hat{B}$, these two possible outcomes will also occur with equal probability. Depending on the outcome of this second measurement, the state of the ancilla qubit is subsequently projected onto one of the associated eigenstates, $|0\rangle$ or $|1\rangle$.

3. If the outcome of the second measurement was '0', then, according to Eq. (4.54), the post-measurement state of the composite system is identical to the original input state, $|0\rangle \otimes |\psi\rangle$. In this case, go back to step 1 and repeat the procedure.

4. If the outcome was '1', on the other hand, then, according to Eq. (4.54), the post-measurement composite state is $|1\rangle \otimes \hat{U} |\psi\rangle$ (up to an irrelevant overall sign). In this case, the physical qubit is in the desired transformed state, $\hat{U} |\psi\rangle$, and we are done.

The particular algorithm described here is probabilistic, in the sense that the desired output is not obtained with certainty after a definite number of steps. Rather, in each iteration it is obtained only with probability one half. In case of failure the procedure is repeated, until eventually it leads to the desired outcome. The probability of success after exactly n iterations is given by

$$\text{prob}(n) = \frac{1}{2^n}.$$

This yields the expected number of iterations,

$$\langle n \rangle = \sum_{n=1}^{\infty} \text{prob}(n)\, n = 2. \qquad (4.55)$$

We discussed our simple protocol for measurement-based computation in the special case where the system to be transformed is a single qubit. Yet it works just as well in Hilbert spaces of arbitrary dimension. In particular, the state $|\psi\rangle$ might pertain to two qubits rather than a single qubit. The above protocol can then be used to implement the unitary transformation associated with a CNOT gate. In this manner we are able to emulate, via measurements only, all members of the universal set of quantum gates. This means, in particular, that measurement-based computation is equally universal. The fact that states may be manipulated, and computations performed, using only measurements is reminiscent of the possibility of 'lossless steering' (Fig. 1.13), that is, our ability to steer an initial state to an arbitrary final state via a sequence of measurements only. Both measurement-based computation and lossless steering exploit the genuine quantum phenomenon that a measurement generally alters a pure state.

4.4 Deutsch–Jozsa Algorithm

A quantum computer can perform some calculations efficiently that a classical computer cannot. I discuss in detail one specific example: the Deutsch–Jozsa algorithm.

Our first, elementary example of a quantum circuit that solves an actual (albeit not particularly useful) computational problem is known as *Deutsch's algorithm*. This algorithm performs a very simple task: it tests whether a Boolean function with a single input bit,

$$f : \{0,1\} \to \{0,1\}, \tag{4.56}$$

is constant, $f(0) = f(1)$, or not.

In the pertinent quantum circuit the Boolean function is implemented by means of a new type of two-qubit gate, called a 'quantum oracle', which acts on the basis states as follows:

$$\begin{array}{c} |x\rangle \quad\boxed{U_f}\quad |x\rangle \\ |y\rangle \qquad\qquad |y \oplus f(x)\rangle \end{array} \quad , \quad x, y \in \{0,1\}. \tag{4.57}$$

Its first input qubit, x, is termed the 'data qubit', while its second input qubit, y, is termed the 'target qubit'. In the standard basis this gate has the matrix representation

$$\mathbf{U}_f = \begin{pmatrix} 1 - f(0) & f(0) & 0 & 0 \\ f(0) & 1 - f(0) & 0 & 0 \\ 0 & 0 & 1 - f(1) & f(1) \\ 0 & 0 & f(1) & 1 - f(1) \end{pmatrix}. \tag{4.58}$$

The matrix is Hermitian and satisfies $\mathbf{U}_f^2 = \mathbf{I}$. Hence it is also unitary, as it should be. In the special case where the data qubit is in one of the basis states, $x \in \{0,1\}$, and the target qubit is in the superposition state $|-\rangle$ defined in Eq. (3.60), the oracle has the effect

$$\hat{U}_f |x,-\rangle = \frac{|x, f(x)\rangle - |x, 1 \oplus f(x)\rangle}{\sqrt{2}}, \tag{4.59}$$

which may be succinctly summarized as

$$\hat{U}_f |x,-\rangle = (-1)^{f(x)} |x,-\rangle. \tag{4.60}$$

In other words, $|x, -\rangle$ is an eigenstate of the oracle, with the associated eigenvalue being either $+1$ (if $f(x) = 0$) or -1 (if $f(x) = 1$).

Deutsch's algorithm does the test with the help of the following quantum circuit:

$$(4.61)$$

In order to understand its inner workings, we track the state of the two qubits at every stage of the process. Initially, the two qubits are in the product state

$$|\psi_0\rangle = |01\rangle. \tag{4.62}$$

The Hadamard gates transform each basis state, individually, into one of the superposition states defined in Eq. (3.60). Consequently, it is

$$|\psi_1\rangle = |+-\rangle = \frac{|0\rangle + |1\rangle}{\sqrt{2}} \otimes |-\rangle. \tag{4.63}$$

Next, thanks to Eq. (4.60), the subsequent action of the quantum oracle will lead to

$$|\psi_2\rangle = \pm \begin{cases} \frac{|0\rangle + |1\rangle}{\sqrt{2}} \otimes |-\rangle : f(0) = f(1) \\ \frac{|0\rangle - |1\rangle}{\sqrt{2}} \otimes |-\rangle : f(0) \neq f(1) \end{cases}. \tag{4.64}$$

While the global sign in front of the curly bracket is not relevant—it has no measurable consequences—the *relative* sign in the state of the first qubit is. Finally, the Hadamard gate for the first qubit rotates its state back to one of the basis states, yielding

$$|\psi_3\rangle = \pm \begin{cases} |0, -\rangle : f(0) = f(1) \\ |1, -\rangle : f(0) \neq f(1) \end{cases}; \tag{4.65}$$

the latter may also be written succinctly as

$$|\psi_3\rangle = \pm |f(0) \oplus f(1)\rangle \otimes |-\rangle. \tag{4.66}$$

The last step in the circuit is a measurement on the first qubit. If the function is constant, this measurement will yield '0'. If the function is not constant, the outcome will be '1'.

Deutsch's algorithm may be extended to an arbitrary number of data qubits; this generalization is known as the *Deutsch–Jozsa algorithm*. In this more general setting, we consider a Boolean function defined not on a single input bit, as in Eq. (4.56), but on an input bit string of arbitrary length, n, $f : \{0,1\}^n \to \{0,1\}$. We know in advance that the Boolean function is either constant (yielding 0 on all inputs or 1 on all inputs) or 'balanced', meaning that it returns 0 for exactly half of its arguments and 1 for the other half. The Deutsch–Jozsa algorithm is designed to find out whether the function is constant or balanced.

Like in Deutsch's algorithm, there exists a unitary gate associated with the function f—another oracle—acting on basis states according to

$$
\begin{array}{c}
|x\rangle \;\overset{/n}{\rule{0.7cm}{0.4pt}}\; \boxed{U_f} \;\overset{/n}{\rule{0.7cm}{0.4pt}}\; |x\rangle \\
|y\rangle \;\rule{1.5cm}{0.4pt}\; |y \oplus f(x)\rangle
\end{array}
\tag{4.67}
$$

The sole difference is that we are now dealing with multiple data qubits, $x \in \{0,1\}^n$, rather than just a single data qubit; this is indicated in the circuit picture by the slash ($/^n$) on the wire. The target is still a single qubit, $y \in \{0,1\}$. Even in this more general case with multiple data qubits, the eigenvalue equation, Eq. (4.60), continues to hold. The circuit used for the Deutsch–Jozsa algorithm is then a straightforward generalization of the corresponding circuit for Deutsch's algorithm:

$$
\tag{4.68}
$$

Here the tensor product of Hadamard operators, $H^{\otimes n}$, means that every single one of the n data qubits is subjected to a Hadamard gate. Similar to Eq. (4.63), the state after the initial Hadamard gates reads

$$
|\psi_1\rangle = \left[\frac{|0\rangle + |1\rangle}{\sqrt{2}}\right]^{\otimes n} \otimes |-\rangle = \frac{1}{2^{n/2}} \sum_{x \in \{0,1\}^n} |x\rangle \otimes |-\rangle .
\tag{4.69}
$$

By virtue of Eq. (4.60), the oracle subsequently produces the state

$$
|\psi_2\rangle = \frac{1}{2^{n/2}} \sum_{x \in \{0,1\}^n} (-1)^{f(x)} |x\rangle \otimes |-\rangle .
\tag{4.70}
$$

Finally, all data qubits are sent once again through Hadamard gates. Generalizing Eq. (4.7), the effect of these Hadamard gates on an n-qubit basis state, $|x\rangle \equiv |x_1 \ldots x_n\rangle$, can be written in the form

$$\hat{H}^{\otimes n} |x\rangle = \frac{1}{2^{n/2}} \sum_{z_i \in \{0,1\}} (-1)^{x_1 z_1 + \ldots + x_n z_n} |z_1 \ldots z_n\rangle = \frac{1}{2^{n/2}} \sum_{z \in \{0,1\}^n} (-1)^{x \cdot z} |z\rangle, \quad (4.71)$$

where $|z\rangle \equiv |z_1 \ldots z_n\rangle$ and $x \cdot z$ denotes the bitwise scalar product of the two bit strings, x and z. Thus, the final quantum state reads

$$|\psi_3\rangle = \frac{1}{2^n} \sum_{z \in \{0,1\}^n} \sum_{x \in \{0,1\}^n} (-1)^{x \cdot z + f(x)} |z\rangle \otimes |-\rangle. \quad (4.72)$$

Subsequent measurements on the data qubits will yield n times '0' with probability

$$\text{prob}(0 \ldots 00 | \psi_3) = \frac{1}{2^{2n}} \left| \sum_{x \in \{0,1\}^n} (-1)^{f(x)} \right|^2. \quad (4.73)$$

If the function is constant, all 2^n summands are equal: they are either all $+1$ or all -1. In both cases, the resultant probability is equal to one; the measurements are certain to yield only '0's. By contrast, if the function is balanced, then half the summands are $+1$ and the other half are -1, exactly cancelling each other. This leads to a probability equal to zero. In other words, the measurements must yield at least one '1'. Thus, by checking whether the measurement results are all '0' or not, we can indeed decide whether the function is constant or balanced.

At first sight, Deutsch's algorithm and the more general Deutsch–Jozsa algorithm might not seem so spectacular. The computational problem which they solve is admittedly an artificial one, which does not have any known practical application. Moreover, a lot of complexity seems to be hidden in the quantum oracle, U_f, and one may wonder how such an oracle would be manufactured in a laboratory. Still, the algorithms offer a first glimpse of why quantum computers might be more powerful than classical computers. The quantum oracle, U_f, needs to be called only once, regardless of the number of data qubits. The number of the other quantum gates needed to solve the problem at hand scales linearly with the number of data qubits. Hence the Deutsch–Jozsa algorithm is efficient. By contrast, obtaining a definitive answer on a classical computer is a hard problem. In the worst case, it requires evaluating the Boolean function for more than half of its possible inputs; that is, $2^n/2 + 1$ times. In other words, the number of function calls on a classical computer grows exponentially with the length of the data bit string. In this sense, the quantum algorithm achieves an exponential speed-up with respect to its classical counterpart. Nowadays there exist many other, more advanced quantum algorithms—such as Shor's factoring algorithm—which achieve a similar speed-up and at the same time solve problems that are actually relevant to the real world. Alas, the hardware implementation of these more interesting algorithms proves extremely challenging and, at the time of writing this book, has not yet reached the point where their theoretical superiority translates into the speed advantage of an actual machine. (This will change, of course, as the development of quantum hardware progresses.) Here we content ourselves with sketching just the basic idea behind the promised 'quantum speed-up' and refrain from delving deeper into those more complex quantum algorithms. Instead, we shall explore more broadly other forms of quantum information processing which have already been, or are very close to being, realized technologically. These include quantum-classical hybrid algorithms for the study of molecules and classical optimization problems (Section 4.5), quantum simulators (Section 4.6), and entanglement-assisted schemes for high-precision measurement (Section 4.7), as well as protocols for quantum communication (Sections 5.3 and 5.4).

4.5   Ground States and Classical Optimization

Many computational tasks related to the study of matter or in classical optimization are framed as minimization problems: for instance, seeking the quantum
state that minimizes the energy, or seeking the classical configuration that
minimizes a particular cost function. I introduce the notions of 'ground state'
and 'Hamiltonian' and show that such minimization problems can often be
mapped to the problem of finding the ground state of a multi-qubit Hamiltonian.
Then I outline a quantum-classical hybrid algorithm—the variational quantum
eigensolver—which solves the latter problem efficiently.

Many important properties of matter are governed by the dynamics of the electrons (rather than of the nuclei) inside the atoms, molecules, or crystals of which it is made. This applies, in particular, to its chemical, electrical, and optical properties. Without external stimuli, the electrons always tend to settle in the state of lowest energy, their *ground state*. This is the eigenstate of the observable of energy with the smallest possible eigenvalue. The observable of energy is commonly called the *Hamiltonian*, and the associated Hermitian operator is denoted by $\hat{H}$. (It should not to be confused with the previously used Hadamard operator.) The Hamiltonian pertains to the totality of all the electrons involved, taking into account both their interactions with the positively charged nuclei and the interactions among themselves. Consequently, it is defined in the composite Hilbert space of all the electrons. The many-electron ground state, $|\psi_0\rangle$, with ground state energy ϵ_0, is the state which minimizes the expectation value of this Hamiltonian:

$$\hat{H}|\psi_0\rangle = \epsilon_0|\psi_0\rangle \quad \Leftrightarrow \quad \langle\psi_0|\hat{H}|\psi_0\rangle \leq \langle\psi|\hat{H}|\psi\rangle \ \forall\psi. \tag{4.74}$$

Any investigation of matter must begin with determining this ground state for its basic building block; for example, a particular molecule. On a classical computer this is a hard problem, because the necessary resources (time, memory) scale polynomially with the dimension of the Hilbert space, which, in turn, scales exponentially with the number of electrons inside the molecule.

Many problems in *classical optimization* also involve a minimization, akin to the minimization of energy. As an example we consider the so-called MaxCut problem (Fig. 4.1). An undirected graph with n nodes is cut into two subsets. In order to describe this cut mathematically, we introduce a two-valued variable, $z_i \in \{+1, -1\}$, for each node, i. We then distinguish the two subsets by setting $z = +1$ in one subset and $z = -1$ in the other. In this way the vector of all n variables, $\vec{z} := (z_1, \ldots, z_n)$, uniquely specifies the cut. Whether two nodes, i and k, belong to the same subset is indicated by the function

$$\frac{1 - z_i z_k}{2} = \begin{cases} 0 : & i, k \text{ in same subset} \\ 1 : & i, k \text{ in different subsets} \end{cases}.$$

Associated with each pair of nodes is a weight, w_{ik}. The ordering of the indices does not matter: $w_{ik} = w_{ki}$. If the two nodes are connected by an edge, this weight is strictly positive, $w_{ik} > 0$; if not, it is $w_{ik} = 0$. The edges which were cut are those whose two ends belong to different subsets. They carry a total weight

$$\sum_{ik} \frac{1 - z_i z_k}{2} w_{ik} = \sum_{i<k} w_{ik} - \sum_{i<k} w_{ik} z_i z_k.$$

One now seeks the cut which maximizes this total weight of the broken edges, or, equivalently, which *minimizes* the function

$$C(\vec{z}) := \sum_{i<k} w_{ik} z_i z_k, \tag{4.75}$$

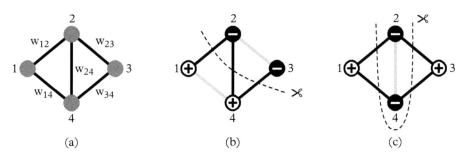

(a)   (b)   (c)

Fig. 4.1 *Classical MaxCut problem. (a) One considers an undirected graph with n nodes, labelled by $i = 1, \ldots, n$ (here: $n = 4$). Not all nodes need to be connected by edges. If there exists an edge connecting nodes i and k, this edge carries a positive weight, $w_{ik} > 0$. (b) A cut divides the graph into two subsets. The nodes in one subset are designated by open circles with a '+', the nodes in the other by black dots with a '−'. The 'value' of the cut is given by the total weight of the edges that were cut (black lines); here, it equals the sum $w_{12} + w_{24} + w_{34}$. (c) We seek the cut which maximizes this value. If we assume that all edges carry equal weight, then in the given example the optimal cut puts nodes 1 and 3 in one subset and nodes 2 and 4 in the other.*

called the *cost function*. Once again, on a classical computer this is a hard problem, in the sense that all known algorithms require resources which scale exponentially with the number of nodes.

The latter problem of minimizing a classical cost function can, in fact, be mapped to the former problem of finding the ground state of a Hamiltonian. In our MaxCut example we convert the cost function into a 'Hamilton operator' of an assembly of n qubits. We do so by replacing the classical variable, z_i, pertaining to the ith node, with the Pauli operator, $\hat{Z}_i$, pertaining to the ith qubit. This yields the n-qubit Hamiltonian

$$\hat{H}_{\mathrm{MC}} := \sum_{i<k} w_{ik} \hat{Z}_i \hat{Z}_k. \tag{4.76}$$

(A Hamiltonian of this form is often referred to as an 'Ising Hamiltonian'.) Furthermore, we establish a one-to-one correspondence between cuts and basis states in the n-qubit Hilbert space. Specifically, we associate with the basis state $|x_1 \ldots x_n\rangle$, $x_i \in \{0,1\}$, the cut $\vec{z}$ where $z_i = (-1)^{x_i}$. Thus, a single qubit in the basis state $|0\rangle$ corresponds to a node in the subgroup where $z = +1$, whereas a qubit in the basis state $|1\rangle$ corresponds to a node in the subgroup where $z = -1$. Each n-qubit basis state is an eigenstate of the Hamiltonian, with the associated eigenvalue equal to the cost of the corresponding cut,

$$\hat{H}_{\mathrm{MC}} |x_1 \ldots x_n\rangle = C(\vec{z}) |x_1 \ldots x_n\rangle, \quad z_i = (-1)^{x_i} \ \forall i. \tag{4.77}$$

Thus, indeed, finding the cut with minimal cost is equivalent to finding the ground state of the Hamiltonian.

A given Hamiltonian does not always relate explicitly to an assembly of qubits like the Ising Hamiltonian did. Rather, the Hamiltonian might pertain to, say, a collection of rotors, harmonic oscillators, or particles of various kinds in a given potential. However, such Hamiltonians can often be mapped to a qubit Hamiltonian. This applies, in particular, to the previously discussed Hamiltonian of the electrons inside a molecule. Electrons are quantum particles and, as such, fundamentally indistinguishable. It does not make sense to say that 'electron A is in the state $|1\rangle$ and electron B is in the state $|2\rangle$' because there is no way to distinguish A and B in the first place. Rather, one may only say, 'There is one electron in the state $|1\rangle$ and one electron in the state $|2\rangle$.' Accordingly, a *many*-electron basis state is denoted by $|n_1 n_2 \ldots\rangle$, where n_i equals the number of electrons occupying the *single*-electron state $|i\rangle$. The $\{|i\rangle\}$, in turn, constitute some orthonormal basis of single-electron states. A further fundamental property of electrons is that they obey the *Pauli principle*: each single-electron state may be occupied at most once, so n_i can only be zero or one. These basic properties of electrons suffice to establish the correspondence with qubits. One associates one qubit with each single-electron state. If the single-electron state is empty, the associated qubit is in the basis state $|0\rangle$; if it is occupied (by exactly one electron), the qubit is in the basis state $|1\rangle$. In the case of a molecule there is a finite number of single-electron states, or 'orbitals'. Accordingly, the electronic states of the molecule can be mapped one to one to states of a finite number of qubits. In a second step, one must map the electronic Hamiltonian to a corresponding qubit Hamiltonian. For this task there exist various mathematical frameworks such as the 'Jordan–Wigner' and 'Bravyi–Kitaev' transformations. However, their discussion is beyond the scope of this introduction.

From now on we shall assume that we have succeeded in mapping the Hamiltonian or cost function at hand to a Hamiltonian of n qubits. An n-qubit Hamiltonian can always be written in a standard form, which we specify below. We begin with the simplest case, $n = 1$. In the space of single-qubit observables the three Pauli operators, $\hat{\sigma}_i$ ($i = 1, 2, 3$), and the unit operator, $\hat{\sigma}_0 \equiv \hat{I}$, constitute a basis; every Hermitian operator in the Hilbert space of a single qubit—like the statistical operator, Eq. (3.65)—can be written as a linear combination of these four operators. Hence, so can the Hamiltonian. This may be generalized to multiple qubits. In the space of n-qubit observables there exists a basis, called the 'Pauli basis', which is comprised of the 4^n distinct n-fold tensor products of these four operators, $\hat{\sigma}_{\mu_1} \otimes \ldots \otimes \hat{\sigma}_{\mu_n}$ ($\mu_k = 0, 1, 2, 3$). (You will verify this in Exercise (4.7).) Consequently, an n-qubit Hamiltonian can always be written as a linear combination of such tensor products,

$$\hat{H} = \sum_{\mu_1, \ldots, \mu_n = 0}^{3} h_{\mu_1 \ldots \mu_n} \hat{\sigma}_{\mu_1} \otimes \ldots \otimes \hat{\sigma}_{\mu_n}. \tag{4.78}$$

This is the standard representation of a multi-qubit Hamiltonian. As a simple example, we consider a two-qubit Hamiltonian with the following matrix representation:

$$\mathbf{H} = \begin{pmatrix} a & & & e \\ & b & e & \\ & e & c & \\ e & & & d \end{pmatrix}.$$

(A Hamiltonian of this form is used in a very simple model of the hydrogen molecule.) Indeed, this Hamiltonian can be written as a linear combination of tensor products,

$$\hat{H} = \frac{a+b+c+d}{4} \hat{I} \otimes \hat{I} + \frac{a-b+c-d}{4} \hat{I} \otimes \hat{Z} + \frac{a+b-c-d}{4} \hat{Z} \otimes \hat{I}$$
$$+ \frac{a-b-c+d}{4} \hat{Z} \otimes \hat{Z} + e \hat{X} \otimes \hat{X}. \tag{4.79}$$

The standard representation of an n-qubit Hamiltonian suggests a simple experimental setup for measuring the energy expectation value. Taking the expectation value on both sides of Eq. (4.78), in an arbitrary n-qubit state $|\psi\rangle$, we find

$$\langle \psi | \hat{H} | \psi \rangle = \sum_{\mu_1, \ldots, \mu_n = 0}^{3} h_{\mu_1 \ldots \mu_n} \langle \psi | \hat{\sigma}_{\mu_1} \otimes \ldots \otimes \hat{\sigma}_{\mu_n} | \psi \rangle. \tag{4.80}$$

Thus, measuring the expectation value of $\hat{H}$ is tantamount to measuring the expectation values of the various tensor products of Pauli operators. The latter is achieved by a simple combination of single-qubit measurements. For instance, in the earlier two-qubit example, Eq. (4.79), we can employ the two measurement setups

$$|\psi\rangle \left\{ \begin{array}{c} \end{array} \right. , \quad |\psi\rangle \left\{ \begin{array}{c} \end{array} \right. . \tag{4.81}$$

The left setup allows us to measure $\hat{I}\otimes\hat{Z}$, $\hat{Z}\otimes\hat{I}$, and $\hat{Z}\otimes\hat{Z}$, whereas the setup on the right, thanks to the Hadamard gates, allows us to measure $\hat{X}\otimes\hat{X}$. We may determine the expectation values of these tensor products in any given two-qubit state, $|\psi\rangle$, by preparing multiple copies of that state and sending them alternately through the two measurement setups, until we have reached a number of measurements large enough to allow us to identify measured averages with expectation values. Forming the appropriate linear combination, Eq. (4.80), then yields the energy expectation value.

Quantum states are specified by some set of parameters. For instance, in the Bloch-sphere representation the pure state of a single qubit is specified by two angles, θ and φ (Eq. (3.59)). More generally, according to Eq. (2.72), an arbitrary pure state in a d-dimensional Hilbert space is specified by $2(d-1)$ parameters. These parameters may be chosen such that they correspond to a specific experimental procedure with which the states are prepared. For example, the choice of the two angles in the case of the single qubit corresponds to an experimental preparation by means of the following circuit:

$$|0\rangle - \boxed{R_y(\theta)} - \boxed{R_z(\varphi)} - |\psi(\theta,\varphi)\rangle, \tag{4.82}$$

where the gate R_i effects the unitary transformation given by Eq. (3.87), to wit, a rotation on the Bloch sphere about the i axis by the given angle (Fig. 3.10). This circuit has a fixed geometry (two gates in a row) and fixed gate types (R_y and R_z, respectively) which are independent of the values of the parameters; it is only the sizes of the respective transformations that vary with the parameters. This idea may be extended to states of n qubits. There we will deal with some larger set of parameters, $\{\theta_i\}$ (in short: $\vec{\theta}$), which can again be chosen such that they correspond to a specific circuit for state preparation. Again, there will be a fixed geometry and fixed gate types, with only the sizes of the individual transformations (or some subset thereof) being controlled by the parameters. Schematically, we will have a circuit of the form

Here θ_i indicates the parameter which controls a given gate (whose type has been left unspecified). While in this conceptual drawing we attributed a parameter dependence to every single gate, a circuit might also feature gates which do not depend on any parameter at all. By combining such a preparation circuit with the measurement of the energy expectation value via Pauli measurements, analogous to Eq. (4.81), we can build a circuit which effectively determines the energy expectation value as a function of $\vec{\theta}$:

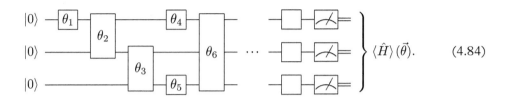

$$\langle \hat{H} \rangle (\vec{\theta}). \qquad (4.84)$$

Here the empty single-qubit gates in the measurement part are placeholders for possible (parameter-independent) rotations of the basis, like the Hadamard gates in Eq. (4.81). In order to generate enough data points for an estimation of the energy expectation value, this circuit must be run repeatedly and with varying settings of the measurement part.

Now we return to our objective of finding the ground state of a given Hamiltonian. We assume that the Hamiltonian has been transformed into an n-qubit Hamiltonian, and that the latter is expressed in the standard form. For the search of its ground state there exists a systematic procedure called the *variational quantum eigensolver* (VQE). This procedure does not calculate the exact ground state as a conventional algorithm would. Rather, it explores the n-qubit Hilbert space in a systematic fashion, converging with high probability to a good enough approximation to the ground state. This is faster than a conventional algorithm and at the same time sufficient for many practical applications, including the study of molecules and the classical optimization problems which we discussed earlier. The VQE is not a pure quantum algorithm but a quantum-classical hybrid; it combines a classical optimization algorithm with a quantum subroutine. In broad terms, here is how its works:

1. Build a quantum circuit like in Eq. (4.83) with adjustable gate or other experimental control parameters, $\{\theta_i\}$, that produces parametrized n-qubit states, $|\psi(\vec{\theta})\rangle$. These are called *ansatz states*.
2. Set up Pauli measurements like in Eq. (4.81) which permit the determination of the energy expectation value.
3. Fix the starting values of the parameters.
4. With the help of the combined setup, like in Eq. (4.84), measure the energy expectation value, $\langle \hat{H} \rangle (\vec{\theta})$. This is the quantum subroutine where the quantum hardware is needed.
5. Use a classical optimizer to determine new parameter values which will likely decrease $\langle \hat{H} \rangle (\vec{\theta})$. Then go back to the previous step.

6. Iterate until the energy expectation value converges. The parameter values at convergence, $\vec{\theta}_*$, then define the state which comes closest to the ground state, $|\psi(\vec{\theta}_*)\rangle$.

The algorithm is probabilistic. It involves measurements with random outcomes, and therefore the path taken in parameter space and the time needed for convergence may vary.

If the goal were to always find the exact ground state, rather than merely an approximation, the search space would have to encompass all conceivable pure states. Accordingly, the number of parameters in the ansatz state would have to coincide with the number of parameters needed to specify an arbitrary pure state. In the case of n qubits this would mean $2(2^n - 1)$ parameters, to wit, a number of parameters which grows exponentially with the number of qubits. As a result, the number of gates needed to prepare the ansatz states would have to grow exponentially, too—a resource requirement which would soon become prohibitive. To make matters worse, even with all these gates an exact result would be far from assured because of the inevitable noise in the circuit. Fortunately, in many practical applications a close approximation to the ground state already suffices. Then the ansatz states may be taken from a restricted family of states, specified by a smaller number of parameters. Still, these restricted states should be spread as uniformly as possible throughout Hilbert space in order to avoid extensive 'blind areas' that remain unexplored. This is achieved by using ansatz states which are highly entangled. Since such states are superpositions of many different basis states, they explore many different directions in Hilbert space simultaneously. Specifically, VQE uses ansatz states that are generated by circuits of the generic form

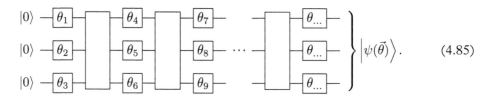

$$\text{(4.85)}$$

The n qubits pass through single- and multi-qubit gates in an alternating fashion. Only the single-qubit gates can be tuned and thus depend on the control parameters. They are used to navigate within the family of ansatz states. By contrast, the multi-qubit gates are hard-wired; they do not depend on any parameter. The generic blocks drawn here indicate unspecified gates involving two, three, or more qubits, or groups of several such gates. In practice, they are often collections of CNOT, controlled-Z, or other controlled unitary transformations of the type of Eq. (4.42). Their purpose is to produce the desired entanglement. The alternation of tunable single-qubit and hard-wired entangling gates is repeated several times, until the ansatz state thus prepared features both a sufficient number of tunable parameters and a sufficient degree of entanglement. Typically, a small, single-digit number of iterations will suffice. Given a fixed number of entangling steps, the total number of parameters in the ansatz state scales only linearly with the number of qubits, in contrast to the exponential growth which would be required for an exact solution. This linear scaling renders the algorithm efficient.

Where the ground state represents the solution to a classical optimization problem, we know that the ground state must be a basis state. In this case the determination of the approximate ground state, $|\psi(\vec{\theta}_*)\rangle$, is followed by one further step. In general, being merely an approximation, $|\psi(\vec{\theta}_*)\rangle$ will still be a superposition of several different basis states. However, we expect that in this superposition the dominant contribution will come from the basis state which represents the true solution. Consequently, in a final step we project the approximate solution onto the basis state with the largest amplitude.

4.6 Simulation

A quantum computer can simulate the dynamics of other quantum many-body systems. I show that such a simulation is not always efficient, but that it is efficient in the important case where the interactions between the constituents are local.

Unless a quantum system has settled in its ground state or some other peculiar stable state, its state will change over time. We have so far steered clear of discussing this time evolution, and will not go into any details now. Suffice it to say that every time evolution exhibits two important properties. First, as long as the quantum system is isolated, it will preserve orthogonality: different initial states which are orthogonal to each other evolve to final states which are orthogonal to each other. Secondly, the state change is a smooth function of the time interval, t, with there being no change at all if $t = 0$. In other words, the transformation of the state is smoothly connected to the identity. Hence, by Wigner's theorem, time evolution is tantamount to a unitary transformation. Whenever one studies time-dependent phenomena, it is of great interest to simulate this unitary transformation on a computer. On a classical computer such simulation is hard because, for the same reasons as in Section 4.5, it requires resources that grow exponentially with the number of particles in the system. As for a quantum computer, we already convinced ourselves in Section 4.2 that any unitary transformation can be implemented in a quantum circuit; so the simulation is possible in principle. The question is whether, or under which circumstances, it can be made efficient.

In fact, it is not possible to simulate *every* time evolution efficiently. To understand why this is so, we recount the various steps of our circuit construction in Section 4.2 and consider the respective resources needed:

1. *Break down the unitary transformation into a sequence of two-level transformations* (Eq. (4.29)). We achieved this by proceeding iteratively: first decompose the original unitary matrix, $\mathbf{U}_{d \times d}$, into two-level transformations and a unitary matrix of reduced dimension, $\mathbf{U}_{(d-1) \times (d-1)}$; then, the latter into two-level transformations and $\mathbf{U}_{(d-2) \times (d-2)}$; and so on. The initial dimensional reduction, from d to $(d-1)$, involves up to $(d-1)$ two-level transformations; the next, from $(d-1)$ to $(d-2)$, up to $(d-2)$ two-level transformations; and so on. Thus the unitary transformation

is broken down into at most

$$(d-1)+(d-2)+\ldots = \frac{d(d-1)}{2} \sim O(d^2)$$

two-level transformations.

2. *Decompose each two-level transformation into two level swaps and one multiply-controlled unitary transformation* (Eq. (4.34)).

3. *Implement each level swap with a sequence of multiply-controlled* NOT *operations.* We did so with the help of a Gray code, a code which connects two arbitrary bit strings via a sequence of nearest-neighbour swaps; one example is shown in Eq. (4.38). Each nearest-neighbour swap corresponds to a multiply-controlled NOT operation. The number of operations scales linearly with the length of the bit strings involved, to wit, the number of qubits. The latter, in turn, scales as $(\log d)$.

4. *Construct each multiply-controlled* NOT *operation or unitary transformation, respectively, from single-qubit gates,* CNOTs, *and (in the latter case) a controlled unitary transformation with a single control qubit.* We achieved this by means of the circuit shown in Eq. (4.41). The circuit features at its centre a single CNOT or controlled unitary transformation (with a single control qubit), respectively. In addition, there are multiple Toffoli gates and ancilla qubits. Their number scales linearly with the number of control qubits, k. The latter is bounded by the total number of qubits in the system and hence scales again as $(\log d)$. Each Toffoli gate, in turn, may be realized with a fixed number of single-qubit and CNOT gates, using the circuit in Eq. (4.26).

5. *Construct each controlled unitary transformation (with a single control qubit) from a fixed number of single-qubit and* CNOT *gates* (Eq. (4.49)).

Altogether, the number of single-qubit and CNOT gates required to implement this scheme scales as $d^2(\log d)^2$. Where the unitary transformation at issue acts on a collection of n qubits, the dimension of the Hilbert space is given by $d=2^n$. So in terms of the number of input qubits—the input length—the requisite number of gates grows exponentially:

$$\#\text{gates} \sim 2^{2n}n^2. \tag{4.86}$$

Thus, indeed, the circuit implementation of an arbitrary unitary transformation is in general *not* efficient.

An efficient circuit implementation does exist, however, for a particular (and particularly important) class of unitary transformations. We explain this for the case where, as before, the unitary transformation acts on a collection of n qubits. A unitary transformation can always be written as the exponential of some anti-Hermitian (or 'skew-Hermitian') operator, $\hat{A}$:

$$\hat{U} = \exp(\hat{A}), \quad \hat{A}^\dagger = -\hat{A}. \tag{4.87}$$

Indeed, the anti-Hermiticity of the exponent guarantees that the transformation is unitary:

$$\hat{U}^\dagger = \exp(\hat{A}^\dagger) = \exp(-\hat{A}) = \hat{U}^{-1}.$$

Conversely, for every unitary transformation such an anti-Hermitian exponent exists. This anti-Hermitian operator is often a sum of distinct terms,

$$\hat{A} = \sum_{i=1}^{l} \hat{A}_i, \tag{4.88}$$

each of which pertains to at most k $(k \leq n)$ out of the n qubits. This means that each individual term is a tensor product of n single-qubit operators,

$$\hat{A}_i = \hat{I} \otimes \ldots \otimes \hat{I} \otimes \hat{A}_i^{(i_1)} \otimes \hat{I} \otimes \ldots \otimes \hat{I} \otimes \hat{A}_i^{(i_2)} \otimes \hat{I} \otimes \ldots,$$

where the number of non-trivial factors never exceeds the given limit, k. One also says that each term 'couples' no more than k qubits. A decomposition of this kind is always possible as long as k is chosen large enough; if need be, one may just choose $k = n$. Now we are interested in the scaling properties when we increase the number of qubits, n. For a generic unitary transformation, the number of non-trivial factors in each summand will then also grow; k will scale with n. As we saw earlier, in this case the complexity of the unitary transformation is such that its circuit implementation requires an exponentially growing number of gates. There is a special class of unitary transformations, however, where k stays constant, independently of the number of qubits. These transformations are called *locally generated*. The terminology stems from the fact that when the unitary transformation describes a time evolution—which is precisely what we want to simulate—the exponent, $\hat{A}$, is proportional to the Hamiltonian, that is, the observable of energy. (Since the former is anti-Hermitian and the latter is Hermitian, the factor of proportionality must be imaginary.) The summands in Eq. (4.88) then correspond to the various contributions to the total energy of the system. They take into account the energies of the qubits on their own, as well as their mutual interactions. The limitation on k corresponds to the requirement that each qubit may interact with at most $(k-1)$ neighbours—that is, 'locally'—regardless of how many other qubits there are. We will show that for locally generated unitary transformations an efficient circuit implementation is possible.

The number of ways to pick k out of n qubits is given by the binomial coefficient, which, at fixed k, scales polynomially with n:

$$\binom{n}{k} \sim \text{poly}(n).$$

Consequently, so does the number of local terms in Eq. (4.88),

$$l \sim \text{poly}(n). \tag{4.89}$$

In general, the exponential of the sum is not equal to the product of exponentials,

$$\exp(\hat{A}) \neq \exp(\hat{A}_1) \dots \exp(\hat{A}_l), \tag{4.90}$$

unless all the $\{\hat{A}_i\}$ commute. However, we can break the unitary transformation down into s small steps,

$$\exp(\hat{A}) = \left[\exp\left(\frac{\hat{A}}{s} \right) \right]^s, \tag{4.91}$$

and then use the so-called 'Suzuki–Trotter decomposition',

$$\left[\exp\left(\frac{\hat{A}}{s} \right) \right]^s = \left[\exp\left(\frac{\hat{A}_1}{s} \right) \dots \exp\left(\frac{\hat{A}_l}{s} \right) \right]^s + \frac{1}{2s} \sum_{i<j} [\hat{A}_i, \hat{A}_j] + O\left(\frac{1}{s^2} \right). \tag{4.92}$$

(You will prove this formula in Exercise (4.8).) The first term on the right-hand side is a product of $s \cdot l$ small locally generated transformations, $\exp(\hat{A}_i/s)$, which involve at most k qubits at a time. As s increases, this product of small local transformations becomes a better and better approximation—the 'Suzuki–Trotter approximation'—to the full unitary transformation (Fig. 4.2). If we want this approximation to be correct to within some error bound, $\epsilon > 0$, we must make the number of steps, s, large enough that the leading-order correction in Eq. (4.92) becomes smaller than ϵ. The leading-order correction is a sum over commutators. The number of distinct commutators scales quadratically with l, which, in turn, scales polynomially with the number of qubits (Eq. (4.89)). So the entire first-order correction term, too, scales polynomially with the number of qubits:

$$\frac{1}{2s} \sum_{i<j} [\hat{A}_i, \hat{A}_j] \sim \frac{\text{poly}(n)}{s}.$$

If this is to be of the order ϵ, then the number of steps, s, must scale as

$$s \sim \frac{\text{poly}(n)}{\epsilon}. \tag{4.93}$$

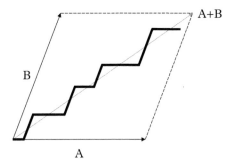

Fig. 4.2 *Schematic illustration of the Suzuki–Trotter approximation. The transformation generated by a sum, $\hat{A} + \hat{B}$, is decomposed into many small increments, each generated by either $\hat{A}$ or $\hat{B}$ alone.*

Hence the total number of local transformations in the Suzuki–Trotter approximation scales as

$$\#\left\{\exp\left(\frac{\hat{A}_i}{s}\right)\right\} = s \cdot l \sim \frac{\text{poly}(n)}{\epsilon}. \tag{4.94}$$

Since a local transformation affects no more than k qubits, each individual local transformation in the Suzuki–Trotter approximation can be implemented with a number of single-qubit and CNOT gates that does not scale with n. Thus, the total number of gates needed for the Suzuki–Trotter approximation also scales as $\text{poly}(n)/\epsilon$. The fact that the number of gates scales only polynomially with the number of qubits means that the circuit implementation is efficient. So indeed, as long as all interactions are local—as is the case in most real systems—an efficient simulation of the dynamics is possible.

Our finding may be generalized in various ways:

1. The quantum system to be simulated need not be a collection of qubits. It may just as well be composed of particles of any kind, or, more generally, constituents of any kind, as long as each of these is described in a finite-dimensional Hilbert space. Under the latter assumption, and provided the constituents interact only locally, the dynamics of any such quantum many-body system can be simulated efficiently on a quantum computer.

2. One can perform several locally generated unitary transformations in sequence. As long as the length of this sequence grows at most polynomially with the number of constituents, the entire sequence, too, may be implemented efficiently. In fact, an efficient quantum circuit is itself such a sequence: each gate is a unitary transformation that is locally generated, because it involves only one or two qubits; and the number of gates—the length of the sequence—grows at most polynomially with the number of input qubits. In this sense, the correspondence between polynomial-length sequences of locally generated transformations and efficient quantum circuits is one to one.

3. Said correspondence is one to one as long as we take into account *all* qubits used in a circuit, including the ancilla qubits. When one disregards the ancilla qubits and focuses on the effective transformation of the input qubits only, the latter might well not be locally generated. (You will encounter a specific example in Exercise (4.9).) Thus, the class of transformations that can be implemented efficiently is actually larger: it includes, but is not limited to, polynomial-length sequences of locally generated transformations.

In all circuit implementations considered so far we have ignored the limited accuracy with which quantum logic gates operate in the laboratory, as well as other sources of noise. It is possible to make a quantum circuit robust against such experimental imperfections by means of so-called 'error-correcting codes'. These codes add redundancy in the form of extra gates and ancilla qubits. While this increases the overall resource requirement, it does so in a manner which maintains the efficiency of the quantum circuit implementation. Unfortunately, a detailed discussion of error correction is beyond the scope of this book. You will encounter a very simple example that conveys the basic idea in Exercise (4.10), and, in case you want to delve deeper, you will find some suggestions for further reading at the end of this chapter.

4.7   High-Precision Measurement

The framework of quantum computation can help us to understand progress in another field of technology: quantum metrology. The latter exploits quantum effects for the purposes of high-precision measurement. I consider one specific measurement and show, using the pictorial language of quantum circuits, how its accuracy may be improved through the use of entangled states.

At first sight it may seem surprising to find a section on high-precision measurement included in the chapter on quantum computation. After all, the development of high-precision measurement techniques falls into the domain of metrology—the science of measurement—which constitutes a fully fledged research area of its own, independent of computation. However, between the two fields there is significant cross-fertilization. First of all, high-precision measurements are critical for realizing, experimentally, the preparation, manipulation, and measurement of quantum states which are required in a working quantum computer. Therefore, progress in the fabrication of quantum computing hardware is inextricably linked to progress in metrology. Conversely, modern metrology makes extensive use of quantum systems and quantum effects. For instance, the sensors used are often finite-dimensional quantum systems like spins, where observables have discrete, clearly resolvable eigenvalues. These sensors are coupled to the entity to be measured; for example, a magnetic field. In this fashion the measurement of the magnetic field, say, is reduced to measurements on the spins—which, thanks to their discreteness, can be done with high precision. In addition, metrology often makes use of the superposition principle of quantum theory; this is the case whenever measurements involve interferometry. Finally, it is possible to enhance the accuracy of measurements even further by exploiting quantum entanglement. How these various quantum features benefit metrology can be conveniently described in the language of quantum computation. In the following, we shall illustrate this with a simple example.

We consider a single-qubit gate, U_φ, which shifts the phase of a basis state by $+\varphi$ or $-\varphi$, respectively:

$$|0\rangle \ -\boxed{U_\varphi}- \ \exp(i\varphi)\,|0\rangle, \qquad |1\rangle \ -\boxed{U_\varphi}- \ \exp(-i\varphi)\,|1\rangle. \qquad (4.95)$$

The associated unitary operator may be written as an exponential,

$$\hat{U}_\varphi = \exp\left(i\varphi\hat{Z}\right), \qquad (4.96)$$

of the Pauli operator $\hat{Z}$. We assume that the phase shift parameter lies somewhere within a given range,

$$\varphi_{\min} \leq \varphi \leq \varphi_{\max}, \qquad (4.97)$$

but is otherwise not known. Our goal is to determine this parameter experimentally.

As *sensors* we use qubits which we can send through the gate. We are allowed to use several sensor qubits in parallel. Moreover, a sensor qubit may pass through the gate multiple times. However, the resources in our laboratory are subject to a number of constraints:

1. The number of sensor qubits which may be used in parallel is limited to some maximum number, s.
2. Each call of the unknown gate costs a certain 'interrogation time', and the total time available for measurement is limited. This means that a sensor qubit can pass through the gate not more than a certain number of times, t.
3. The overall experiment may be repeated at most n times.

One rather straightforward circuit which makes maximal use of the available resources is the following:

$$(4.98)$$

Here the s sensor qubits run in parallel and independently of each other through the unspecified phase shift gate t times. As the effects of the t sequential U_φ gates accumulate, they act like a single gate, $U_{t\varphi}$, which shifts the phases of basis states by $\pm t\varphi$ rather than $\pm\varphi$. The phase shifters are sandwiched between Hadamard gates in order to create a superposition state. Indeed, for each individual qubit the net effect of all gates, including the two Hadamard gates, is to transform the initial basis state, $|0\rangle$, into a superposition state,

$$
\begin{aligned}
\hat{H}\hat{U}_{t\varphi}\hat{H}\,|0\rangle &= \hat{H}\hat{U}_{t\varphi}\frac{|0\rangle + |1\rangle}{\sqrt{2}} \\
&= \hat{H}\frac{\exp(it\varphi)\,|0\rangle + \exp(-it\varphi)\,|1\rangle}{\sqrt{2}} \\
&= \frac{\exp(it\varphi)\,|0\rangle + \exp(it\varphi)\,|1\rangle + \exp(-it\varphi)\,|0\rangle - \exp(-it\varphi)\,|1\rangle}{2} \\
&= \cos(t\varphi)\,|0\rangle + i\sin(t\varphi)\,|1\rangle .
\end{aligned}
\tag{4.99}
$$

Given φ and t, subsequent measurement on the single qubit will yield '0' with probability

$$
p_{\varphi,t} := \mathrm{prob}(0|\varphi,t) = \cos^2(t\varphi).
\tag{4.100}
$$

When s sensor qubits are used in parallel and the entire experiment is repeated n times, a total of sn independent single-qubit measurements are performed. The likelihood that these yield the result '0' k times is given by the binomial distribution,

$$
\mathrm{prob}(k|\varphi,t,s,n) = \binom{sn}{k} p_{\varphi,t}{}^{k}\,(1 - p_{\varphi,t})^{sn-k}.
\tag{4.101}
$$

For sufficiently large n, this binomial distribution approaches a Gaussian,

$$
\mathrm{prob}(f|\varphi,t,s,n) \sim \frac{1}{sn}\frac{1}{\sqrt{2\pi}\sigma_f}\exp\left[-\frac{(f - p_{\varphi,t})^2}{2\sigma_f^2}\right],
\tag{4.102}
$$

where, instead of the absolute frequency, k, we denoted the relative frequency, $f := k/(sn)$. The Gaussian is peaked at $f = p_{\varphi,t}$ and has width

$$
\sigma_f := \sqrt{\frac{p_{\varphi,t}(1 - p_{\varphi,t})}{sn}} = \left|\frac{\sin(2t\varphi)}{2}\right|\frac{1}{\sqrt{sn}}.
\tag{4.103}
$$

As $n \to \infty$, this width tends towards zero, $\sigma_f \propto 1/\sqrt{n} \to 0$; so in this limit it becomes quasi certain that the observed relative frequency will coincide with the probability, $p_{\varphi,t}$. Thus, the observed relative frequency will allow us to infer $p_{\varphi,t}$. The latter, in turn, allows us to infer the phase shift, φ. Ultimately, this method of estimating the phase shift rests on the law of large numbers of classical statistics.

We delve a bit deeper into the details of this inference. In particular, we wish to understand how the accuracy of our estimate scales with the parameters that specify our experiment: t, s, and n. We assume that prior to the experiment nothing is known about the phase shift, except that it must lie in the interval $[\varphi_{min}, \varphi_{max}]$. This initial ignorance is modelled by a uniform prior probability density function,

$$\text{pdf}(\varphi) = \begin{cases} 1/(\varphi_{max} - \varphi_{min}) & : \varphi_{min} \leq \varphi \leq \varphi_{max} \\ 0 & : \text{else} \end{cases}. \tag{4.104}$$

In light of the experimental data, this prior is updated according to Bayes' rule, Eq. (1.39):

$$\text{pdf}(\varphi|f,t,s,n) \propto \text{prob}(f|\varphi,t,s,n)\,\text{pdf}(\varphi) \propto \left| \frac{1}{\sin(2t\varphi)} \right| \exp\left[-\frac{(p_{\varphi,t} - f)^2}{2\sigma_f^2} \right]. \tag{4.105}$$

Provided n is sufficiently large, and therefore σ_f sufficiently small, this posterior distribution is dominated by the exponential. The latter is peaked at those phase shifts, $\{\varphi_i\}$, which give $p_{\varphi_i,t} = f$; to wit, for which it is

$$\cos^2(t\varphi_i) = f. \tag{4.106}$$

We assume that one of these lies inside the a priori allowed range between φ_{min} and φ_{max}. This particular phase shift becomes our *point estimate*, φ_*. (In case there are several candidates within the allowed range, we assume that we know from other considerations which one to pick.) Taylor-expanding the exponent around this point estimate, up to second order,

$$\frac{(p_{\varphi,t} - f)^2}{2\sigma_f^2} = 2sn \frac{(\cos^2(t\varphi) - \cos^2(t\varphi_*))^2}{\sin^2(2t\varphi)} \approx 2snt^2(\varphi - \varphi_*)^2, \tag{4.107}$$

then yields a Gaussian posterior for φ,

$$\text{pdf}(\varphi|f,t,s,n) \propto \exp\left[-\frac{(\varphi - \varphi_*)^2}{2\sigma_\varphi^2} \right], \tag{4.108}$$

of width

$$\sigma_\varphi = \frac{1}{2t\sqrt{sn}}. \tag{4.109}$$

The non-zero width of the posterior shows that the accuracy with which we can estimate the phase shift is limited. This residual uncertainty about the phase shift is known as 'shot noise' or 'projection noise'. It is only due to the limited resources in the experiment, to wit, the finite number of 'shots' available for our sensor qubits. It does not take into account other potential sources of noise, such as a loss of coherence in the course of the experiment. Reducing the noise down to this inevitable shot noise is sometimes referred to as the *standard quantum limit*.

It turns out that we can do even better when we harness entanglement. Rather than using the s sensor qubits independently, we can entangle them prior to sending them through the unspecified phase shift gates. The following circuit does so via a sequence of CNOT gates:

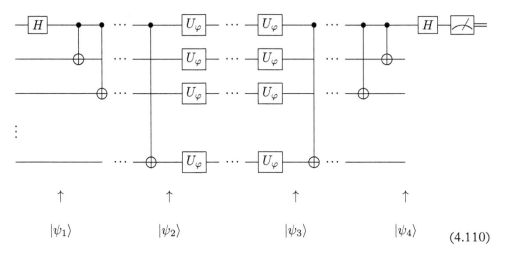

$$ (4.110) $$

Again the s sensor qubits are initially prepared in the basis state $|0\rangle$ (not shown due to lack of space). Yet only the first qubit passes through Hadamard gates and is eventually subjected to a measurement; the others play the role of ancilla qubits. After the initial Hadamard gate, the sensor qubits are in the state

$$ |\psi_1\rangle = \frac{|00\ldots0\rangle + |10\ldots0\rangle}{\sqrt{2}}. \qquad (4.111) $$

Then the CNOT gates successively flip the basis states in the second term, yielding

$$ |\psi_2\rangle = \frac{|00\ldots0\rangle + |11\ldots1\rangle}{\sqrt{2}}. \qquad (4.112) $$

A state of this form is highly entangled. It constitutes a generalization of the Greenberger–Horne–Zeilinger state which you encountered in Exercise (3.20), and is sometimes referred to as a *cat state*. Next, as in Eq. (4.99) but now for all sensor qubits in concert, the phase shift gates alter the phases of the two terms in the superposition, in opposite directions:

$$ |\psi_3\rangle = \frac{\exp(ist\varphi)\,|00\ldots0\rangle + \exp(-ist\varphi)\,|11\ldots1\rangle}{\sqrt{2}}. \qquad (4.113) $$

The subsequent CNOT gates undo the entanglement of the qubits, yielding

$$|\psi_4\rangle = \frac{\exp(ist\varphi)|0\rangle + \exp(-ist\varphi)|1\rangle}{\sqrt{2}} \otimes |0\ldots0\rangle. \tag{4.114}$$

At this point the ancilla qubits have done their part, and we may focus on the first qubit only. In the final step, its state is transformed by the Hadamard gate into

$$|\psi_{\text{final}}\rangle = \cos(st\varphi)|0\rangle + i\sin(st\varphi)|1\rangle. \tag{4.115}$$

This looks just like the final state in Eq. (4.99), only with t replaced by st; it is as if the first sensor qubit had run through the phase shift gate not t times but st times.

We can now repeat our statistical analysis, Eqs (4.100) to (4.109), with two alterations. First, as already noted, t is replaced by st. Secondly, in each run of the experiment we perform a measurement on just the first of the s sensor qubits rather than on all of them. This means that in n runs the total number of independent measurements no longer equals sn but only n. With these modifications, $t \rightarrow st$ and $sn \rightarrow n$, we obtain a Gaussian posterior for φ of width

$$\sigma'_\varphi = \frac{1}{2st\sqrt{n}}. \tag{4.116}$$

Compared to our previous result, Eq. (4.109), this width is smaller—and thus the accuracy is improved—by a factor $\sqrt{s}$. Indeed, the clever use of entanglement has allowed us to beat the standard quantum limit.

In terms of its scaling with the number of sensor qubits, $\sigma'_\varphi \sim 1/s$ rather than $1/\sqrt{s}$, the accuracy of this entanglement-assisted scheme is the best we can hope for. Indeed, there is a fundamental limit on the resolution with which we can determine a phase shift. This is due to the fact that the observed relative frequency, f, may only take discrete values. After n runs of the experiment, these possible values of f are spaced by $\Delta f = 1/n$. The necessary condition for the point estimate,

$$\cos^2(st\varphi_*) = f$$

(identical to Eq. (4.106), yet with t replaced by st), then leads to a corresponding spacing of possible point estimates, $\Delta\varphi_*$. To first order, it is

$$\left| \Delta \left[\cos^2(st\varphi_*) \right] \right| \approx st \left| \sin(2st\varphi_*) \right| \Delta\varphi_* = 2st\sqrt{f(1-f)}\Delta\varphi_*$$

and hence

$$\Delta\varphi_* \approx \frac{1}{2st\sqrt{f(1-f)}}\Delta f = \frac{1}{2\sqrt{f(1-f)}}\frac{1}{stn}. \tag{4.117}$$

As long as we exclude the special cases $f = 0$ and $f = 1$, this spacing is finite. It constitutes the minimal uncertainty which cannot be undercut by any measurement scheme. This fundamental limit is sometimes referred to as the *Heisenberg limit*. It implies, in particular, that as a function of the number of qubit sensors, s, the accuracy of a measurement cannot improve faster than $1/s$. In this sense, our entanglement-assisted scheme makes optimal use of the available sensor qubits, whereas the previous scheme did not.

The improvement thanks to entanglement must be taken with a grain of salt. In our analysis we assumed that the added complexity of the CNOT gates came at no extra cost. In practice, however, the CNOT operations do cost time, reducing the time available for the phase shift gates and, thus, their maximum number, t. This has an adverse effect on the accuracy. Moreover, and even more seriously, the entangled states which the CNOT gates produce can be very unstable, which further reduces the available measurement time and resulting accuracy. Still, the basic concept remains true that accuracy can be gained by working with sensor qubits in some non-trivial, entangled state rather than with uncorrelated qubits. One example is so-called 'squeezed states' of light, which are entangled photon states. Such states are being used, for instance, in Advanced LIGO, a gravitational wave detector based on ultra-high precision laser interferometry.

Chapter Summary

- A computer is a machine which receives an input, processes it according to some set of rules, and produces an output. A classical computer processes classical bits. It can be constructed from elementary logic gates, each acting on one or two bits at a time.

- In a quantum computer the elementary carriers of information are two-level systems, or 'qubits'.

- The input of a computation is encoded in the initial state of the qubits. The computation proper consists in a unitary transformation in the composite Hilbert space of these qubits.

- The unitary transformation is effected by means of a quantum circuit. The latter is composed of elementary quantum logic gates that each act on one or two qubits at a time, possibly involving ancilla qubits.

- Frequently used single-qubit gates are the Pauli, Hadamard, phase, and $\pi/8$ gates; the most important two-qubit gate is the CNOT gate. These form a universal set; that is, they can implement any multi-qubit unitary transformation.

- In contrast to classical computing, where copying a classical bit poses no problem, the no-cloning theorem forbids the duplication of an arbitrary qubit state.

- To read out a classical result, measurements are performed by means of measurement gates. Unless the qubits are in basis states, these measurements have various possible outcomes, of which only one represents the correct solution. Therefore, the typical quantum algorithm works in a probabilistic fashion and requires multiple runs of the circuit.

- In principle, it is possible to do without unitary gates and implement any quantum algorithm with measurements only. This is termed measurement-based computation. It has no classical analogue.

- If the requisite amount of resources—gates, ancilla qubits, reruns—scales at most polynomially with the number of input qubits, an algorithm is deemed efficient. For some computational problems which cannot be solved efficiently on a classical computer, efficient algorithms do exist on a quantum computer. A simple example is the Deutsch–Jozsa algorithm.

- The study of chemical, electrical, or optical properties of matter requires the determination of the electron state which minimizes the energy. Similarly, many classical optimization problems amount to finding the configuration that minimizes some cost function.

- The electronic states in a molecule, and often the possible configurations in a classical optimization problem, can be mapped to states of qubits, and the pertinent energy or cost function, to the energy observable—the 'Hamiltonian'—of these qubits.

- The variational quantum eigensolver uses entangled states to systematically search the multi-qubit Hilbert space for the lowest-energy state. It converges with high probability in polynomial time.

- A quantum computer can efficiently simulate the dynamics of a quantum many-body system, provided its constituents interact locally.

- Quantum circuits can be made robust against noise by means of error-correcting codes, without compromising their efficiency.

- Given finite resources, there is a fundamental limit to the accuracy that any measurement can achieve. Clever use of entanglement can bring this limit from the standard quantum limit down to the Heisenberg limit.

Further Reading

As a gentle start, the informal and playful (yet scientifically sound) books by Aaronson (2013) and Rudolph (2017) might help you warm to quantum computation without too much mathematics. The former also features an introduction to the theory of computational complexity. Some basic components of a quantum computer and simple circuits are already covered in the previously mentioned text by Schumacher and Westmoreland (2010). The standard reference in the field of quantum computation and quantum information is the classic by Nielsen and Chuang (2000), where you find a wealth of rigorous results, many additional algorithms, a discussion of experimental implementations, and a comprehensive list of references to original research papers. A shorter, somewhat older, but very readable introduction is the review by Steane (1998). In addition to textbooks and reviews, there are a number of good lecture notes available online with varying focuses and levels of difficulty; for example, those by Preskill (2018a), Watrous (2006), and Vazirani (2007). As an introduction to quantum computing that is aimed explicitly at non-physicists, I recommend the books by Mermin (2007) and Lipton and Regan (2014). These also contain a detailed analysis of Shor's factorization algorithm along with the requisite background in number theory. If you set out to build a real quantum computer, you will have to confront error correction, a topic which I touch upon only briefly in Exercise (4.10). Many of the books and lecture notes just mentioned discuss error correction in much greater detail than I do, and are therefore a good place to start. Measurement-based quantum computation was introduced (in a form different from the toy example I gave here) by Raussendorf and Briegel (2001); additional details can be found in Briegel *et al.* (2009), and a somewhat different approach can be seen in Perdrix (2007). For a list of classical optimization problems and how they may be recast as cost or energy minimization problems, you may consult, for instance, Lucas (2014). Mapping atomic or molecular Hamiltonians to qubit Hamiltonians, as well as the variational quantum eigensolver algorithm, is described in Peruzzo *et al.* (2014) and McClean *et al.* (2016). The idea of simulating quantum physics on a quantum computer goes back to Feynman (1982). That this simulation can be made efficient if the quantum system features only local interactions was shown by Lloyd (1996). Finally, to learn more about high-precision measurements in general, and quantum-enhanced measurements in particular, I suggest the article by Giovannetti *et al.* (2004) and the review by Degen *et al.* (2017). How all these ideas are implemented in actual hardware is a vast and rapidly evolving area of—mainly experimental—research, well beyond the scope of this book. For a succinct overview of the current status, you may consult regularly updated reports by various research and cybersecurity agencies; for example, the consensus study report by the National Academies of Sciences, Engineering, and Medicine (2019). In the near to medium term, one will have to deal with noisy intermediate-scale quantum (NISQ) technology, which already allows for many interesting applications; these are reviewed in Preskill (2018b).

······················

EXERCISES

4.1. Single-qubit gates

(a) Which rotations do the Pauli (X, Y, Z), Hadamard (H), phase (S), $\pi/8$ (T), and phase shift (U_φ) gates effect on the Bloch sphere?

(b) Show the following relationship between single-qubit gates:

$$-\boxed{H}-\boxed{S}-\boxed{Y}-\boxed{S^\dagger}-\boxed{H}- \;=\; -\boxed{Z}-$$

4.2. Controlled-Z gate

Show that

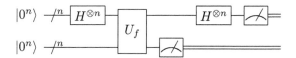

That is to say, the roles of control and target qubit may be interchanged.

4.3. Measurement process revisited

Consider the measurement process discussed in Exercise (3.22), involving either two (AB) or three (ABC) systems. For both scenarios, represent the measurement process as a quantum circuit.

4.4. Simon's algorithm

Consider a function that maps bit strings of length n to bit strings of length n, $f : \{0,1\}^n \to \{0,1\}^n$. You are told that there exists some unknown bit string $s \in \{0,1\}^n$ such that

$$f(x) = f(y) \quad \Leftrightarrow \quad y \in \{x, x \oplus s\},$$

where $\oplus$ means bitwise addition modulo two. This makes f a two-to-one function (unless $s = 0^n$, in which case it is one to one). You are allowed to consult an oracle: you submit x and the oracle tells you $f(x)$. With the help of such oracle queries, you must find the unknown s.

(a) Show that on a classical computer this problem is hard: the number of oracle queries scales exponentially with the length of the bit string, n.

(b) On a quantum computer the oracle is realized as a unitary gate, U_f, like in the Deutsch–Jozsa algorithm (Eq. (4.67)). The sole difference is that now the second input and the second output also consist of n qubits, $y, f(x) \in \{0,1\}^n$. Verify that this quantum oracle performs a unitary transformation.

(c) The circuit, too, is similar to that of the Deutsch–Jozsa algorithm, Eq. (4.68), but with a few twists:

$$
\begin{array}{l}
|0^n\rangle \;\;/^n\; \boxed{H^{\otimes n}} \quad\quad \boxed{H^{\otimes n}} \;\; \measuredangle \\
\quad\quad\quad\quad\quad\quad\; \boxed{U_f} \\
|0^n\rangle \;\;/^n \quad\quad\quad\quad \measuredangle
\end{array}
$$

As we did for the Deutsch–Jozsa algorithm, determine the state after each step in the circuit. (After the first measurement you may focus on the remaining data qubits only.) In particular, show that the state of the data qubits just before the final measurement is given by

$$|\psi\rangle \propto \sum_{z \in \{0,1\}^n} \left[(-1)^{x \cdot z} + (-1)^{(x \oplus s) \cdot z} \right] |z\rangle.$$

Hint: Use Eq. (4.71).

(d) Show that the final measurement will yield some random $z \in \{0,1\}^n$ for which $s \cdot z = 0$ modulo two, where $s \cdot z$ denotes the bitwise scalar product. When you repeat this protocol sufficiently many times, you will obtain $(n-1)$ independent such equations, $s \cdot z_i = 0$ modulo two $(i = 1, \ldots, n-1)$, for the n unknown digits of s; an nth condition is provided by the fact that the digits must be either zero or one. These equations can be efficiently solved for s using classical methods. Since the outcome of each individual run, z, is random, the algorithm is probabilistic: the number of runs needed to obtain $(n-1)$ independent equations may vary. However, on average, it scales only linearly with n rather than exponentially, rendering the algorithm efficient.

Simon's algorithm solves a problem which in itself is of little practical use. However, it represented an important milestone in the development of more complex, and potentially more useful, algorithms such as Shor's factorization algorithm.

4.5. Quantum search

For many mathematical problems it is much easier to recognize the solution than to find it. For example, the problem of decomposing a large integer into two prime factors is hard; but if someone proposes a solution, it is easy to check whether this solution is correct. One (inefficient) way of finding the solution is to go through all possibilities and check them one by one, until you have found the one which is correct. Consider the simple example where there are four possible solutions, and the task for a (classical or quantum) computer is to find the correct one. All the computer may do is consult an 'oracle', which will tell it whether a proposed solution is correct.

(a) How often does a classical computer have to consult the oracle, on average, before it finds the correct solution?

(b) Label the four possible solutions by $0, 1, 2, 3$ or, in binary form, $00, 01, 10, 11$. Let the correct solution be the one with binary label jk $(j, k \in \{0,1\})$. A quantum computer may consult a quantum version of the oracle. This is a three-qubit gate with two control (or data) qubits and one target qubit, akin to the Toffoli gate, Eq. (4.25). It preserves basis states of the data qubits, and it flips basis states of the target qubit if and only if the data qubits are in the basis state $|jk\rangle$, that is, the state which represents the correct solution:

$$
\begin{array}{c}
|x\rangle \\
|y\rangle \\
|t\rangle
\end{array}
\quad \boxed{O} \quad
\begin{array}{c}
|x\rangle \\
|y\rangle \\
|t \oplus \delta_{xj}\delta_{yk}\rangle
\end{array}
\qquad , \qquad x, y, t \in \{0,1\}.
$$

Verify that this quantum oracle performs a unitary transformation. Relate it to other three-qubit gates that you have encountered in Chapter 4. Discuss the similarities and differences to the quantum oracle used in the Deutsch and Deutsch–Jozsa algorithms.

(c) Show that the two-qubit gate

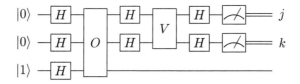

implements the unitary transformation

$$\hat{V} = \hat{I} - 2\,|00\rangle\langle00|$$

in the Hilbert space of two qubits.

(d) Show that the circuit

$$
\begin{array}{c}
|0\rangle - H - \cdots - H - \cdots - H - \measuredangle = j \\
|0\rangle - H - O - H - V - H - \measuredangle = k \\
|1\rangle - H - \cdots
\end{array}
$$

yields the correct solution with only a single oracle query. This simple example illustrates that, given an oracle which can check the correctness of a proposed solution, a quantum computer can search for the correct solution faster (that is, with fewer oracle queries) than its classical counterpart. There exists a well-known quantum algorithm, *Grover's algorithm*, which extends the basic idea presented here to the general situation where one searches for the correct solution among an arbitrary number, N, of possibilities. While a classical computer would have to call the oracle $O(N)$ times, Grover's algorithm requires only $O(\sqrt{N})$ oracle queries.

4.6. Knapsack problem

You have N objects, labelled by $i = 1,\ldots,N$, with integer weight $w_i \in \mathbb{N}$ and (say, monetary) value $v_i > 0$, and you have a knapsack which can only carry weight $W_{\max}$. If x_i is a binary variable indicating whether object i is contained ($x_i = 1$) or not ($x_i = 0$) in the knapsack, the total weight in the knapsack equals

$$W(\vec{x}) = \sum_{i=1}^{N} w_i x_i,$$

and the total value is

$$V(\vec{x}) = \sum_{i=1}^{N} v_i x_i.$$

You want to maximize $V(\vec{x})$, subject to the constraint that $W(\vec{x}) \leq W_{\max}$. This problem has many practical applications, particularly in economics and finance, and is hard on a classical computer. It can be mapped to the problem of minimizing a cost function with the help of the following trick. For simplicity, assume that $W_{\max} = 2^M - 1$ for some integer $M \in \mathbb{N}$. Then the total weight in the knapsack, being an integer, too, may be represented as a binary number with M digits. Let $y_k \in \{0,1\}$ be the kth digit from the right of that binary number. One treats the $\{x_i\}$ and $\{y_k\}$ as $(N+M)$ formally independent binary variables (even though in reality they are not) and defines the cost function

$$C(\vec{x}, \vec{y}) := A \left(\sum_{i=1}^{N} w_i x_i - \sum_{k=1}^{M} 2^{k-1} y_k \right)^2 - B \sum_{i=1}^{N} v_i x_i$$

with $A, B > 0$. The first term penalizes any deviation of the actual weight from the weight indicated by the binary number and thus, in effect, enforces that the two are equal; this re-establishes the factual dependence of $\vec{y}$ on $\vec{x}$. The second term rewards any increase in total value.

(a) Explain why solving the original problem is indeed equivalent to the minimization of this cost function. Show that in order to ensure that the weight in the knapsack never exceeds the upper limit, $W_{\max}$, it must be

$$A > B \cdot \max_i \{v_i\}.$$

(b) Convert the cost function into a Hamiltonian of an assembly of $(N+M)$ qubits.

4.7. Pauli basis

Verify that the 4^n distinct n-fold tensor products $\hat{\sigma}_{\mu_1} \otimes \dots \otimes \hat{\sigma}_{\mu_n}$ $(\mu_k = 0,1,2,3)$ constitute a basis in the space of n-qubit observables, and hence that every n-qubit Hamiltonian can be represented in the form of Eq. (4.78). As in the main text, $\hat{\sigma}_0$ is the unit operator and $\hat{\sigma}_i$ $(i=1,2,3)$ are the three Pauli operators.

4.8. Suzuki–Trotter decomposition

Prove the Suzuki–Trotter decomposition, Eq. (4.92).

4.9. Multi-qubit transformations

(a) Consider the following circuit with three input qubits and one ancilla qubit, with the single-qubit gate U_φ defined as in Eq. (4.95):

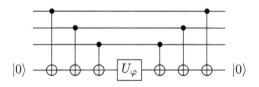

Which unitary transformation does this circuit effect on the three input qubits? Write this transformation in exponential form as in Eq. (4.87) and determine the exponent.

(b) In the following circuit two Hadamard gates have been added:

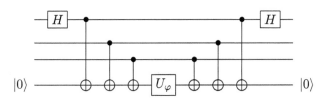

Which unitary transformation does this modified circuit effect on the three input qubits? Write it again in exponential form.

(c) Draw a circuit that implements the unitary transformation $\exp(i\varphi \hat{X} \otimes \hat{Y} \otimes \hat{Z})$ on the three input qubits.
Hint: Exercise (4.1).

(d) Generalize to n input qubits and unitary transformations of the form

$$\hat{U} = \exp(i\varphi \hat{\sigma}_{\mu_1} \otimes \ldots \otimes \hat{\sigma}_{\mu_n}),$$

where each $\hat{\sigma}_{\mu_k}$ ($\mu_k = 0, 1, 2, 3$) is either the unit operator, $\hat{\sigma}_0 \equiv \hat{I}$, or one of the three Pauli operators, $\hat{\sigma}_i$ ($i = 1, 2, 3$). In other words, the exponent is proportional to a member of the Pauli basis (see Exercise (4.7)). Is such a transformation locally generated? Sketch a circuit implementation. How does the requisite number of gates scale with n? Is the circuit implementation efficient? Discuss.

(e) Show that the above results, combined with the general considerations of Section 4.6, imply that an efficient circuit implementation exists for any unitary transformation of the form

$$\hat{U} = \exp(it\hat{H}),$$

where t is a real parameter and $\hat{H}$ is an n-qubit Hamiltonian in the standard representation (Eq. (4.78)), *provided* the number of non-zero coefficients, $h_{\mu_1 \ldots \mu_n} \neq 0$, scales at most polynomially with n.

4.10. Error correction

Quantum error correction is more challenging than classical error correction, for several reasons: the no-cloning theorem prevents us from adding redundancy by simply copying qubits; the fact that measurements generally alter a quantum state complicates the detection and correction of errors; and there is a continuum of possible errors, corresponding to the continuum of possible unitary transformations. Here we consider a very simple example which leaves out many of these complexities. Imagine that a qubit is being sent from one place to another through a quantum channel—say, a photon through an optical fibre—and that, with some probability p, the state of the qubit is corrupted by noise in that channel. In practice, such noise can cause an arbitrary unitary transformation. Here, however, we limit ourselves to the special case where the noise merely triggers a random flip of the basis state $|0\rangle$ to the basis state $|1\rangle$ and vice versa. In other words, the noise consists in a random application of the Pauli-X gate. The following protocol, called the 'three-qubit bit flip code', corrects this type of error:

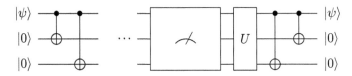

In addition to the qubit which carries the state of interest, $|\psi\rangle$, there are two extra qubits, initially in the basis state $|0\rangle$. The three qubits are entangled via two CNOT gates and then enter the noisy channel (indicated by the dotted line). The error probability, p, is assumed to be so small that a random flip, if it occurs at all, affects at most one of the three qubits. Upon receipt at the other end, all three qubits are subjected to a joint measurement called 'syndrome measurement'. This measurement establishes whether a flip occurred at all and, if so, on which of the three qubits. It has four possible outcomes, labelled $0,1,2,3$ and corresponding to 'no flip', 'flip on qubit 1', 'flip on qubit 2', et cetera. Depending on the outcome (called the 'error syndrome'), a 'recovery' operation, U, is performed to correct the error. If the error syndrome was zero, no recovery is needed; if the syndrome was one, a Pauli-X gate is applied to the first qubit; and so on. (The dependency of U on the error syndrome is not shown in the circuit.) Finally, two CNOT gates disentangle the three qubits and restore their initial state.

(a) Suppose that the state to be transmitted, $|\psi\rangle$, is a superposition state,

$$|\psi\rangle = a\,|0\rangle + b\,|1\rangle\,.$$

What is the state of the three qubits before entering the quantum channel? Which are the possible states of the three qubits when they exit the channel?

(b) Show that the projectors

$$\hat{P}_0 := |000\rangle\langle000| + |111\rangle\langle111|$$
$$\hat{P}_1 := |100\rangle\langle100| + |011\rangle\langle011|$$
$$\hat{P}_2 := |010\rangle\langle010| + |101\rangle\langle101|$$
$$\hat{P}_3 := |001\rangle\langle001| + |110\rangle\langle110|$$

in three-qubit Hilbert space project onto the eigenspaces associated with the four potential outcomes of the syndrome measurement. Verify that these projectors are mutually orthogonal and add up to the unit operator.

    (c) Does the syndrome measurement alter the state of the three qubits? Explain.

    (d) Does the syndrome measurement reveal anything about the transmitted state, $|\psi\rangle$? Explain.

    (e) Show that subsequent recovery and the two CNOTs will indeed restore the correct state.

4.11. Michelson interferometer

A Michelson interferometer (Fig. 4.3) allows one to measure small variations in distance with high precision. An interferometer of this kind was employed in the famous Michelson–Morley experiment, which yielded the first evidence against the existence of the aether and had a strong influence on the development of the special theory of relativity; and its basic design is still in use today for the detection of gravitational waves. Its mathematical description is similar to that of the Mach–Zehnder interferometer (Fig. 3.15) which you encountered in Exercise (3.14). Again, a photon can be in one of two orthogonal kinetic states, $|0\rangle$ and $|1\rangle$, corresponding to the horizontal and vertical beam directions, respectively. Assume that when a photon passes the beam splitter for the first time, its effect is described by the same unitary matrix, $\mathbf{U}_{BS}$, as in Exercise (3.14), whereas when the photon passes the splitter for the second time, in reverse direction, it is described by the inverse, $\mathbf{U}_{BS}^{\dagger}$. The two mirrors effect no unitary transformation at all.

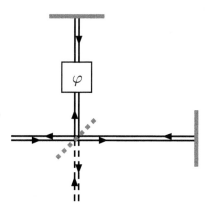

Fig. 4.3 *Michelson interferometer. The beam splitter at the centre splits an incoming laser beam (or beams) in two. The two outgoing beams are reflected by mirrors at the top and to the right and then pass through the splitter a second time. Below and to the left of the splitter are light sources, detectors, and possibly other optical components (not shown here) to prepare the incoming and measure the outgoing light, respectively. Photons travelling to the upper mirror acquire an extra phase, φ, relative to those travelling to the right. This happens both on their way up and on their way down, so the total phase difference equals 2φ. Such a phase difference might be due to a medium placed in the vertical beam, or it might be due to the fact that the distances of the two mirrors from the splitter are slightly different. In the latter case the phase difference is given by $\varphi = 2\pi\Delta l/\lambda$, where Δl denotes that length difference and λ denotes the wavelength of the used light.*

(a) Photons travelling to and from the top mirror undergo an extra phase shift, φ, relative to those travelling to and from the mirror on the right. That is to say, their state is multiplied—twice—by an extra phase factor, $\exp(i\varphi)$. Show that the net effect of beam splitter, mirrors, and this phase shift is (up to a global phase factor) the unitary transformation

$$\hat{V}_\varphi = \exp\left(i\varphi\hat{Y}\right),$$

where $\hat{Y}$ is one of the Pauli operators. Describe the associated rotation on the Bloch sphere.

(b) Suppose that there is a detector below the splitter. A photon comes in from the left, that is, in the state $|0\rangle$. What is the probability that the detector will click?

(c) In order to estimate the phase shift, φ, you send in s independent photons in the state $|0\rangle$ and count the detector clicks. Calculate the accuracy, σ_φ, of the resultant estimate. How does it scale with s?

(d) Suppose you want the accuracy of your measurement to reach the Heisenberg limit. To this end, you want to employ a circuit akin to Eq. (4.110), with $t = 1$ and with U_φ replaced by V_φ. In order for this circuit to work, which further changes do you have to make to its design? (Never mind whether this can be done in practice.) With these modifications, write down the s-photon state that enters the beam splitter.

Hint: Use the circuit identity from Exercise (4.9).

5

Communication

5.1 Classical Information

I explain how classical information is defined operationally and introduce the entropy as its quantitative measure. I consider different variants of entropy—the Shannon, conditional, and relative entropies, as well as the mutual information—and discuss their interrelationships and basic properties. I show that classical information can be encoded in quantum states, yet that the extent to which this is possible is limited by a fundamental bound: the Holevo bound.

Communication is the exchange of information. We must begin, therefore, by defining 'information' and devising a way to quantify it. In this section we shall discuss how this is done classically; in Section 5.2 we will transfer the idea to the quantum realm. In classical theory, 'information' is defined operationally by considering the communication between two protagonists; for instance, our two familiar characters, Alice and Bob (Fig. 5.1). Alice performs N runs—called *trials*—of some random experiment, each with d possible results, $i = 1, \ldots, d$. For instance, she might roll a die ($d = 6$) a hundred times ($N = 100$). The trials are statistically independent, in the sense that the outcome probabilities of every trial are unaffected by the results of the other trials. After her experiment, Alice wants to communicate to her partner, Bob, the outcomes of all N trials, including the order in which these results occurred; she wants to send a message of the form, say, 'The first roll of the die yielded a "5", the second roll yielded a "3", …'. However, she is constrained to use a communication channel that allows only classical bits ('0' and '1'). In other words, she is forced to encode her message in a bit string. In this idealized setup there is no noise in the channel, so Alice need not worry about redundant coding. At the receiving end, Bob knows the probability distribution for Alice's experiment beforehand. In accordance with our discussion in Section 1.3, this distribution constitutes the 'state' associated with each trial; we shall denote it by $\rho = \{p_i\}_{i=1}^d$. In addition, Bob knows the number of trials, N. What he does *not* know, however, is the specific sequence of results that Alice has obtained. This is the information that he misses, and that has yet to be communicated in the form of a bit string.

Alice and Bob have agreed in advance on a coding algorithm. They want this algorithm to provide for maximal compression, in the sense that the bit string becomes as short as

Quantum Theory: An Information Processing Approach. Jochen Rau, Oxford University Press (2021). © Jochen Rau.
DOI: 10.1093/oso/9780192896308.003.0005

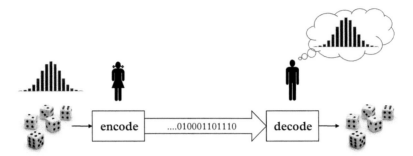

Fig. 5.1 *Setup for the definition of classical information.*

possible. Let S_1 denote the minimum length of the bit string needed to communicate the result of a single trial, and S_N the minimum length needed to communicate the results of N trials. Clearly, it is possible to communicate the results of N trials with NS_1 bits. However, Alice might do better than that if, rather than encoding all results separately, she encodes the totality of the N results *en bloc*. For example, if Alice rolls a perfect die once, she will need at least $S_1 = 3$ bits to encode the result of this single roll. Three bits can accommodate up to $2^3 = 8$ different outcomes, enough for a single die roll with its six possible outcomes; two bits, on the other hand, would not suffice. Consequently, if Alice rolls the die three times and encodes each result separately, she will need $3S_1 = 9$ bits. But if she encodes the results of all three rolls *en bloc*, she can do with only $S_3 = 8$ bits; for there are in total $6^3 = 216$ possibilities, which is less than $2^8 = 256$. So in general it is $S_N \leq NS_1$. As the number of trials becomes very large, the average number of bits needed per trial approaches a well-defined limit,

$$S := \lim_{N \to \infty} \frac{S_N}{N} \leq S_1. \tag{5.1}$$

In classical information theory, this asymptotic number of bits per trial defines the *Shannon entropy*. It quantifies the *information*, in units of bits, which is contained in the outcome of one trial; or, equivalently, Bob's initial ignorance, or subsequent surprise, about that outcome.

The Shannon entropy is some as yet unknown function of the state, ρ. A few special cases are straightforward:

- The outcome of the random experiment is certain if one of the results occurs with probability one, $\rho = \{0,\ldots,0,1,0,\ldots\}$. In this case the outcome is known to Bob beforehand, and the actual run of the experiment will not yield any new information. Therefore, there is no need to send any bits; it is $S = 0$.

- If each trial yields one out of two possible results, $d = 2$, and both results are equally likely, $p_1 = p_2 = 1/2$—such as Heads or Tails when flipping a fair coin—then the optimal encoding requires one bit per trial; say, '0' for Heads and '1' for Tails. Consequently, it is $S = 1$.

- If in each trial we flip r fair coins simultaneously, there are $d = 2^r$ possible outcomes that are equally likely, $p_i = 1/2^r$ for all i. In this case the number of bits needed per trial equals the number of bits needed to encode the results of r coin flips, $S = r$. This may also be written as the binary logarithm, $S = \log_2(2^r)$, of the number of outcomes per trial.

- The latter result can be generalized: whenever the d possible results of a trial are equally likely, $p_i = 1/d$ for all i, the average number of bits needed to encode the result of one trial equals the binary logarithm of d, $S = \log_2 d$.

For an arbitrary state, Shannon derived the formula

$$S(\rho) = -\sum_{i=1}^{d} p_i \log_2 p_i, \quad \rho = \{p_i\}_{i=1}^{d};$$ (5.2)

this is known as 'Shannon's noiseless channel coding theorem'. You will prove this theorem in Exercise (5.1). The general formula is consistent with all the special cases discussed earlier. (In the case of an experiment whose outcome is certain, one must use the limit $\lim_{p\to 0} p\log_2 p = 0$.) The Shannon entropy is closely related to the Boltzmann entropy known in statistical mechanics; the two differ only by a multiplicative constant. For brevity, we shall henceforth omit 'Shannon' and refer to S simply as the entropy, and we shall omit the base of the logarithm and simply write 'log' rather than 'log$_2$'.

The entropy exhibits three important properties:

1. The entropy is non-negative, $S(\rho) \geq 0$. It vanishes if and only if the outcome of the experiment is known beforehand:

$$S(\rho) = 0 \quad \Leftrightarrow \quad \rho = \{0, \ldots, 0, 1, 0, \ldots\}.$$ (5.3)

2. It is invariant under arbitrary permutations, π, of the probabilities $\{p_i\}$,

$$S(\pi(\rho)) = S(\rho),$$ (5.4)

where $\pi(\rho) := \{p_{\pi(i)}\}$. In other words, the entropy is unaffected by a mere relabelling of the results.

3. The entropy is additive, in the following sense. When two random experiments, A and B, with possible outcomes $\{a_i\}$ and $\{b_j\}$, respectively, are performed jointly, with possible joint outcomes $\{a_i \otimes b_j\}$, the number of bits needed—on average— to encode the result of this joint experiment is the sum of two terms: the average number of bits needed to encode the result of A and the average number of bits needed to encode the result of B, *given the result of* A. Correspondingly, the *joint entropy* of A and B is a sum of two entropies,

$$S(\rho_{AB}) = S(\rho_A) + S(B|A), \tag{5.5}$$

where $\rho_{AB} := \{\mathrm{prob}(a_i \otimes b_j)\}$, $\rho_A := \{\mathrm{prob}(a_i)\}$ and

$$S(B|A) := \sum_i \mathrm{prob}(a_i) \, S(\rho_{B|a_i}), \tag{5.6}$$

with $\rho_{B|a_i} := \{\mathrm{prob}(b_j|a_i)\}$. The latter entropy is called the *conditional entropy* of B, given A. You will prove this additivity in Exercise (5.2). The caveat 'given the result of A' is necessary because the two experiments might be correlated. For example, consider a single die roll where the first random experiment, A, consists in determining whether the number shown is odd or even; and the second random experiment, B, whether it is a prime number. Since there are only two non-prime numbers on a die, '4' and '6', which are both even, the two random experiments are correlated. If Alice communicates to Bob that the first experiment yielded 'odd', Bob already knows that the second experiment must have yielded 'prime', so in this case no further bits need to be sent concerning the outcome of B. Whenever there are such correlations, fewer bits are needed to encode the result of the joint experiment than to encode the results of both experiments separately. Hence, in general, it is

$$S(\rho_{AB}) \le S(\rho_A) + S(\rho_B). \tag{5.7}$$

Equality holds if and only if the two experiments are uncorrelated, that is, the joint state is a product state:

$$S(\rho_{AB}) = S(\rho_A) + S(\rho_B) \quad \Leftrightarrow \quad \rho_{AB} = \rho_A \otimes \rho_B. \tag{5.8}$$

The above three properties—non-negativity, permutation invariance, and additivity—are properties which one would intuitively expect from any information measure. It turns out that, up to a multiplicative constant, the entropy defined in Eq. (5.2) is in fact the only function that satisfies all these properties.

When two random experiments are correlated, learning about the result of one already provides some information about the result of the other. If, say, we learn the result of A, then on average our ignorance about the result of B is reduced from $S(\rho_B)$ to $S(B|A)$. The difference,

$$I(A:B) := S(\rho_B) - S(B|A), \tag{5.9}$$

is the average amount of information provided by A about B; it is called the *mutual information* of A and B. By virtue of Eq. (5.5), this mutual information may also be written in the form

$$I(A:B) = S(\rho_A) + S(\rho_B) - S(\rho_{AB}), \tag{5.10}$$

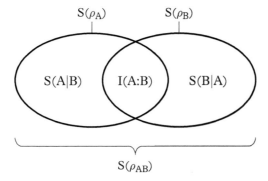

Fig. 5.2 *Pictorial representation of the relationships between various classical entropies. The letters A and B refer to two (possibly correlated) random experiments. The entropies $S(\rho_A)$ and $S(\rho_B)$ pertain to the individual experiments, $S(\rho_{AB})$ to the joint experiment. $S(A|B)$ is the conditional entropy of A, given the result of B (likewise for $S(B|A)$); and $I(A:B)$ is the mutual information of A and B. The relationships between the sets indicate the relationships between the entropies. For instance, it is $S(\rho_{AB}) = S(\rho_A) + S(\rho_B) - I(A:B)$.*

which shows that it is in fact symmetric in A and B:

$$I(A:B) = I(B:A). \tag{5.11}$$

The relationships between the various entropies and the mutual information are conveniently summarized in Fig. 5.2. There exists yet another way to express the mutual information, which will prove helpful later. Given two classical distributions, $\rho = \{p_i\}$ and $\sigma = \{q_i\}$, we define their *relative entropy* as

$$S(\rho\|\sigma) := \sum_i p_i(\log p_i - \log q_i). \tag{5.12}$$

It vanishes if and only if the two states are identical, and is positive otherwise:

$$S(\rho\|\sigma) \begin{cases} = 0 : \rho = \sigma \\ > 0 : \text{otherwise} \end{cases}. \tag{5.13}$$

(You will prove this in Exercise (5.3).) In this sense, the relative entropy quantifies the degree of distinguishability of two states. It does not qualify as a proper metric, however, because it is in general not symmetric: $S(\rho\|\sigma) \neq S(\sigma\|\rho)$. As you will verify in Exercise (5.4), the mutual information of A and B can be written as a weighted average of relative entropies:

$$I(A:B) = \sum_i \text{prob}(a_i) \, S(\rho_{B|a_i} \| \rho_B). \tag{5.14}$$

This form, too, makes clear that the mutual information measures the strength of correlations between the two random experiments. If they are uncorrelated, ascertaining the outcome of A will not affect our expectations as to the outcome of B, so $\rho_{B|a_i} = \rho_B$ for all i. The two states being equal, their relative entropy vanishes, and so does the mutual information. If, on the other hand, the two experiments are correlated, the relative entropy becomes greater than zero for at least some i, and so does the mutual information. The stronger the correlations, the more $\rho_{B|a_i}$ differs from ρ_B, the larger their relative entropy, and thus the larger the mutual information.

In the context of quantum information processing, it is interesting to know if, and to what extent, classical information can be encoded in a quantum state. In order to address this question, we modify our communication setup: rather than via a string of bits, Alice now conveys the classical information to Bob via a sequence of quantum systems prepared in various 'coding states', $\{\rho_i\}$ (Fig. 5.3). Naively, one might think that a single quantum system can carry an infinite amount of classical information. For instance, a qubit has a continuum of pure states, corresponding to the surface of the Bloch sphere. In order to specify a particular point on this surface with arbitrary precision, one needs an infinite amount of classical information. So why not, conversely, encode an infinite amount of classical information in a point on the Bloch sphere—that is, by preparing the qubit in the associated pure state? Alas, due to the limitations on quantum measurements, Bob would have no chance of retrieving this information. As for information that can actually be retrieved, it turns out that an individual qubit can carry no more than one classical bit. In the following, we shall provide a short proof of that assertion, building on the framework of information theory that we laid out earlier.

Alice's classical random experiment has several possible outcomes, $\{a_i\}$, with probability distribution $\rho_A = \{\mathrm{prob}(a_i)\}$. She encodes each outcome, a_i, in a different coding state. In the Hilbert space of the carrier system, this state is represented by a statistical

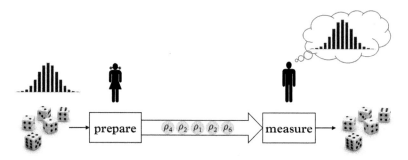

Fig. 5.3 *Quantum systems as carriers of classical information. Alice (left) performs classical random experiments as before, but this time her results are encoded in quantum states: depending on the results of her experiments, she prepares quantum systems in various coding states, $\{\rho_i\}$, and sends these to Bob. Like in the classical setup, Bob knows in advance the classical probability distribution of Alice's experiment, the number of trials, and the coding algorithm. Bob extracts information about the specific sequence of Alice's results by performing measurements on the quantum systems he receives.*

operator, $\hat{\rho}_i$. Bob's subsequent measurement may be regarded as a second random experiment, correlated to that of Alice. Associated with the possible outcomes of Bob's measurement are mutually exclusive propositions, $\{b_j\}$. These, in turn, are represented in the Hilbert space of the quantum carrier by some set of mutually orthogonal projectors, $\{\hat{P}_j\}$. (A more general setup allows for the coupling of the carrier to some quantum device, or 'environment', and a measurement on the composite system including the environment. However, this will not affect the final result.) Given a particular outcome of Alice's experiment, a_i, the outcome probabilities of Bob's experiment read

$$\rho_{\mathrm{B}|a_i} = \{\mathrm{prob}(b_j|a_i)\} = \{\mathrm{tr}(\hat{\rho}_i\hat{P}_j)\}. \tag{5.15}$$

If, on the other hand, the outcome of Alice's experiment is not known, then we must take the weighted average, yielding

$$\rho_{\mathrm{B}} = \{\mathrm{prob}(b_j)\} = \{\mathrm{tr}(\hat{\rho}\hat{P}_j)\}, \quad \hat{\rho} := \sum_i \mathrm{prob}(a_i)\,\hat{\rho}_i. \tag{5.16}$$

The average amount of information that Bob's measurement provides about the result of Alice's experiment is given by the mutual information, $I(\mathrm{A}:\mathrm{B})$. Since we know $\rho_{\mathrm{B}|a_i}$ and ρ_{B}, as well as the probability distribution of Alice's experiment, we can calculate this mutual information with the help of Eq. (5.14). The latter features a relative entropy of two classical states, which may be related to the relative entropy of two *quantum* states as follows. In general, the relative entropy of two quantum states, represented by statistical operators $\hat{\rho}$ and $\hat{\sigma}$, is defined as

$$S(\hat{\rho}\|\hat{\sigma}) := \mathrm{tr}(\hat{\rho}\log\hat{\rho} - \hat{\rho}\log\hat{\sigma}). \tag{5.17}$$

It has properties very similar to its classical counterpart, Eq. (5.12), which you will investigate in Exercise (5.5); in particular, it also vanishes if and only if the two states are identical, and is positive otherwise. In the present context, you will show in Exercise (5.11) that the weighted average featuring in Eq. (5.14) is bounded from above by the weighted average of the quantum relative entropy of $\hat{\rho}_i$ and $\hat{\rho}$,

$$\sum_i \mathrm{prob}(a_i)\,S(\rho_{\mathrm{B}|a_i}\|\rho_{\mathrm{B}}) \leq \sum_i \mathrm{prob}(a_i)\,S(\hat{\rho}_i\|\hat{\rho}). \tag{5.18}$$

This implies an upper bound for the mutual information,

$$I(\mathrm{A}:\mathrm{B}) \leq H[\{\mathrm{prob}(a_i),\hat{\rho}_i\}], \tag{5.19}$$

the so-called *Holevo bound*:

$$H[\{\mathrm{prob}(a_i),\hat{\rho}_i\}] := \sum_i \mathrm{prob}(a_i)\,S(\hat{\rho}_i\|\hat{\rho}). \tag{5.20}$$

In the same exercise you will prove that this upper bound is saturated if the coding states commute, $[\hat{\rho}_i, \hat{\rho}_k] = 0$. The Holevo bound depends only on the characteristics of Alice's experiment and on the states she has chosen to encode the results. It no longer depends on the projectors $\{\hat{P}_j\}$, that is, on the characteristics of Bob's measurement. In Exercise (5.11) you will also show that the Holevo bound, in turn, is bounded from above by the entropy associated with Alice's experiment, $S(\rho_A)$:

$$H[\{\text{prob}(a_i), \hat{\rho}_i\}] \leq S(\rho_A). \tag{5.21}$$

You will verify that the latter bound is saturated whenever the coding states are mutually orthogonal, $\hat{\rho}_i \hat{\rho}_k = 0$ for all $i \neq k$.

Equipped with the above mathematical results, we can now assess the suitability of quantum systems as carriers of classical information. The classical information about the outcomes of Alice's trials has been transmitted successfully if and only if Bob is able to retrieve it in full via suitable measurements. He can do so if and only if the amount of information that he gleans from a measurement, $I(A : B)$, equals the amount of information that Alice has encoded, $S(\rho_A)$. This means that we must have equality in both Eqs (5.19) and (5.21). The former presupposes that the coding states commute, whereas the latter presupposes, more strongly, that they are mutually orthogonal. So we conclude that the transmission of classical information requires orthogonal coding states. In the case of a qubit, at most two states can be mutually orthogonal; for instance, the two standard basis states, $|0\rangle$ and $|1\rangle$. This is no more than the two states of a classical bit. Thus, indeed, an individual qubit cannot carry more classical information than a classical bit.

Despite our sobering conclusion, the use of quantum systems can aid classical communication in various other ways. For instance, the exchange of quantum systems plays a pivotal role in secure protocols for the distribution of cryptographic keys. These protocols are 'physically' secure in that they are protected against eavesdropping by the basic laws of physics. We shall discuss a very simple such protocol in Section 5.3. Another example is the artificial enhancement of the information-carrying capacity of a qubit, circumventing—sort of—the above limitation of one classical bit per qubit with a trick called 'superdense coding'. This enhancement presupposes that Alice and Bob have previously shared a pair of entangled qubits, which they can then employ as an additional communication resource. We will outline the basic idea behind this scheme in Section 5.4.

5.2 Quantum Information

I discuss the operational definition of quantum information and the pertinent measure, the von Neumann entropy. I sketch the derivation of the formula for the von Neumann entropy (Schumacher's theorem) and consider its most important properties. In particular, I point out where these differ from the classical case and what this may entail for quantum communication.

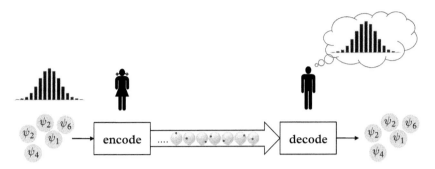

Fig. 5.4 *Setup for the definition of quantum information.*

Quantum information is defined in a manner analogous to classical information. The pertinent setup is shown in Fig. 5.4, which in many ways mirrors that in Fig. 5.1. Again, Alice performs N independent trials of a random experiment. Each trial produces one of n quantum states, $\{|\psi_i\rangle\}_{i=1}^{n}$—termed 'signal states'—according to some probability distribution, $\{p_i\}_{i=1}^{n}$. These signal states are required to be pure, but not necessarily mutually orthogonal. (This contrasts with the coding states used previously for the transmission of classical information (Fig. 5.3), which, for perfect retrieval, had to be orthogonal but not necessarily pure.) For instance, Alice might send photons one by one through a polarization filter. Some will be absorbed, others will pass; each photon that passes counts as a trial. The filter has n different settings, which are applied randomly and independently to each passing photon. Behind the filter there will then be photons in n different polarization states, occurring with respective probabilities that depend on the probabilities of the various polarizer settings and possibly on the prior polarization, if any, of the incoming beam. Afterwards, Alice wants to convey to Bob all the quantum states which were generated in this way, including the order in which they occurred. This is tantamount to transmitting to Bob the product state $|\psi_{i_1}\rangle \otimes \ldots \otimes |\psi_{i_N}\rangle$. She has at her disposal a quantum communication channel that allows her to send information exclusively in the form of qubits, symbolized in the figure by little Bloch spheres. This contrasts with the setup in Fig. 5.3, where arbitrary quantum systems were allowed. Also in contrast to Fig. 5.3, Alice does the encoding—the conversion of said product state into a state of carrier qubits—by means of a unitary transformation, rather than state preparation. Likewise, at the receiving end Bob must decode the incoming qubit state—that is, convert it back into the original product state—by means of unitaries only; he does not perform any measurements. Bob knows in advance the set of possible signal states, $\{|\psi_i\rangle\}$, their respective probabilities, $\{p_i\}$, and the number of trials, N, as well as Alice's coding algorithm. Just as in the classical case, Alice and Bob want their communication to be as compressed as possible, requiring the least number of qubits. Once again, as the number of trials increases, the minimum number of qubits divided by the number of trials approaches a well-defined limit. According to *Schumacher's theorem*, this limit depends only on the weighted average of the signal states—that is, the convex combination

$$\hat{\rho} = \sum_{i=1}^{n} p_i |\psi_i\rangle\langle\psi_i| \tag{5.22}$$

—and is given by the latter's *von Neumann entropy,*

$$S(\hat{\rho}) := -\operatorname{tr}(\hat{\rho}\log\hat{\rho}). \tag{5.23}$$

It quantifies the *quantum information,* in units of qubits, contained (on average) in a signal state; or, from Bob's perspective, the quantum information he still needs—beyond his advance knowledge—in order to reconstruct a signal state. The formula for the von Neumann entropy has a mathematical structure which is similar to its classical counterpart, Eq. (5.2). Indeed, you will prove in Exercise (5.10) that in the special case where the signal states are mutually orthogonal, $\langle\psi_i|\psi_k\rangle = \delta_{ik}$, the von Neumann entropy coincides with the Shannon entropy of the classical distribution $\{p_i\}$. If the signal states are not orthogonal, however, the former is strictly smaller: $S(\hat{\rho}) < S(\{p_i\})$.

The proof of Schumacher's theorem is worth understanding, and we shall briefly sketch the basic idea. It proceeds along the same lines as that of Shannon's noiseless channel coding theorem (Exercise (5.1)), albeit with a few twists due to the possibility that the signal states might not be orthogonal. To begin with, the mixture of signal states, Eq. (5.22), possesses a spectral decomposition,

$$\hat{\rho} = \sum_k \rho_k \hat{P}_k, \quad \operatorname{tr}(\hat{P}_k) = 1, \quad \rho_k \geq 0, \quad \sum_k \rho_k = 1. \tag{5.24}$$

In contrast to Eq. (3.49), we assume here that all projectors are one-dimensional. Should one of the eigenvalues be degenerate, we take that into account by having the same eigenvalue appear multiple times in the sum. The equality on the right-hand side expresses the fact that the state is normalized, $\operatorname{tr}(\hat{\rho}) = 1$. Unless the signal states are mutually orthogonal, the eigenvalues, $\{\rho_k\}$, and associated projectors, $\{\hat{P}_k\}$, differ from the probabilities and projectors of the signal states, $\{p_i\}$ and $\{|\psi_i\rangle\langle\psi_i|\}$. In particular, the projectors $\{\hat{P}_k\}$ are always mutually orthogonal, even if the signal states are not. After Alice has produced N signal states, as yet unknown to Bob, the latter's knowledge about the composite system is described by the N-fold product state

$$\hat{\rho}^{\otimes N} = \left(\sum_k \rho_k \hat{P}_k\right)^{\otimes N}.$$

Multiplying out the brackets in this N-fold tensor product yields a convex combination of tensor products of projectors. The weight of a particular tensor product of projectors depends only on the frequencies, $\{N_k\}$, $\sum_k N_k = N$, with which the various projectors occur in that product, not on their order. Thus, we may group together tensor products which differ only by a permutation, π, of the projectors, yielding

$$\hat{\rho}^{\otimes N} = \sum_{\{N_i\}} \left(\prod_k \rho_k^{N_k} \right) \hat{P}(\{N_i\}), \tag{5.25}$$

where

$$\hat{P}(\{N_i\}) := \sum_\pi \pi \left[\bigotimes_i \hat{P}_i^{\otimes N_i} \right]. \tag{5.26}$$

The latter is a projector onto a higher-dimensional subspace of the composite Hilbert space. Its dimension equals the number of inequivalent permutations, or combinations, of the constituent projectors:

$$\mathrm{tr}(\hat{P}(\{N_i\})) = \binom{N}{N_1, N_2, \ldots} = \frac{N!}{N_1! N_2! \cdots}. \tag{5.27}$$

Expanding by this dimension, we can cast our formula for the product state in the alternative form

$$\hat{\rho}^{\otimes N} = \sum_{\{N_i\}} \left(\mathrm{tr}(\hat{P}(\{N_i\})) \prod_k \rho_k^{N_k} \right) \hat{\rho}(\{N_i\}), \tag{5.28}$$

where

$$\hat{\rho}(\{N_i\}) := \frac{\hat{P}(\{N_i\})}{\mathrm{tr}(\hat{P}(\{N_i\}))} \tag{5.29}$$

is now a statistical operator. The latter only has support on the subspace associated with $\hat{P}(\{N_i\})$ and, within that subspace, represents total ignorance. Written in this form, the product state appears as a mixture of states of total ignorance defined on varying—in fact, mutually orthogonal—subspaces. When the number of trials becomes large, we may approximate the factorials in Eq. (5.27) by Stirling's formula, Eq. (2.46). This gives the asymptotic behaviour of the weights,

$$\mathrm{tr}(\hat{P}(\{N_i\})) \prod_k \rho_k^{N_k} \sim \sqrt{\frac{2\pi N}{\prod_k (2\pi N_k)}} \prod_i \left(\frac{\rho_i N}{N_i} \right)^{N_i},$$

which, for very large $\{N_i\}$, is dominated by that factor which is exponential in N,

$$\prod_i \left(\frac{\rho_i N}{N_i} \right)^{N_i} = 2^{-NS(\omega\|\rho)}.$$

The last expression features the classical relative entropy, Eq. (5.12), of the relative frequency distribution, $w := \{N_i/N\}$, and the eigenvalue distribution, $\rho := \{\rho_i\}$. In the mixture, Eq. (5.28), only those frequency distributions contribute significantly for which that exponential is of the order one. This only happens when the relative entropy is at most of the order $1/N$ and hence, in view of Eq. (5.13), the relative frequencies are (almost) identical with the eigenvalues, $w \approx \rho$. (A more rigorous analysis reveals that the deviation may be at most of the order $1/\sqrt{N}$.) Such relative frequencies are said to be 'typical'. In the original decomposition, Eq. (5.25), all terms with typical frequencies have practically the same probability. Therefore, the associated projectors can be grouped together once again, this time to a single projector onto the *typical subspace*,

$$\hat{P}_{\text{typ}} := \sum_{S(w\|\rho)\sim 1/N} \hat{P}(\{N_i\}). \tag{5.30}$$

As $N \to \infty$, the product state effectively has support only on this typical subspace and, within that subspace, represents total ignorance:

$$\hat{\rho}^{\otimes N} \sim \frac{\hat{P}_{\text{typ}}}{\text{tr}(\hat{P}_{\text{typ}})}. \tag{5.31}$$

This result is, in essence, the statement of the *typical subspace theorem*. (I am omitting all the technical bells and whistles here.) It might look somewhat familiar to readers who know statistical mechanics: there, *cum grano salis*, it reflects the equivalence of the canonical (left) and microcanonical (right) ensembles. For the purposes of encoding, the typical subspace theorem implies that the dimension of the composite Hilbert space can be effectively reduced to that of the typical subspace. Combinations of signal states that lie outside the typical subspace are so rare that they may be safely ignored; their omission produces a negligible error. For the transmission over the quantum channel, therefore, it is sufficient to map only the typical subspace to the Hilbert space of qubits. We achieve maximal compression if we make the latter as small as possible; that is to say, if we make its dimension just equal to the dimension of the typical subspace. The minimum number of qubits is then given by the logarithm of that dimension,

$$\log[\text{tr}(\hat{P}_{\text{typ}})] = -\text{tr}[\hat{\rho}^{\otimes N}\log(\hat{\rho}^{\otimes N})] = -\text{tr}[(\hat{\rho}\log\hat{\rho})^{\otimes N}].$$

Thanks to the factorization of the trace, Eq. (3.96), we can write this as

$$\log[\text{tr}(\hat{P}_{\text{typ}})] = -N\,\text{tr}(\hat{\rho}\log\hat{\rho}) = NS(\hat{\rho}).$$

Dividing by the number of trials, or signal states, we obtain the average number of qubits required to encode a signal state. Indeed, this number equals the von Neumann entropy.

Like its classical counterpart, the von Neumann entropy exhibits properties that conform with our expectations about a reasonable information measure:

1. The von Neumann entropy is non-negative, $S(\hat{\rho}) \geq 0$. It vanishes if and only if the state is pure,

$$S(\hat{\rho}) = 0 \quad \Leftrightarrow \quad \hat{\rho} \text{ pure}, \tag{5.32}$$

which is the analogue of Eq. (5.3). Purity—that is, maximal knowledge about a quantum system—is the counterpart of certainty about the outcome of a classical random experiment.

2. It is invariant under unitary transformations:

$$S(\hat{U}\hat{\rho}\hat{U}^{\dagger}) = S(\hat{\rho}). \tag{5.33}$$

This is the quantum analogue of the classical permutation invariance, Eq. (5.4).

3. For a bipartite system composed of constituents A and B, the entropy is subadditive,

$$S(\hat{\rho}_{\mathrm{AB}}) \leq S(\hat{\rho}_{\mathrm{A}}) + S(\hat{\rho}_{\mathrm{B}}). \tag{5.34}$$

Here $\hat{\rho}_{\mathrm{AB}}$ denotes the state of the composite system, and $\hat{\rho}_{\mathrm{A}}$ and $\hat{\rho}_{\mathrm{B}}$ denote the reduced states of the respective constituents. Subadditivity reflects the fact that going from the composite state to the reduced states generally entails a loss of information, namely the information about correlations. This inequality is the analogue of Eq. (5.7) for the Shannon entropy. And parallel to Eq. (5.8), equality holds if and only if the two constituents are uncorrelated:

$$S(\hat{\rho}_{\mathrm{AB}}) = S(\hat{\rho}_{\mathrm{A}}) + S(\hat{\rho}_{\mathrm{B}}) \quad \Leftrightarrow \quad \hat{\rho}_{\mathrm{AB}} = \hat{\rho}_{\mathrm{A}} \otimes \hat{\rho}_{\mathrm{B}}. \tag{5.35}$$

Once again, the above three properties (plus mathematical assumptions of a more technical nature) specify the function $S(\hat{\rho})$ uniquely, up to a multiplicative constant. In this sense, the von Neumann entropy is the unique measure of quantum information.

Despite the many similarities between the classical and quantum entropies, there are also some important differences. One crucial difference concerns the relationship between the entropy of a composite system and that of its individual constituents. While in both the classical and the quantum case we have the same upper bound for the joint entropy, Eqs (5.7) and (5.34), there is a characteristic difference in its lower bound. In the classical case, the additivity of the entropy, Eq. (5.5), and the fact that the conditional entropy, Eq. (5.6), is never negative imply that the joint entropy cannot be smaller than that of any constituent:

$$S_{\mathrm{class}}(\rho_{\mathrm{AB}}) \geq S_{\mathrm{class}}(\rho_{\mathrm{A}}). \tag{5.36}$$

Here I added the subscript 'classical' to emphasize that the ρ's (without hat) represent classical distributions rather than quantum states. By contrast, in the quantum case a non-negative conditional entropy cannot be defined and an analogous argument cannot be made. Indeed, there are instances where the joint entropy is smaller than that of an individual constituent. A case in point is the Bell states, Eq. (3.99), which are pure composite states with zero entropy, $S(\hat{\rho}_{AB}) = 0$. The reduced states of the individual constituents, on the other hand, are totally mixed (Eq. (3.100)) and have an entropy equal to one, $S(\hat{\rho}_A) = S(\hat{\rho}_B) = 1$. Thus, the above classical inequality is violated. In fact, by definition, this happens whenever a pure composite state is entangled. For the von Neumann entropy of a composite quantum system there exists only a weaker lower bound, given by the triangle inequality, or *Araki–Lieb inequality*,

$$S(\hat{\rho}_{AB}) \geq |S(\hat{\rho}_A) - S(\hat{\rho}_B)|. \tag{5.37}$$

You will prove this inequality in Exercise (5.10). In the case of pure entangled states like the Bell states, it implies that the entropies of the individual constituents, albeit not zero, must be equal. Indeed, in a Bell state the reduced states of both qubits are totally mixed, with identical non-zero entropies.

The properties of the von Neumann entropy can be useful to derive very elegantly, for instance, constraints on the correlations among three or more quantum systems. One simple argument goes as follows. We consider three systems, A, B, and C, and assume that the first two together are in some pure composite state, $\hat{\rho}_{AB}$; this pure state has zero entropy, $S(\hat{\rho}_{AB}) = 0$. The joint entropy of all three systems, $S(\hat{\rho}_{ABC})$, must satisfy both the Araki–Lieb inequality and subadditivity,

$$|S(\hat{\rho}_{AB}) - S(\hat{\rho}_C)| \leq S(\hat{\rho}_{ABC}) \leq S(\hat{\rho}_{AB}) + S(\hat{\rho}_C).$$

Since the entropy for the pair AB vanishes, this implies $S(\hat{\rho}_{ABC}) = S(\hat{\rho}_C) = S(\hat{\rho}_{AB}) + S(\hat{\rho}_C)$, which, by Eq. (5.35), further implies that the tripartite state factorizes, $\hat{\rho}_{ABC} = \hat{\rho}_{AB} \otimes \hat{\rho}_C$. In sum, we have

$$S(\hat{\rho}_{AB}) = 0 \quad \Rightarrow \quad \hat{\rho}_{ABC} = \hat{\rho}_{AB} \otimes \hat{\rho}_C. \tag{5.38}$$

(You already anticipated this result, using different means, in Exercise (3.21).) This innocent-looking conclusion means, in particular, that whenever two systems A and B are in a pure entangled state like a Bell state, they cannot be correlated in any way with a third system, C. In other words, strong quantum correlations between two systems preclude correlations of any kind with a third system. In this sense, entanglement is *monogamous*. Monogamy is exploited, for instance, in the design of secure quantum communication protocols for the distribution of cryptographic keys. Such protocols allow two legitimate parties to share information through quantum correlations, and at the same time prevent an illegitimate third party from eavesdropping on that exchange—which would require establishing correlations, forbidden by monogamy, with the legitimate parties.

There are many more practical applications of quantum information theory, including its extension to noisy channels and methods for error correction, which are of paramount importance for the development of real-world quantum hard- and software. These are, however, beyond the scope of this book, and we shall limit ourselves to just a few simple protocols that illustrate the basic principles. They include the quantum key distribution and superdense coding protocols, already alluded to in Section 5.1, which we will explain in Sections 5.3 and 5.4, as well as the protocol for 'teleportation': the transmission of a quantum state over a classical channel, assisted by previously shared entanglement. The latter will also be discussed in Section 5.4.

5.3 Cryptography

Communication that is provably secure requires private cryptographic keys, which must be generated and shared between the parties in advance. It is imperative that this advance distribution of keys be protected against eaves-dropping. I show that such protection can be achieved with a suitable quantum communication protocol, the BB84 protocol.

Secure communication requires the use of cryptographic keys. With the help of such a key—and the longer it is, the more secure—a message is encrypted by the sender; and with the help of another key, it is decrypted by the recipient. There are two basic forms of cryptography: private-key cryptography and public-key cryptography. In the former, the sender and the receiver use the same private key for encryption and decryption. This form of cryptography poses two practical challenges: the sharing of the private key must occur in perfect secrecy; and each pair of communication partners must have its own set of private keys. The number of keys needed thus quickly becomes very large; in a network where arbitrary pairs of participants should be able to communicate with each other in private, the total number of keys grows quadratically with the size of the network. Both issues are resolved by public-key cryptography, which is today the most widely used form of cryptography. In this form of cryptography, there are two different keys used to encrypt and decrypt a message. The key used to encrypt a message is the recipient's *public* key; whereas the key used to decrypt the message is the recipient's *private* key. There is no longer a need to share a private key with others; a party who wishes to join a secure conversation simply posts his or her public key. All other parties who wish to send encrypted messages to the new participant then use this same public key. Thus, the total number of keys in the network grows only linearly with the number of its members.

The public and the private key must be related to each other in order to ensure that a message encrypted with the former can actually be decrypted with the latter. Thus, in principle, it is possible to infer the secret, private key from the public key. However, the algorithm used to create a pair of public and private keys is such that it would require enormous computational resources, growing exponentially with the length of the keys, to do so. Specifically, in the widely used RSA cryptosystem, inferring the private key amounts to solving a mathematical problem known as factoring, that is, finding the prime factors of a large non-prime integer. It is widely believed, albeit not known with absolute certainty, that this problem is computationally hard—at least on a classical computer. One of the milestones in the development of quantum computing has been the discovery that factorization is among the tasks that a quantum computer can perform efficiently, in contrast to a classical computer. So far, the pertinent quantum algorithm—Shor's algorithm—exists only on paper; it is not yet possible to build the hardware to actually implement it, beyond trivial factorizations like $21 = 3 \cdot 7$. But if one day a sufficiently large quantum computer can be built, this will spell the end of secure public-key encryption based on the RSA cryptosystem. (Moreover, there is still the possibility that an efficient classical factorization algorithm can be found.)

Considerations of this kind have led to renewed interest in private-key encryption. If the private keys are not just exclusive to each pair of communication partners but, moreover, randomly changed every time a communication takes place, then private-key encryption can actually be *proven* to be secure. However, if one wants to return to private-key encryption, one must find ways to solve the associated practical problems which we mentioned earlier. From a security perspective, the most pressing of these is to find a secure method of sharing a private key, especially if the parties are not physically in the same room. It turns out that this problem can be tackled by clever exploitation of some fundamental features of quantum theory; the pertinent protocols are known as *quantum key distribution*. Here we shall discuss a basic variant of such a protocol, the BB84 protocol, named after its inventors, Charles Bennett and Gilles Brassard, and the year of its publication, 1984. In Exercise (5.16) you will encounter an alternative protocol, the EPR protocol, whose physical security is based on the monogamy of entanglement.

The basic idea behind the BB84 protocol is that while an eavesdropper might be able to access the communication channel between two legitimate parties, she would not be able to read out data without causing a disturbance, which, in turn, would be detected by the legitimate parties. Once alerted that their communication has been compromised, the legitimate parties would immediately terminate their communication and discard the—now insecure—shared key. The protocol works with the following quantum circuit:

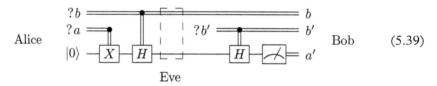

$$(5.39)$$

The two parties, Alice and Bob, are connected by two channels, one classical (indicated by the uninterrupted double line at the top) and one quantum (single line at the bottom). Between Alice and Bob there might be a third party, Eve, eavesdropping on their classical and quantum communication. The first part of the protocol proceeds as follows:

1. Alice randomly generates two classical bits, a and b. (Their randomness is indicated by the question marks, and their classicality by the use of double lines.) These two random experiments are independent, and their possible outcomes, '0' or '1', occur with equal probability one half. In addition, Alice has at her disposal a qubit in the basis state $|0\rangle$.

2. The bit a controls a NOT (that is, Pauli-X) operation on the qubit: depending on whether a is '0' or '1', the qubit is either left in the state $|0\rangle$ or flipped to the orthogonal basis state, $|1\rangle$. That the Pauli-X gate is controlled by a classical

bit, rather than by a control qubit, is represented graphically by the double vertical line.

3. The second bit, b, controls a subsequent Hadamard gate. If b is '0', the qubit state is left in the basis state $|0\rangle$ or $|1\rangle$, respectively. If b is '1', on the other hand, then, according to Eq. (4.6), the Hadamard gate rotates these basis states to $|+\rangle$ or $|-\rangle$, respectively. Altogether, depending on the values of a and b, the two consecutive controlled operations yield one out of four possible qubit states, $|\psi\rangle$, as determined by the following table:

$$
\begin{array}{cc|c}
a & b & |\psi\rangle \\
\hline
0 & 0 & |0\rangle \\
0 & 1 & |+\rangle \\
1 & 0 & |1\rangle \\
1 & 1 & |-\rangle
\end{array}
\tag{5.40}
$$

Each of these possibilities occurs with probability one quarter.

4. Alice sends the qubit to Bob through the quantum channel.

5. At the receiving end, Bob randomly generates a classical bit, b', which is used to control a Hadamard gate on the incoming qubit. After this controlled operation there are the following possibilities for the final state of the qubit, $|\psi'\rangle$:

$$
\begin{array}{ccc|c}
a & b & b' & |\psi'\rangle \\
\hline
0 & 0 & 0 & |0\rangle \\
0 & 0 & 1 & |+\rangle \\
0 & 1 & 0 & |+\rangle \\
0 & 1 & 1 & |0\rangle \\
1 & 0 & 0 & |1\rangle \\
1 & 0 & 1 & |-\rangle \\
1 & 1 & 0 & |-\rangle \\
1 & 1 & 1 & |1\rangle
\end{array}
\tag{5.41}
$$

Once again, there are four distinct possibilities for the qubit state, each occurring with probability one quarter.

6. Bob performs on the qubit a measurement in the standard basis, $\{|0\rangle, |1\rangle\}$. Its outcome, '0' or '1', is another classical bit, a'.

7. Bob confirms to Alice (over the phone, say) the receipt of the qubit. Only then does Alice send to Bob, via the classical channel, her classical bit b.

8. Bob compares Alice's b to his own b' and announces whether they agree. There are two cases:

   a. If they do agree, then, as the table in Eq. (5.41) shows, the state of the qubit just before Bob's measurement must have been one of the standard basis states, $|0\rangle$ or $|1\rangle$; more precisely, it was $|0\rangle$ if $a = 0$ and $|1\rangle$ if $a = 1$. Hence whenever

$b = b'$, it is also $a = a'$. In this case Alice and Bob retain a or a', respectively, as part of their shared key.

    b. If b is different from b', on the other hand, the qubit must have been in one of the two rotated basis states, $|+\rangle$ or $|-\rangle$, prior to measurement. In this case there is no certainty about whether a and a' agree. Consequently, Alice and Bob discard this bit.

9. The procedure is repeated until Alice and Bob have generated a key which is long enough.

Up to this point the protocol has clearly succeeded in producing a shared key, but what about its security? Since the information transmitted through the classical channel, b, is of no use, an eavesdropper would have to perform measurements on the qubits which are transmitted through the quantum channel. The eavesdropper, Eve, knows that the qubit is in one of four possible quantum states, $|0\rangle$, $|1\rangle$, $|+\rangle$, or $|-\rangle$. Consequently, she will perform her measurements in either the $\{|0\rangle, |1\rangle\}$ or the $\{|+\rangle, |-\rangle\}$ basis. For any given qubit passing through the channel, the probability that Eve has chosen the 'correct' measurement basis, in the sense that it includes the state of the passing qubit, is equal to one half. Then, and only then, the state of the qubit is not altered by the measurement. Otherwise, with the remaining probability one half, the measurement does disturb the state of the qubit: after the measurement, the state changes to one of the basis states in Eve's 'wrong' basis. Due to this disturbance, it is no longer guaranteed that whenever $b = b'$, it is also $a = a'$. Rather, there will be instances where, even though $b = b'$, it is $a \neq a'$. (In Exercise (5.15) you will show that the latter occurs with probability one quarter.) This forms the basis for the second part of the BB84 protocol:

10. Alice draws a random sample from the bits in her key string, and tells Bob which bits she selected. Bob selects the corresponding bits from his key string.

11. Alice and Bob compare the values of their selected bits. Provided there was no interference (and no noise) in their communication, these values should all agree. If Alice and Bob find that they do indeed agree (up to some allowance for noise in the channel), they discard the check bits and use the remaining bit string as a secure private key.

12. If, on the other hand, Alice and Bob find significant disagreement, there must have been an eavesdropper. In this case, the protocol is aborted.

Like many other quantum algorithms, the BB84 protocol is probabilistic; there is randomness involved at various stages of the protocol. As a consequence, the time needed to generate a key of a prescribed length may vary.

The complete protocol stipulates two further steps, called 'information reconciliation' and 'privacy amplification'. However, these do not alter the key physical principle behind the security of the BB84 protocol—namely, that no measurement on an individual quantum system is capable of clearly distinguishing non-orthogonal quantum states; for example, $|0\rangle$ from $|+\rangle$. Any attempt by an eavesdropper to do so regardless causes a disturbance which the legitimate parties can detect.

5.4 Superdense Coding and Teleportation

I present two examples of how communication may be enhanced with the help of previously shared entanglement. In the first example, known as superdense coding, pre-shared entanglement is used to effectively double the classical information-carrying capacity of a qubit. In the second example, teleportation, it is exploited to transmit a quantum state with just two classical bits. I describe the respective protocols and show that they are optimal.

The two protocols in this section serve to illustrate how the clever use of entanglement can enhance the communication between two parties, via both classical and quantum channels. In both protocols entanglement comes in the form of *Bell pairs*; that is, pairs of qubits in one of the four Bell states, Eq. (3.99). (They are sometimes also called 'EPR pairs', named after Albert Einstein, Boris Podolsky, and Nathan Rosen, the authors of an influential critique of quantum mechanics where these pairs played a central role.) Such Bell pairs may be prepared, for instance, by means of the following simple circuit:

$$
\left.\begin{array}{c} |j\rangle \ -\boxed{H}\!\!-\!\!\bullet\!\!- \\[4pt] |k\rangle \ -\!\!\!\!-\!\!\oplus\!\!- \end{array}\right\} |\beta_{jk}\rangle, \quad j,k \in \{0,1\}. \tag{5.42}
$$

The communication partners—our familiar characters Alice and Bob—do this preparation and share the Bell pair in advance, in the sense that each party gets to keep one of the two entangled qubits. They keep the shared Bell pair in storage until it is needed for one of the communication protocols.

We begin with the protocol for *superdense coding* (sometimes simply referred to as 'dense coding'), which exploits entanglement to enhance the transmission of classical information over a quantum channel. Specifically, while we learnt in Section 5.1 that an individual qubit can carry only one bit of classical information (encoded in a basis state, $|0\rangle$ or $|1\rangle$), superdense coding allows a single qubit to effectively carry *two* bits of classical information. In the process, the superdense coding protocol 'consumes' entanglement in a sense which will be explained below. The pertinent quantum circuit looks like this:

$$\tag{5.43}$$

This circuit features two qubits, marked by single lines. At the beginning, they are the two members of the Bell pair which Alice and Bob have prepared and shared in advance. One of the members (top) is initially in the possession of Alice, while the other (bottom) is in the possession of Bob. Later in the process, Alice will pass her member of the pair on to Bob. The pair has been prepared in the Bell state

$$|\psi_0\rangle = |\beta_{00}\rangle = \frac{|00\rangle + |11\rangle}{\sqrt{2}}. \tag{5.44}$$

Alice wants to send two classical bits, x and y, but has only one qubit (her member of the Bell pair) at her disposal. She succeeds, nevertheless, with the help of the following protocol:

1. Alice lets the second classical bit, y, control a Pauli-X operation on her member of the Bell pair. Afterwards, the two qubits are in the state

$$|\psi_1\rangle = \frac{|y0\rangle + |1 \oplus y, 1\rangle}{\sqrt{2}}. \tag{5.45}$$

2. Alice lets the first classical bit, x, control another operation on her member of the Bell pair, this time a Pauli-Z. This yields

$$|\psi_2\rangle = \pm\frac{1}{\sqrt{2}} \begin{cases} |y0\rangle + |1 \oplus y, 1\rangle : x = 0 \\ |y0\rangle - |1 \oplus y, 1\rangle : x = 1 \end{cases}. \tag{5.46}$$

The overall sign is not important.
3. Alice sends a *single* qubit—namely, her member of the Bell pair—to Bob.
4. Bob couples the incoming qubit to his own member of the Bell pair via a CNOT gate, yielding

$$|\psi_3\rangle = \begin{cases} (|yy\rangle + |1 \oplus y, y\rangle)/\sqrt{2} = |+, y\rangle & : x = 0 \\ (|yy\rangle - |1 \oplus y, y\rangle)/\sqrt{2} = \pm|-, y\rangle & : x = 1 \end{cases}. \tag{5.47}$$

If $x = 1$, there is an overall sign depending on the value of y; but once again, this is not important.
5. Finally, Bob sends the incoming qubit through a Hadamard gate (Eq. (4.6)), which leads to the final state

$$|\psi_4\rangle = \begin{cases} |0y\rangle : x = 0 \\ |1y\rangle : x = 1 \end{cases} = |xy\rangle. \tag{5.48}$$

6. Bob performs measurements on both qubits in the standard basis, yielding the classical bits x and y.

Thus, indeed, Bob retrieves two bits of classical information, even though he received just a single qubit from Alice. In this sense, the classical information-carrying capacity of the qubit has been doubled, from one to two classical bits. That only worked because Alice and Bob had previously shared a qubit pair in an entangled state. The entanglement served as an additional communication 'resource', which the protocol subsequently 'used up'. Indeed, at the end of the process the two qubits are in a product state; the pair cannot be reused to pull the same trick a second time.

There exists another well-known protocol, *teleportation*, which is in a sense the mirror image of superdense coding. It also exploits entanglement to accomplish an otherwise impossible feat: the transmission of a quantum state over a classical channel. Under normal circumstances, if Alice is only allowed to send a finite amount of classical information to Bob, it is impossible for her to convey to Bob an arbitrary quantum state of a qubit, $|\psi\rangle$. In principle, she might represent this state in the standard basis,

$$|\psi\rangle = \alpha_0 |0\rangle + \alpha_1 |1\rangle, \tag{5.49}$$

and then relay to Bob the coefficients α_0 and α_1—but doing so with perfect accuracy would require an infinite number of classical bits. The situation changes when, in addition to the classical communication channel, Alice and Bob have at their disposal a previously shared Bell pair. With its help, Alice can convey to Bob any single-qubit state, $|\psi\rangle$, by sending just two classical bits. The pertinent quantum circuit looks like this:

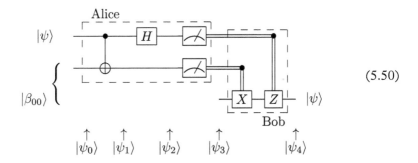

$$(5.50)$$

It comprises a total of three qubits: the qubit whose state, $|\psi\rangle$, Alice wants to communicate to Bob through the classical channel; and the two members of the previously shared qubit pair, prepared—like in the case of superdense coding—in the Bell state $|\beta_{00}\rangle$. One member of this pair is in the possession of Alice, whereas the other member resides with Bob. Assuming that $|\psi\rangle$ is represented as in Eq. (5.49), these three qubits are initially in the state

$$|\psi_0\rangle = |\psi\rangle \otimes |\beta_{00}\rangle = \frac{1}{\sqrt{2}} \Big[\alpha_0 |0\rangle \otimes (|00\rangle + |11\rangle) + \alpha_1 |1\rangle \otimes (|00\rangle + |11\rangle) \Big]. \tag{5.51}$$

The protocol now proceeds in the following steps:

1. Alice couples the qubit of interest, which is in the state $|\psi\rangle$, via a CNOT gate to her member of the Bell pair. Then the state of the three qubits changes to

$$|\psi_1\rangle = \frac{1}{\sqrt{2}}\left[\alpha_0\,|0\rangle \otimes (|00\rangle + |11\rangle) + \alpha_1\,|1\rangle \otimes (|10\rangle + |01\rangle)\right]. \qquad (5.52)$$

2. Alice sends the first qubit through a Hadamard gate. This entails a further change of the state, to

$$
\begin{aligned}
|\psi_2\rangle &= \frac{1}{2}\left[\alpha_0(|0\rangle + |1\rangle) \otimes (|00\rangle + |11\rangle) + \alpha_1(|0\rangle - |1\rangle) \otimes (|10\rangle + |01\rangle)\right] \\
&= \frac{1}{2}\left[|00\rangle \otimes (\alpha_0\,|0\rangle + \alpha_1\,|1\rangle) + |01\rangle \otimes (\alpha_0\,|1\rangle + \alpha_1\,|0\rangle)\right. \\
&\quad \left. + |10\rangle \otimes (\alpha_0\,|0\rangle - \alpha_1\,|1\rangle) + |11\rangle \otimes (\alpha_0\,|1\rangle - \alpha_1\,|0\rangle)\right].
\end{aligned}
\qquad (5.53)
$$

3. Alice performs measurements (in the standard basis) both on the qubit of interest and on her member of the Bell pair. Depending on the outcomes of these measurements, the remaining qubit—Bob's member of the Bell pair—is in one of the following four states:

$$|\psi_3\rangle = \begin{cases} \alpha_0\,|0\rangle + \alpha_1\,|1\rangle : 00 \\ \alpha_0\,|1\rangle + \alpha_1\,|0\rangle : 01 \\ \alpha_0\,|0\rangle - \alpha_1\,|1\rangle : 10 \\ \alpha_0\,|1\rangle - \alpha_1\,|0\rangle : 11 \end{cases} \qquad (5.54)$$

4. Alice sends the results of her two measurements to Bob. This requires two classical bits.

5. Depending on the measurement results received from Alice, Bob performs one out of four possible transformations on his qubit:

results received	operation on qubit
00	none
01	Pauli-X
10	Pauli-Z
11	first Pauli-X, then Pauli-Z

$$(5.55)$$

This is indicated in the circuit by the two classically controlled gates. In all four cases Bob's qubit ends up in the original state of Alice's qubit,

$$|\psi_4\rangle = |\psi\rangle. \qquad (5.56)$$

After the protocol, Bob possesses a qubit in precisely the state that Alice wanted to transmit. Indeed, this was achieved without sending a physical qubit; sending two classical bits was enough.

Like in the case of superdense coding, the crucial ingredient here is the entanglement of the previously shared Bell pair. As before, this entanglement serves as a resource which is being used up during the process. By the end, it is gone; it is no longer available to teleport a second qubit. Moreover, on Alice's side the measurements have eliminated all traces of the original state of interest, $|\psi\rangle$. The qubit which initially carried this state is now in one of the two basis states. Loosely speaking, the state $|\psi\rangle$ has 'disappeared' on Alice's side and 'reappeared' on Bob's side. Its disappearance on Alice's side is inevitable, for otherwise the circuit could serve as a device to clone a state, which is forbidden by the no-cloning theorem.

Could we do even better than the superdense coding and teleportation protocols? Might there be a way to use pre-shared entanglement even more cleverly—for instance, involving entangled states of more than two qubits—that would allow a qubit to carry *more* than two classical bits, or allow the transmission of a qubit state with *less* than two classical bits, respectively? The answer is no, for the following reason. Let us say that in an optimal world it is possible for a qubit to encode s bits of classical information, and for a qubit state to be teleported with t classical bits. Then we could combine the respective protocols as follows: teleport a qubit with t classical bits; and let that teleported qubit, in turn, encode s classical bits. In effect, this combination conveys s bits of classical information by sending t physical bits. If s were strictly greater than t, we would arrive at a logical contradiction. In that case we could iterate the procedure, leveraging the $s > t$ transmitted bits of classical information to teleport (on average) more than one qubit state, which, in turn, would encode $s' > s$ classical bits; and so on, until, in the end, we would be able to send an infinite amount of classical information with a finite number of physical bits. This is clearly impossible. Hence we conclude that it must be $s \leq t$. (You will weigh this argument more carefully and respond to a possible criticism in Exercise (5.17).) From the above protocols we already know that $s \geq 2$ and $t \leq 2$. So altogether, we obtain the sequence of inequalities $2 \leq s \leq t \leq 2$, which can only be satisfied if $s = t = 2$. Thus, indeed, the superdense coding and teleportation protocols are already optimal.

Chapter Summary

- Classical information is quantified by the minimum number of bits needed to communicate it. When the information pertains to the outcome of a classical random experiment, embedded in a sequence of many trials, this number is given by the Shannon entropy.
- Classical information may be encoded in quantum states. In order for the information to be retrievable, the coding states must be orthogonal. As a consequence, a qubit cannot carry more classical information than a classical bit.

- Quantum information is quantified by the minimum number of qubits needed to communicate it. When the information pertains to a pure quantum state, drawn at random from a set of possible 'signal states' and embedded in a long sequence of such states, this number is given by the von Neumann entropy. It does not depend on the individual signal states but only on their weighted average.

- Classically, the entropy of a composite system can never be smaller than that of a constituent. By contrast, this may happen in the quantum realm when the composite state is entangled.

- Entanglement is monogamous, in the sense that entanglement between two systems precludes correlations of any kind with a third system.

- Cryptographic keys can be generated and shared between distant parties in a manner that is protected against eavesdropping by the laws of quantum physics.

- One specific realization of quantum key distribution is the BB84 protocol. It exploits the fact that measurements on an individual quantum system cannot resolve non-orthogonal states. Any attempt by an eavesdropper to do so regardless causes a disturbance which will alert the legitimate parties.

- Entanglement, in the form of previously shared Bell pairs, can be harnessed to enhance communication via both classical and quantum channels. The entanglement is consumed in the process; it cannot be reused.

- Superdense coding exploits entanglement to enhance the transmission of classical information over a quantum channel. Specifically, it allows a single qubit to effectively carry two bits, rather than just one bit, of classical information.

- Teleportation uses entanglement to transmit a quantum state with just two classical bits.

- The superdense coding and teleportation protocols are optimal in the sense that there exist no other protocols that would allow a qubit to carry more than two classical bits, or allow the transmission of a qubit state with less than two classical bits, respectively.

Further Reading

A standard reference for classical information theory is the book by Cover and Thomas (2006). For more details about quantum information theory you can turn once again to the classic by Nielsen and Chuang (2000) or to the equally comprehensive book by Wilde (2013). As a careful and pedagogical introduction to both classical and quantum information I recommend the text by Barnett (2009). The central mathematical objects in information theory—entropy and relative entropy—exhibit a wealth of interesting and useful properties, which are reviewed in Wehrl (1978), Vedral (2002), Ruskai (2002), and in the book by Petz (2008). Information and entropy also play an important role in statistical mechanics, as explained, for instance, in the texts by Balian (1991) and Rau (2017). The BB84 protocol for quantum key distribution was devised by Bennett and Brassard (1984), the EPR protocol in Exercise (5.16) by Ekert (1991). Superdense coding is due to Bennett and Wiesner (1992), and the teleportation protocol to Bennett *et al.* (1993). These protocols are also discussed in several of the books and lecture notes mentioned in Chapter 4. The hardware implementation of quantum communication, especially quantum key distribution, is pretty advanced and beginning to be commercialized. One example of recent progress is the entanglement-based secure quantum cryptography over a distance of more than 1,000 km described in Yin *et al.* (2020).

EXERCISES

5.1. Shannon's noiseless channel coding theorem

Consider a random experiment with d possible outcomes. The probability of outcome i equals p_i, $i = 1, \ldots, d$. When you repeat the experiment N times, you obtain an ordered sequence of N outcomes. This sequence is called 'typical' if the relative frequency of outcome i, f_i, coincides with its probability, p_i. By the law of large numbers, for $N \to \infty$ virtually all sequences are typical.

(a) For a given (large) N and probability distribution $\{p_i\}$, how many typical sequences are there?

(b) In the communication setup shown in Fig. 5.1, Alice and Bob might compile beforehand a numbered list of all typical sequences. Then, in order to communicate the results of her N trials, which for $N \to \infty$ will almost certainly form a typical sequence, Alice just has to communicate the associated item number on the pre-compiled list. Argue that this encoding is optimal. Show that the average number of bits required per trial equals the Shannon entropy, Eq. (5.2).

5.2. Additivity of classical information

Prove the additivity of the Shannon entropy, Eq. (5.5).

5.3. Non-negativity of classical relative entropy

Verify that the classical relative entropy, defined by Eq. (5.12), is non-negative, and that it vanishes if and only if the two distributions are identical.

Hint: Combine the two logarithms into a single one and use $\log x = \ln x / \ln 2 \leq (x - 1)/\ln 2$.

5.4. Mutual information and relative entropy

Verify that the mutual information can be expressed as a weighted average of relative entropies, Eq. (5.14).

5.5. Quantum relative entropy

The quantum relative entropy, defined by Eq. (5.17), exhibits many interesting and useful properties. Show:

(a) The quantum relative entropy is invariant under simultaneous unitary transformations of both arguments.

(b) If the two statistical operators share a common eigenbasis, $\{|i\rangle\}$, their quantum relative entropy equals the relative entropy of the classical distributions $\rho :=\{\langle i|\hat{\rho}|i\rangle\}$ and $\sigma := \{\langle i|\hat{\sigma}|i\rangle\}$:

$$S(\hat{\rho}\|\hat{\sigma}) = S(\rho\|\sigma).$$

(c) For an arbitrary classical probability distribution $\{p_i\}$ and normalized statistical operators $\{\hat{\rho}_i\}$ and $\hat{\sigma}$, it is

$$\sum_i p_i S(\hat{\rho}_i\|\hat{\sigma}) = \sum_i p_i S(\hat{\rho}_i\|\hat{\rho}) + S(\hat{\rho}\|\hat{\sigma})$$

with

$$\hat{\rho} := \sum_i p_i \hat{\rho}_i.$$

(d) When the arguments are swapped, it is

$$\sum_i p_i S(\hat{\sigma}\|\hat{\rho}_i) = S(\hat{\sigma}\|\hat{\rho}') + \sum_i p_i S(\hat{\rho}'\|\hat{\rho}_i)$$

with

$$\log \hat{\rho}' := \sum_i p_i \log \hat{\rho}_i + c,$$

where c is adjusted such that $\hat{\rho}'$ is normalized.

(e) The quantum relative entropy can be written as a classical relative entropy even when the two statistical operators do not share a common eigenbasis. In this case, given their spectral decompositions,

$$\hat{\rho} = \sum_i \rho_i \hat{P}_i, \quad \hat{\sigma} = \sum_k \sigma_k \hat{Q}_k$$

with $\mathrm{tr}(\hat{P}_i) = \mathrm{tr}(\hat{Q}_k) = 1$, their quantum relative entropy is given by

$$S(\hat{\rho}\|\hat{\sigma}) = S(P\|\Sigma),$$

where $P := \{P_{ik}\}$ (uppercase 'rho') and $\Sigma := \{\Sigma_{ik}\}$ are the classical distributions (with two indices)

$$P_{ik} := \text{tr}(\hat{P}_i \hat{Q}_k)\rho_i, \quad \Sigma_{ik} := \text{tr}(\hat{P}_i \hat{Q}_k)\sigma_k.$$

This result implies, in particular, that the quantum relative entropy shares with its classical counterpart two important properties: it is non-negative and it vanishes if and only if the two states are identical (prove the latter). *Hint*: Use (5.5c).

(f) The quantum relative entropy is convex in its first argument,

$$\sum_i p_i S(\hat{\rho}_i \| \hat{\sigma}) \geq S(\hat{\rho} \| \hat{\sigma}),$$

with $\hat{\rho}$ defined as before.

(g) More strongly, the quantum relative entropy is *jointly convex*,

$$\sum_i p_i S(\hat{\rho}_i \| \hat{\sigma}_i) \geq S(\hat{\rho} \| \hat{\sigma}),$$

with $\hat{\rho}$ defined as before and

$$\hat{\sigma} := \sum_i p_i \hat{\sigma}_i.$$

This includes the convexity in the first argument as a special case. (Project) *Hint*: Show that the relative entropy may be written in the form

$$S(\hat{\rho} \| \hat{\sigma}) = -\lim_{p \to 0^+} \frac{1}{p} \left[\text{tr}(\hat{\rho}^{1-p} \hat{\sigma}^p) - \text{tr}(\hat{\rho}) \right]$$

and prove that the function

$$f_p(\hat{\rho}, \hat{\sigma}) := \text{tr}(\hat{\rho}^{1-p} \hat{\sigma}^p), \quad 0 < p < 1,$$

is jointly concave. (This is a special instance of 'Lieb's concavity theorem'.) The latter looks plausible because the analogous function of two variables,

$$f_p(x, y) := x^{1-p} y^p, \quad 0 < p < 1, \quad x \geq 0, \quad y \geq 0,$$

is jointly concave; the challenge is to generalize this to operators.

5.6. Relative entropy for composite systems

Consider a system composed of two constituents, A and B. Show that the quantum relative entropy is:

(a) *superadditive,*

$$S(\hat{\rho}_{AB}\|\hat{\sigma}_A \otimes \hat{\sigma}_B) \geq S(\hat{\rho}_A\|\hat{\sigma}_A) + S(\hat{\rho}_B\|\hat{\sigma}_B),$$

with equality if and only if $\hat{\rho}_{AB}$ factorizes, $\hat{\rho}_{AB} = \hat{\rho}_A \otimes \hat{\rho}_B$

(b) *monotonic,*

$$S(\hat{\rho}_{AB}\|\hat{\sigma}_{AB}) \geq S(\hat{\rho}_A\|\hat{\sigma}_A).$$

The latter reflects the intuitive expectation that the distinguishability of two states can only decrease when a part of the system (here: constituent B) is ignored.

Hint: Use the joint convexity of the relative entropy (Exercise (5.5)), with $p_i = 1/d_B$, $\hat{\rho}_i = \hat{U}_i\hat{\rho}_{AB}\hat{U}_i^\dagger$, $\hat{\sigma}_i = \hat{U}_i\hat{\sigma}_{AB}\hat{U}_i^\dagger$, and $\hat{U}_i$ defined as in Exercise (3.20).

5.7. von Neumann entropy

The von Neumann entropy exhibits many more interesting properties than those listed in Section 5.2. In this exercise you will investigate a few of them. Show that:

(a) The von Neumann entropy can be expressed in terms of the quantum relative entropy as

$$S(\hat{\rho}) = \log d - S(\hat{\rho}\|\hat{\iota}),$$

where d is the dimension of the Hilbert space and $\hat{\iota} := \hat{I}/d$ is the statistical operator representing total ignorance.

(b) For an arbitrary orthonormal basis, $\{|i\rangle\}$, it is

$$S(\hat{\rho}) \leq S_{\text{class}}(\{\langle i|\hat{\rho}|i\rangle\}),$$

with equality if and only if $\{|i\rangle\}$ is an eigenbasis of $\hat{\rho}$.

(c) The von Neumann entropy is *concave*: for an arbitrary classical probability distribution $\{p_i\}$ and normalized statistical operators $\{\hat{\rho}_i\}$, it is

$$S\left(\sum_i p_i\hat{\rho}_i\right) \geq \sum_i p_i S(\hat{\rho}_i).$$

Concavity reflects the fact that any 'mixing' of quantum states,

$$\hat{\rho}_1, \hat{\rho}_2, \dots \rightarrow \sum_i p_i\hat{\rho}_i,$$

generally entails a loss of information.

Hint: Express the von Neumann entropy as a quantum relative entropy and use results of Exercise (5.5).

5.8. **Entropy of a two-level system**

(a) Consider a normalized statistical operator for a two-level system in the form of Eq. (3.65), with $\mathrm{tr}(\hat\rho) = 1$. Calculate its von Neumann entropy as a function of the 'position vector' $\vec{r}$.

(b) Calculate the von Neumann entropy of the statistical operators with the following matrix representations:

$$\rho = \begin{pmatrix} 1 & 0 \\ 0 & 0 \end{pmatrix}$$

$$\rho = \frac{1}{2}\begin{pmatrix} 1 & 1 \\ 1 & 1 \end{pmatrix}$$

$$\rho = \frac{1}{3}\begin{pmatrix} 2 & 1 \\ 1 & 1 \end{pmatrix}.$$

5.9. **Losing and gaining information during the measurement process**

Consider a measurement whose potential outcomes are represented by mutually orthogonal projectors, $\{\hat{P}_i\}$, which add up to the unit operator.

(a) After performing this measurement, but *before* learning the specific result, the statistical operator, $\hat\rho$, is updated to

$$\mathcal{D}(\hat\rho) := \sum_i \hat{P}_i \hat\rho \hat{P}_i.$$

Show that the von Neumann entropy of this updated state is at least as large as the original von Neumann entropy,

$$S(\mathcal{D}(\hat\rho)) \geq S(\hat\rho),$$

with equality if and only if $\mathcal{D}(\hat\rho) = \hat\rho$. Thus, performing a measurement without reading the result generally leads to a *loss* of information.

Hint: Non-negativity of the quantum relative entropy.

(b) Consider a system composed of two constituents. The first constituent, A, is an auxiliary system which will prove helpful for your calculations; the second constituent, B, is the one on which the actual measurement is performed. The Hilbert space of A has dimension equal to the number of different outcomes of our measurement, and has some orthonormal basis, $\{|i\rangle\}$. On the composite Hilbert space, define the operator

$$\hat\rho_{\mathrm{AB}} := \sum_i |i\rangle\langle i| \otimes \sqrt{\hat\rho}\,\hat{P}_i\sqrt{\hat\rho}.$$

Show:

i. This is a valid statistical operator for the composite system.

ii. The reduced states of A and B are given by

$$\hat{\rho}_A = \sum_i \text{tr}(\hat{\rho}\hat{P}_i)|i\rangle\langle i|, \quad \hat{\rho}_B = \hat{\rho}.$$

iii. The joint entropy equals the entropy of the updated state prior to learning the result,

$$S(\hat{\rho}_{AB}) = S(\mathcal{D}(\hat{\rho})).$$

Hint: Verify that $\text{tr}(\hat{\rho}_{AB}{}^n) = \text{tr}(\mathcal{D}(\hat{\rho})^n)$ for all $n \geq 1$.

iv. The latter, in turn, is the sum of two terms,

$$S(\mathcal{D}(\hat{\rho})) = \sum_i \text{tr}(\hat{\rho}\hat{P}_i) S(\hat{\rho}_{|i}) + S(\hat{\rho}_A),$$

where $\hat{\rho}_{|i}$ is the post-measurement state given by Lüders' rule, Eq. (3.56).

(c) Show that the expected von Neumann entropy *after* reading the result is smaller than or equal to the original entropy,

$$\sum_i \text{tr}(\hat{\rho}\hat{P}_i) S(\hat{\rho}_{|i}) \leq S(\hat{\rho}),$$

with equality if and only if $S(\hat{\rho}_{|i}) = S(\hat{\rho})$ for all i. So—at least on average— learning a specific outcome generally leads to a *gain* of information.
Hint: Use the previous result and the subadditivity of the entropy.

5.10. Purification reloaded

Purification (Exercise (3.23)) is a powerful tool that allows one to derive further interesting properties of the von Neumann entropy:

(a) Use purification to show directly that

$$S(\hat{\rho}_{AB}) = 0 \quad \Rightarrow \quad S(\hat{\rho}_A) = S(\hat{\rho}_B).$$

(b) Verify that the von Neumann entropy of the mixture defined by Eq. (5.22) is smaller than, or at most equal to, the Shannon entropy of the classical distribution $\{p_i\}$,

$$S(\hat{\rho}) \leq S(\{p_i\}),$$

with equality if and only if the signal states are mutually orthogonal, $\langle \psi_i | \psi_k \rangle = \delta_{ik}$.

Hint: Consider a composite system AB in the pure state

$$|\psi_{AB}\rangle = \sum_i \sqrt{p_i}\,|i\rangle \otimes |\psi_i\rangle,$$

where the $\{|i\rangle\}$ constitute an orthonormal basis in the Hilbert space of constituent A and the $\{|\psi_i\rangle\}$ pertain to constituent B. Exploit the fact that $S(\hat{\rho}_{AB}) = 0$. Then use for constituent A the first result of Exercise (5.9), with $\hat{P}_i = |i\rangle\langle i|$.

(c) Prove the Araki–Lieb inequality, Eq. (5.37).
   Hint: Purify AB by adding a third system, C. Then apply the subadditivity property, Eq. (5.34), to $S(\hat{\rho}_{AC})$ and $S(\hat{\rho}_{BC})$.

5.11. Holevo bound

(a) Show that if $\hat{\rho}$ is a normalized statistical operator, then so is

$$\mathcal{E}(\hat{\rho}) := \sum_k \frac{\mathrm{tr}(\hat{\rho}\hat{P}_k)}{\mathrm{tr}(\hat{P}_k)}\,\hat{P}_k.$$

Show that, with the help of this map $\mathcal{E}$, the classical relative entropy on the left-hand side of Eq. (5.18) can be written as a quantum relative entropy,

$$S(\rho_{B|a_i}\|\rho_B) = S(\mathcal{E}(\hat{\rho}_i)\|\mathcal{E}(\hat{\rho})).$$

(b) Show that

$$S(\hat{\rho}_i\|\mathcal{E}(\hat{\rho})) = S(\hat{\rho}_i\|\mathcal{E}(\hat{\rho}_i)) + S(\mathcal{E}(\hat{\rho}_i)\|\mathcal{E}(\hat{\rho})).$$

This is a special case of a more general 'Pythagorean theorem' for the relative entropy which you will prove in Exercise (5.13).

(c) Show that, with $p_i := \mathrm{prob}(a_i)$, it is

$$\sum_i p_i S(\hat{\rho}_i\|\mathcal{E}(\hat{\rho}_i)) - S(\hat{\rho}\|\mathcal{E}(\hat{\rho})) \geq 0.$$

Hint: Use the joint convexity of the quantum relative entropy (Exercise (5.5)).

(d) Prove Eq. (5.18). Show that the upper bound is saturated if the coding states commute.
   Hint: Use all previous results and the result of Exercise (5.5c).

(e) Show that the Holevo bound can be expressed in terms of von Neumann entropies:

$$H[\{\mathrm{prob}(a_i), \hat{\rho}_i\}] = S(\hat{\rho}) - \sum_i \mathrm{prob}(a_i)\,S(\hat{\rho}_i).$$

In particular, it is itself bounded from above by the von Neumann entropy of the weighted average of the coding states,

$$H[\{\text{prob}(a_i), \hat{\rho}_i\}] \leq S(\hat{\rho}).$$

Show that this upper bound is saturated if and only if all coding states $\hat{\rho}_i$ are pure.

(f) Verify that the Holevo bound is bounded from above by the entropy associated with Alice's experiment (Eq. (5.21)), with equality if and only if the coding states are mutually orthogonal, $\hat{\rho}_i \hat{\rho}_k = 0$ for all $i \neq k$.
Hint: As in the preceding exercise, express the Holevo bound in terms of von Neumann entropies. For each coding state write its spectral decomposition,

$$\hat{\rho}_i = \sum_k \rho_{ik} |\psi_{ik}\rangle\langle\psi_{ik}|.$$

Consider a composite system AB in the pure state

$$|\psi_{AB}\rangle = \sum_{ik} \sqrt{\text{prob}(a_i)\rho_{ik}} |ik\rangle \otimes |\psi_{ik}\rangle,$$

where the $\{|ik\rangle\}$ constitute an orthonormal basis in the Hilbert space of constituent A and the $\{|\psi_{ik}\rangle\}$ pertain to constituent B. Then repeat the analysis of Exercise (5.10).

5.12. Quantum-state tomography

The relative entropy is a useful concept not just in the theory of information and communication but also for state tomography, whose basic idea we outlined in Section 2.7. One takes a sample of size M from an exchangeable assembly and subjects the members of that sample to various measurements, yielding data, D. These data trigger an update of the probability density function on the manifold of single-constituent states, in accordance with Bayes' rule, Eq. (2.52). If the various observables measured on (different) members of the sample are informationally complete, the collected sample means determine a unique 'tomographic image', μ: it is the state in which the expectation values of the measured observables coincide with the observed sample means. The data, D, are then tantamount to the specification of that state, μ. According to the *quantum Stein lemma*, for large sample sizes the log-likelihood of finding μ is proportional to a relative entropy:

$$\log \text{prob}(\mu | \rho^{\otimes M}) \approx -MS(\hat{\mu} \| \hat{\rho}).$$

Show that for the purposes of Bayesian updating, via Eq. (2.52), obtaining first a tomographic image μ from a sample of size M and subsequently a tomographic image μ' from another sample of size M' is tantamount to obtaining the weighted

average of μ and μ' from the combined sample of size $(M + M')$. In other words, it does not matter whether data from various samples are processed sequentially or lumped together and processed jointly.
Hint: Exercise (5.5c).

5.13. Pythagorean theorem

Prove that for an arbitrary set of observables, $\{G_a\}$, the quantum relative entropy, Eq. (5.17), satisfies the 'Pythagorean theorem'

$$S(\hat{\rho}\|\hat{\sigma}) = S(\hat{\rho}\|\mathcal{P}^{\sigma}_{\{G_a\}}(\hat{\rho})) + S(\mathcal{P}^{\sigma}_{\{G_a\}}(\hat{\rho})\|\hat{\sigma}).$$

Here $\mathcal{P}^{\sigma}_{\{G_a\}}(\hat{\rho})$ is the statistical operator defined as

$$\mathcal{P}^{\sigma}_{\{G_a\}}(\hat{\rho}) := \arg\min_{\rho'} \left\{ S(\hat{\rho}'\|\hat{\sigma}) \,\middle|\, \langle G_a \rangle_{\rho'} = \langle G_a \rangle_{\rho} \,\forall a \right\}.$$

Among all statistical operators $\hat{\rho}'$ which yield the same expectation values for the observables $\{G_a\}$ as $\hat{\rho}$, this is the one which comes 'closest' to $\hat{\sigma}$, in the sense that it minimizes the quantum relative entropy with respect to $\hat{\sigma}$. The map $\hat{\rho} \mapsto \mathcal{P}^{\sigma}_{\{G_a\}}(\hat{\rho})$ is idempotent, $\mathcal{P}^{\sigma}_{\{G_a\}} \circ \mathcal{P}^{\sigma}_{\{G_a\}} = \mathcal{P}^{\sigma}_{\{G_a\}}$, and may therefore be considered a projection; however, it is in general not linear. (Project)

5.14. Gibbs states

In quantum-state tomography (see Exercise (5.12)) it may happen that the measurement data are incomplete, in the sense that they do not amount to the specification of a unique tomographic image. Rather, they might consist of sample means pertaining to just a small set of observables, $\{G_a\}$, which is not informationally complete. In that case, for large sample sizes the log-likelihood attains the form

$$\log \mathrm{prob}(\{g_a\}|\rho^{\otimes M}) \approx -MS(\mathcal{P}^{\rho}_{\{G_a\}}(\hat{\mu})\|\hat{\rho}),$$

where $\{g_a\}$ denotes the observed sample means, μ is *any* state in which the expectation values of the $\{G_a\}$ coincide with these sample means, and $\mathcal{P}^{\rho}_{\{G_a\}}$ is the projector introduced in Exercise (5.13). (This is due to a quantum version of 'Sanov's theorem'.)

(a) Suppose that you do have complete data, specifying a unique tomographic image, μ. But in the Bayesian update, Eq. (2.52), you process them in two steps: first the data pertaining to the observables $\{G_a\}$, and then the rest. This rest is said to be 'irrelevant' for the update if the probability density function on the manifold of single-constituent states changes only in the first step but not the second. Show that this is the case if the prior, $\mathrm{pdf}(\rho)$, has support only on the submanifold of single-constituent states for which

$$\mathcal{P}^\sigma_{\{G_a\}}(\hat\rho) = \hat\rho.$$

Here σ is some 'reference state' which also lies in this submanifold. (Project)
Hint: Pythagorean theorem for the quantum relative entropy (Exercise (5.13)).
(b) Assuming that the submanifold contains the state of total ignorance, show that the states on the submanifold satisfy

$$\hat\rho = \arg\max_{\rho'}\left\{ S(\hat\rho') \,\middle|\, \langle G_a\rangle_{\rho'} = \langle G_a\rangle_\rho \,\forall a \right\};$$

that is to say, they are states which maximize the von Neumann entropy, given the expectation values of the $\{G_a\}$. (Project)

Maximum-entropy states play a central role in statistical mechanics, where they are known as *Gibbs states*. The above results mean that a theory which a priori considers only a limited set of 'relevant' observables, $\{G_a\}$, and a priori allows only the associated Gibbs states, is closed under Bayesian updating in the following sense. Any Bayesian update of a probability density function over the Gibbs states will yield another probability density function over the Gibbs states. This update will depend solely on data pertaining to the relevant observables. It will not depend on any other, irrelevant data which one may have obtained contemporaneously.

5.15. BB84 protocol

(a) When Eve eavesdrops on the communication between Alice and Bob, it may be that $a \neq a'$ even though $b = b'$. Verify that this occurs with probability one quarter.
(b) How many qubits have to be transmitted, on average, in order to produce a shared key of length l?
(c) Imagine that the random bit generators which generate a, b, and b' are not perfect but biased, all three in the same way: they generate '0' with probability p and '1' with probability $(1-p)$. How does that change the number of qubits which have to be transmitted in order to produce a key of the same length? How about the number of qubits needed to produce a key that allows the same strength of encryption? (Do not worry about eavesdropping during the key distribution.)
Hint: Shannon entropy.

5.16. EPR protocol

In addition to the BB84 protocol, there exist various other quantum protocols for the secure sharing of a cryptographic key. One of these is the EPR protocol, which, like the superdense coding and teleportation protocols, makes use of previously shared Bell (or 'EPR') pairs. Its basic idea is the following. Alice and Bob share many Bell pairs. These pairs have been prepared in the state $|\beta_{11}\rangle$ (Eq. (3.102)), to wit, the same Bell state which we used to show the violation of the CHSH inequality (Eq. (3.111)). Randomly and independently of each other, Alice and Bob perform measurements on each qubit in their possession—like in the Bell experiment depicted in Fig. 3.13. In contrast to the latter, however, each party may choose from one of *three* (rather than two) measurements: Alice may measure Q, R, or S, whereas Bob may measure S, T, or Q. There will be instances, therefore, where Alice and Bob measure on their respective members of a Bell pair the same observable (both Q or both S). After the measurements have been completed, Alice and Bob tell each other which observable they measured on which qubit (but not the measurement results). Then they divide the Bell pairs into two groups: a first group where their measurements were different, and a second group where they were identical. Subsequently, Alice and Bob share the results of their measurements for the first group only. They use these data to determine the average value of the observable A and check whether it agrees with its expected value, $2\sqrt{2}$ (Eq. (3.111)). Provided it does, Alice and Bob convert their measurement results for the second group into a shared secret key.

What does the final conversion step look like? Why does it result in a shared key? Why is the protocol secure? Explain.
Hint: Exercise (3.25).

5.17. No-communication theorem

Alice and Bob share a quantum system composed of two constituents (say, a Bell pair), of which one is in the possession of Alice and the other in the possession of Bob. The initial state of the composite system is described by the statistical operator $\hat{\rho}_{AB}$.

(a) Consider the reduced state of Alice's constituent, $\hat{\rho}_A$, given by Eq. (3.105). Show that this reduced state is invariant under

  i. an arbitrary unitary transformation which pertains to Bob's constituent only,

$$\hat{\rho}_{AB} \to (\hat{I}_A \otimes \hat{U})\hat{\rho}_{AB}(\hat{I}_A \otimes \hat{U}^\dagger)$$

  ii. an arbitrary measurement on Bob's constituent only, *before* its outcome is revealed:

$$\hat{\rho}_{AB} \to \sum_i (\hat{I}_A \otimes \hat{P}_i)\hat{\rho}_{AB}(\hat{I}_A \otimes \hat{P}_i).$$

The projectors, $\{\hat{P}_i\}$, represent the potential outcomes of Bob's measurement. They are mutually orthogonal and add up to the unit operator.

Thus, in the absence of an extra communication channel over which Bob may reveal to Alice the outcome of his measurement, whatever he does on his side will not affect in any way Alice's expectations. In other words, a previously shared quantum system alone cannot be used to send information between the parties. Communication always requires an additional classical or quantum channel, to wit, the physical transmission of some other system. This result is known as the 'no-communication theorem' or 'no-signalling principle'.

(b) In light of the no-communication theorem, revisit Exercises (3.19) and (3.21) and interpret the results.

(c) Reconsider the argument we gave in Section 5.4 for the optimality of the superdense coding and teleportation protocols. A clever student might object that it is not fair to say that in our hypothetical scenario we convey an infinite amount of classical information simply with 'a finite number of physical bits'; after all, we have to burn an infinite number of Bell pairs in the process. Can they not contribute to the transmission of information? Respond.

5.18. Remote state preparation

Consider the protocol represented by the following circuit:

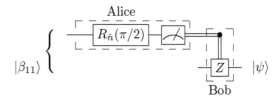

Alice and Bob have shared in advance a qubit pair in the Bell state $|\beta_{11}\rangle$. One member of the pair is in the possession of Alice, the other in the possession of Bob. Alice applies to her qubit a Bloch sphere rotation by $\pi/2$ about the axis $\hat{n} = (-\sin\varphi, \cos\varphi, 0)$. Then she performs a measurement in the standard basis and sends the result to Bob. Depending on the result received, Bob either does nothing or applies a Pauli-Z gate to his qubit.

(a) Show that after this protocol the state of Bob's qubit is (up to some irrelevant global phase factor)

$$|\psi\rangle = |0\rangle + e^{i\varphi}|1\rangle.$$

This state lies on the equator of the Bloch sphere; for this reason it is called an 'equatorial' state. Thus, by performing a suitable unitary transformation on her own qubit and sending just a *single* classical bit, Alice can remotely prepare equatorial states on Bob's side with arbitrary values of φ.

Hint: Exploit the invariance of the Bell state, Eq. (3.103), and replace the standard basis by the basis $\{|\bar{\psi}\rangle, |\psi\rangle\}$, where $|\psi\rangle$ is the state you wish to prepare and $|\bar{\psi}\rangle$ is the state orthogonal to it.

(b) How does the circuit have to be modified if the final state of Bob's qubit ought to lie in the x–z plane rather than on the equator?

(c) Is it possible with a circuit of this kind to remotely prepare arbitrary states on Bob's side that are not a priori constrained to some two-dimensional plane?

(d) Compare this protocol to the teleportation protocol and discuss the similarities and differences.

5.19. Local operations and classical communication

Teleportation and remote state preparation are examples of protocols where only local operations (unitary transformations or measurements on one constituent at a time) and classical communication—in short: LOCC—are allowed. The local operations on one constituent may depend on the classical information received from the other. More generally, the local operations may be probabilistic; that is, they are performed according to some classical probability distribution, which, in turn, may depend on the information received.

(a) Show that LOCC will map any initial state of the form

$$\rho_{AB} = \sum_i p_i \, \rho_i \otimes \sigma_i$$

to a final state of the same form. Here $\{p_i\}$ is a classical probability distribution, and $\{\rho_i\}$ and $\{\sigma_i\}$ are single-constituent states. States of this form are called 'separable'.

(b) Show that in classical probability theory every composite state is separable. Show that, by contrast, in quantum theory there exist composite states which are not. In particular, show that pure entangled states are not separable.

(c) For two qubits, prove that every separable state can be created by means of LOCC, starting from the initial state $|0\rangle\langle 0| \otimes |0\rangle\langle 0|$. More generally, all separable states can be created by means of LOCC starting from an initial pure product state.

5.20. Entangled versus separable states

Since an initial pure product state is not correlated at all, and, as you learnt in Exercise (5.19), subsequent LOCC create at most correlations of the kind also seen in classical probability theory, the resultant separable state will exhibit at most classical correlations; it will not exhibit quantum correlations, to wit, entanglement. This opens a way to extend the definition of entanglement from pure to mixed states: a mixed state is entangled if and only if it is *not* separable. The preparation of an entangled state out of an initial product state can never be achieved by LOCC alone but requires either quantum communication or a supply of preexisting entangled states, like previously shared Bell pairs, that are consumed in the process.

(a) Verify that this definition of entanglement includes our earlier definition of pure entangled states as a special case.

(b) Prove that if the state of a bipartite system is separable, then the joint entropy respects the classical lower bound, Eq. (5.36). This is a stronger bound than the

Araki–Lieb inequality, Eq. (5.37). Conversely, if the classical bound is violated, there must be entanglement. (The latter is a sufficient but not a necessary condition for entanglement.)

Hint: Show

$$S(\hat{\rho}_{AB}) = S(\hat{\rho}_A \otimes \hat{\iota}_B) - S(\hat{\rho}_{AB} \| \hat{\rho}_A \otimes \hat{\iota}_B),$$

where $\hat{\iota}_B := \hat{I}_B / d_B$, and use the joint convexity of the relative entropy (Exercise (5.5)).

References

Aaronson, S. (2013). *Quantum Computing Since Democritus*. Cambridge University Press.

Abraham, R., Marsden, J. E., and Ratiu, T. (1988). *Manifolds, Tensor Analysis, and Applications* (2nd edn). Springer.

Arnold, V. I. (1997). *Mathematical Methods of Classical Mechanics* (2nd edn). Springer.

Balian, R. (1991). *From Microphysics to Macrophysics*. Volume 1. Springer.

Ballentine, L. E. (1998). *Quantum Mechanics: A Modern Development* (2nd edn). World Scientific.

Banaszek, K., Cramer, M., and Gross, D. (2013). Focus on quantum tomography. *New J. Phys.*, 15, 125020.

Barnett, S. (2009). *Quantum Information*. Oxford University Press.

Barrett, J. and Leifer, M. (2009). The de Finetti theorem for test spaces. *New J. Phys.*, 11, 033024.

Bell, J. S. (2004). *Speakable and Unspeakable in Quantum Mechanics* (2nd edn). Cambridge University Press.

Bennett, C. H. and Brassard, G. (1984). Quantum cryptography: Public key distribution and coin tossing. In *Proceedings of IEEE International Conference on Computers, Systems and Signal Processing*, pp. 175–179. IEEE, New York.

Bennett, C. H., Brassard, G., Crépeau, C., Jozsa, R., Peres, A., and Wootters, W. (1993). Teleporting an unknown quantum state via dual classical and Einstein-Podolsky-Rosen channels. *Phys. Rev. Lett.*, 70, 1895.

Bennett, C. H. and Wiesner, S. J. (1992). Communication via one- and two-particle operators on Einstein-Podolsky-Rosen states. *Phys. Rev. Lett.*, 69, 2881.

Birkhoff, G. (1967). *Lattice Theory* (3rd edn). Volume 25, Colloquium Publications. American Mathematical Society.

Birkhoff, G. and v. Neumann, J. (1936). The logic of quantum mechanics. *Ann. Math.*, 37, 823.

Bohm, A. (1993). *Quantum Mechanics: Foundations and Applications* (3rd edn). Springer.

Bohr, N. (1987). *The Philosophical Writings of Niels Bohr*. Volumes 1–3. Ox Bow Press.

Bridgman, P. W. (1927). *The Logic of Modern Physics*. Macmillan.

Briegel, H. J., Browne, D. E., Dür, W., Raussendorf, R., and Van den Nest, M. (2009). Measurement-based quantum computation. *Nature Physics*, 5, 19.

Caves, C. M., Fuchs, C. A., and Schack, R. (2002). Unknown quantum states: The quantum de Finetti representation. *J. Math. Phys.*, 43, 4537.

Chevalier, G. (2007). Wigner's theorem and its generalizations. In *Handbook of Quantum Logic and Quantum Structures* (ed. K. Engesser, D. M. Gabbay, and D. Lehmann), pp. 429–475. Elsevier.

Chiribella, G. and Spekkens, R. W. (eds) (2016). *Quantum Theory: Informational Foundations and Foils*. Springer.

Cohen-Tannoudji, C., Diu, B., and Laloë, F. (1991). *Quantum Mechanics*. Volumes 1–2. Wiley.

Cover, T. M. and Thomas, J. A. (2006). *Elements of Information Theory* (2nd edn). Wiley.

Cox, R. T. (1946). Probability, frequency and reasonable expectation. *Am. J. Phys.*, 14, 1.

D'Ariano, G. M., Paris, M. G. A., and Sacchi, M. F. (2004). Quantum tomographic methods. *Lect. Notes Phys.*, 649, 7.

de Finetti, B. (1990). *Theory of Probability*. Wiley.

Degen, C. L., Reinhard, F., and Cappellaro, P. (2017). Quantum sensing. *Rev. Mod. Phys.*, **89**, 035002.

Ekert, A. K. (1991). Quantum cryptography based on Bell's theorem. *Phys. Rev. Lett.*, **67**, 661.

Elitzur, A. C. and Vaidman, L. (1993). Quantum mechanical interaction-free measurements. *Found. Phys.*, **23**, 987.

Feynman, R. P. (1982). Simulating physics with computers. *Int. J. Theor. Phys.*, **21**, 467.

Foulis, D. J., Greechie, R. J., and Rüttimann, G. T. (1992). Filters and supports in orthoalgebras. *Int. J. Theor. Phys.*, **31**, 789.

Foulis, D. J., Greechie, R. J., and Rüttimann, G. T. (1993). Logicoalgebraic structures II: Supports in test spaces. *Int. J. Theor. Phys.*, **32**, 1675.

Foulis, D. J. and Randall, C. H. (1972). Operational statistics I: Basic concepts. *J. Math. Phys.*, **13**, 1667.

Foulis, D. J. and Randall, C. H. (1981). What are quantum logics and what ought they to be? In *Current Issues in Quantum Logic* (ed. E. G. Beltrametti and B. C. van Fraassen), Volume 8, Ettore Majorana International Science, Physical Sciences, pp. 35–52. Plenum Press.

Giovannetti, V., Lloyd, S., and Maccone, L. (2004). Quantum-enhanced measurements: Beating the standard quantum limit. *Science*, **306**, 1330.

Gudder, S. (1988). *Quantum Probability*. Academic Press.

Jauch, J. M. (1968). *Foundations of Quantum Mechanics*. Addison-Wesley.

Jaynes, E. T. (2003). *Probability Theory: The Logic of Science*. Cambridge University Press.

Kirkpatrick, K. A. (2003). 'Quantal' behavior in classical probability. *Found. Phys. Lett.*, **16**, 199.

Kraus, K. (1983). *States, Effects and Operations: Fundamental Notions of Quantum Theory*. Volume 190, Lecture Notes in Physics. Springer.

Kwiat, P. G. and Hardy, L. (2000). The mystery of the quantum cakes. *Am. J. Phys.*, **68**, 33.

Liang, Y.-C., Spekkens, R. W., and Wiseman, H. M. (2011). Specker's parable of the overprotective seer: A road to contextuality, nonlocality and complementarity. *Phys. Rep.*, **506**, 1.

Lipton, R. J. and Regan, K. W. (2014). *Quantum Algorithms via Linear Algebra: A Primer*. MIT Press.

Lloyd, S. (1996). Universal quantum simulators. *Science*, **273**, 1073.

Lucas, A. (2014). Ising formulations of many NP problems. *Frontiers in Physics*, **2**, 5.

MacKay, D. J. C. (1992). Bayesian interpolation. *Neural Comp.*, **4**, 415.

MacKay, D. J. C. (2003). *Information Theory, Inference and Learning Algorithms*. Cambridge University Press.

Mackey, G. (1963). *The Mathematical Foundations of Quantum Mechanics*. Benjamin.

McClean, J. R., Romero, J., Babbush, R., and Aspuru-Guzik, A. (2016). The theory of variational hybrid quantum-classical algorithms. *New J. Phys.*, **18**, 023023.

Mermin, N. D. (1990). Simple unified form for the major no-hidden-variables theorems. *Phys. Rev. Lett.*, **65**, 3373.

Mermin, N. D. (1993). Hidden variables and the two theorems of John Bell. *Rev. Mod. Phys.*, **65**, 803.

Mermin, N. D. (1995). Limits to quantum mechanics as a source of magic tricks: Retrodiction and the Bell-Kochen-Specker theorem. *Phys. Rev. Lett.*, **74**, 831.

Mermin, N. D. (2007). *Quantum Computer Science*. Cambridge University Press.

Misra, B. and Sudarshan, E. C. G. (1977). The Zeno's paradox in quantum theory. *J. Math. Phys.*, **18**, 756.

National Academies of Sciences, Engineering, and Medicine (2019). *Quantum Computing: Progress and Prospects*. The National Academies Press.

Nielsen, M. A. and Chuang, I. L. (2000). *Quantum Computation and Quantum Information*. Cambridge University Press.

Öttinger, H. C. (2005). *Beyond Equilibrium Thermodynamics*. Wiley.

Perdrix, S. (2007). Towards minimal resources of measurement-based quantum computation. *New J. Phys.*, **9**, 206.

Peres, A. (1978). Unperformed experiments have no results. *Am. J. Phys.*, **46**, 745.

Peres, A. (1990). Incompatible results of quantum measurements. *Phys. Lett. A*, **151**, 107.

Peres, A. (1995). *Quantum Theory: Concepts and Methods*. Kluwer Academic Publishers.

Peruzzo, A., McClean, J., Shadbolt, P., Yung, M.-H., Zhou, X.-Q., Love, P. J., Aspuru-Guzik, A., and O'Brien, J. L. (2014). A variational eigenvalue solver on a photonic quantum processor. *Nature Comm.*, **5**, 4213.

Petz, D. (2008). *Quantum Information Theory and Quantum Statistics*. Springer.

Piron, C. (1964). Axiomatique quantique. *Helv. Phys. Acta*, **37**, 439.

Piron, C. (1976). *Foundations of Quantum Physics*. Benjamin.

Preskill, J. (1997–2018a). Lecture notes on quantum computation. Caltech Particle Theory Group http://www.theory.caltech.edu/~preskill/ph219/index.html accessed 12 January 2021.

Preskill, J. (2018b). Quantum computing in the NISQ era and beyond. *Quantum*, **2**, 79.

Randall, C. H. and Foulis, D. J. (1973). Operational statistics II: Manuals of operations and their logics. *J. Math. Phys.*, **14**, 1472.

Rau, J. (2017). *Statistical Physics and Thermodynamics: An Introduction to Key Concepts*. Oxford University Press.

Raussendorf, R. and Briegel, H. J. (2001). A one-way quantum computer. *Phys. Rev. Lett.*, **86**, 5188.

Renner, R. (2007). Symmetry of large physical systems implies independence of subsystems. *Nature Physics*, **3**, 645.

Rudolph, T. (2017). *Q is for Quantum*. Terence Rudolph.

Ruskai, M. B. (2002). Inequalities for quantum entropy: A review with conditions for equality. *J. Math. Phys.*, **43**, 4358. Erratum 46:019901 (2005).

Sakurai, J. J. (1985). *Modern Quantum Mechanics*. Benjamin Cummings.

Scarani, V., Chua, L., and Liu, S. Y. (2010). *Six Quantum Pieces: A First Course in Quantum Physics*. World Scientific.

Schack, R., Brun, T. A., and Caves, C. M. (2001). Quantum Bayes rule. *Phys. Rev. A*, **64**, 014305.

Schumacher, B. and Westmoreland, M. (2010). *Quantum Processes Systems, and Information*. Cambridge University Press.

Sivia, D. S. (1996). *Data Analysis: A Bayesian Tutorial*. Oxford University Press.

Specker, E. (1960). Die Logik nicht gleichzeitig entscheidbarer Aussagen. *Dialectica*, **14**, 239.

Spekkens, R. W. (2007). Evidence for the epistemic view of quantum states: A toy theory. *Phys. Rev. A*, **75**, 032110.

Steane, A. (1998). Quantum computing. *Rept. Prog. Phys.*, **61**, 117.

Susskind, L. and Friedman, A. (2014). *Quantum Mechanics: The Theoretical Minimum* (2nd edn). Basic Books.

Varadarajan, V. S. (1985). *Geometry of Quantum Theory* (2nd edn). Springer.

Vazirani, U. (2007). Lecture notes on quantum computation. Berkeley EECS https://people.eecs.berkeley.edu/~vazirani/quantum.html accessed 12 January 2021.

Vedral, V. (2002). The role of relative entropy in quantum information theory. *Rev. Mod. Phys.*, **74**, 197.

Watrous, J. (2006). Lecture notes on quantum computing. University of Waterloo: Cheriton School of Computer Science https://cs.uwaterloo.ca/~watrous/ accessed 12 January 2021.

Wehrl, A. (1978). General properties of entropy. *Rev. Mod. Phys.*, **50**, 221.

Wheeler, J. A. (1983). Law without law. In *Quantum Theory and Measurement* (ed. J. A. Wheeler and W. H. Zurek), Princeton Series in Physics, pp. 182–213. Princeton University Press.

Wilce, A. (2000). Test spaces and orthoalgebras. In *Current Research in Operational Quantum Logic: Algebras, Categories, Languages* (ed. B. Coecke, D. Moore, and A. Wilce), Volume 111, Fundamental Theories of Physics, pp. 81–114. Kluwer.

Wilde, M. M. (2013). *Quantum Information Theory*. Cambridge University Press.

Yin, J., Li, Y.-H., Liao, S.-K., Yang, M., Cao, Y., Zhang, L., Ren, J.-G., Cai, W.-Q., Liu, W.-Y., Li, S.-L., Shu, R., Huang, Y.-M., Deng, L., Li, L., Zhang, Q., Liu, N.-L., Chen, Y.-A., Lu, C.-Y., Wang, X.-B., Xu, F., Wang, J.-Y., Peng, C.-Z., Ekert, A. K., and Pan, J.-W. (2020). Entanglement-based secure quantum cryptography over 1,120 kilometres. *Nature*, **582**, 501.

Index